Imaging Brain Function With EEG

Walter J. Freeman • Rodrigo Quian Quiroga

Imaging Brain Function With EEG

Advanced Temporal and Spatial Analysis of Electroencephalographic Signals

Springer

Walter J. Freeman
Molecular and Cell Biology
University of California at Berkeley
Berkeley, CA, USA

Rodrigo Quian Quiroga
Centre for Systems Neuroscience
University of Leicester
Leicester, UK

ISBN 978-1-4614-4983-6 ISBN 978-1-4614-4984-3 (eBook)
DOI 10.1007/978-1-4614-4984-3
Springer New York Heidelberg Dordrecht London

Library of Congress Control Number: 2012948983

Printed on acid-free paper

Springer is part of Springer Science+Business Media (www.springer.com)

To our wives, who have inspired us,
and to our children and grandchildren,
whom we aim to inspire.

Preface

This book is about temporal and spatial patterns that we find in the electric fields on the scalp (electroencephalogram, EEG) and cerebral cortex (electrocorticogram, ECoG) (Lopes da Silva 1993; Basar 1998). The patterns are enigmatic, ephemeral, easily dismissed as noise, and by most accounts epiphenomenal (Freeman and Baird 1989). Yet, some of the patterns are neural correlates of intentional actions, specifically the perception and discrimination of sensory stimuli by alert, aroused human and animal subjects. For this reason, they have become a focus of our experimental and theoretical investigations. What can they tell us about how brains work? What tools do we need to record and analyze them? Which related disciplines of science, mathematics, and engineering do we turn to for guidance in simulating them with computational models of cortical dynamics?

We begin with a brief overview of electroencephalography. Temporal analysis predominated in the first two decades after the discovery of EEG by Hans Berger (1873–1941). Thereafter, two main breakthroughs advanced the analysis of temporal signals. The first was the use of ensemble averaging – that is, the average over several stimulus presentations – to better visualize evoked responses (Dawson 1954), and the second was the introduction of personal computers in the analysis of EEG signals, especially after the implementation of the fast Fourier transform (FFT) by Cooley and Tukey (1965), which enabled a rapid and reliable representation of the frequencies in EEG signals. More recent advances include the study of time-frequency patterns and the introduction of wavelets in the analysis of EEGs and evoked responses (Quian Quiroga et al. 2001; Majumdar et al. 2006). Pioneers undertook spatial analysis with racks of primitive amplifiers. W. Gray Walter (1953) focused on the "toposcopy" of alpha waves in the EEG. John Lilly (Lilly 1954) recorded spontaneous and evoked potentials in the ECoG. Large, dense electrode arrays for scalp EEG were introduced by Donald W. DeMott (1970), Dietrich Lehmann (1971), M. N. Livanov (1977), and Konrad Maurer (1989), mainly to analyze the topography of alpha waves. The systematic study of beta and gamma patterns in the ECoG at high spatial resolution in the ECoG was introduced by Freeman and Schneider (1982). High temporal and spectral resolutions were later achieved by introduction to brain studies of the Hilbert transform (Freeman and Rogers 2002).

In this book, we pursue EEG and ECoG patterns as we would study the natural history of a new species, like searching for elusive forest animals, trying to catch and hold them for description without damaging or distorting them. We ask the following questions: Where are they found in brains? What behaviors are they correlated with and when? How large are they? How long do they last? Can we group them into recognizable categories? How often do samples that can be categorized recur? What are their internal structures and textures that constitute their features? What frequencies appear in their temporal and spatial spectra? Beyond empirical description, how do they form? Are they transmitted? If so, where do they go, by what means, and with what delays? Are they epiphenomenal or do they play an active role in the genesis and control of behavior? Do other parts of the brain detect and respond to them, and if so, how? Can we find meanings in the patterns? Are the meanings only for objective observers like ourselves or do the patterns reflect the construction and deployment of subjective meanings within the brains for the subjects?

In our book, we propose answers to these questions by showing examples of the textured patterns both in time and space and the contexts of recording. We describe the optimal conditions and methods for their measurement and present hypotheses on how they form and why they are significant. Our results give first glimpses of these patterns, which may already seem primitive but nevertherless provide prescriptions on how our results can be replicated, improved, and extended. What makes the work so difficult is that the electric potential differences we observe are samples from extracellular fields of very weak electric energy. They are signs of the transmembrane electric currents that give shape and texture to great clouds of cortical action potentials. The fields emerge because every neuron interacts with many thousand others in the cortical tissue that anatomists call laminar neuropil (from a Greek word for *felt*). It is the textured fabric of axons (Gr., *axis*), dendrites (Gr., *tree*), glia (Gr., *glue*), and capillaries (Latin, *hairs*) that generates and regulates its own spontaneous background activity. The pulse clouds emerging from the neuropil do the work of cortex (Gr., *tree bark*) by forming vector fields, which are manifested in scalar fields of electric potential. We cannot at present record enough pulse trains simultaneously to see the textures directly; so we infer them through the potentials that we can record, and confirm them when we can by simultaneously recording spikes from representative single cells in the population (Sect. 3.3.3 and Sect. 4.5 in Freeman 1975). The mixed activity of axons and dendrites is robust and resilient; yet, it resembles the bubbling of a pan of boiling water. Finding and extracting self-organized patterns emerging in such noise is not a trivial undertaking.

Laminar neuropil, in other words, is an active medium that embeds the sensory, cognitive, and motor systems and serves as a massive axodendritic channel of communication among them. We propose that the spatiotemporal patterns manifest the forms taken by *macroscopic* perceptual and cognitive information, carried by dense pulse clouds in the neuropil, in parallel with the *microscopic* sensory and motor information, carried by sparse pulse trains of neurons singly in local networks. We suggest that the large-scale patterns, which are the focus of our book, can convey the relevant context and meaning of the information, in a word, the knowledge that the subject has about the received information, because the laminar neuropil provides the neural mechanisms for constructing and storing knowledge during sensation, and for mobilizing the knowledge for transmission during perception. It is the massive quantity of integrated information that supports our experience and feeling of recognition in perception, variously described as "metastable coordination dynamics" (Kelso and Tognoli 2006), "virtual associative networks" (Yufik 1998), "mind force" (Orsucci 1998, 2009), "holographic brain" (Pribram 1999), "global work space" (Baars et al. 2003), and others (Jordan 2008; Koch and Tononi 2008; Tononi 2008; Seth 2009; Tallon-Baudry 2009).

The large-scale patterns from the neuropil, measured using EEG, ECoG, or local field potentials (LFP), constitute the first method of imaging brain activity in awake subjects, going back to Berger in the 1920s (see Chap. 1). Our knowledge about brain function has been greatly increased by the introduction of single-cell recordings in the 1950s, now advanced to high level of sophistication with the identification of *concept cells* (Quian Quiroga 2012) and, more recently, by imaging techniques of MEG, PET, fMRI and BOLD (blood oxygen level dependent), and fMRI. Why then do we focus on the predecessor of these methods? This is because EEG, ECoG, and LFP signals are the most challenging in terms of data processing, and in spite of being known for nearly a century, we still learn a lot from their analysis, especially when we use advanced signal-processing methods and bold experimental designs. Most prior research with EEG signals has been constrained to variations of a couple basic paradigms: the study of evoked responses and the study of EEG oscillations in given frequency bands in single channels or a judicious sample of channels. In our book, we provide some tools to go beyond these standard analyses and experimental designs. In particular, we propose two radical paradigm shifts. First, we argue that the ensemble averaging that is typically used to observe evoked responses imposes a large loss of information of systematic and nonsystematic changes of the trial-by-trial responses (Quian Quiroga 2000; Quian Quiroga and Garcia 2003). New powerful signal-processing tools, like wavelet denoising, indeed allow the visualization of the single-trial responses, thus opening a window to new types of analyses and experiments (Quian Quiroga et al. 2007). In fact, some of the most interesting cognitive processes (e.g., learning) are revealed by changes during an experimental session. The use of these new techniques requires new experimental designs, where trial-by-trial changes are sought in order to study their correlation with different cognitive processes, instead of being avoided in order to get cleaner averages. It is the tracking of this variability that allows us to study different cognitive processes and merge the spatial and temporal information from fMRI and EEG (Eichele et al. 2005; Eichele et al. 2008; Freeman et al. 2009). The second main paradigm shift we propose is to study the dynamics and propagation of spatial patterns of field potentials, as one can study the continuous evolution of waves in fields. So far, the information from different EEG or ECoG channels has been studied independently or at most by the use of topographic plots at precise times. However, both the single-channel temporal analysis and the multiple-channel topographic analysis are too limited because they do not display how spatial patterns of brain activity evolve in time.

Without getting into deep philosophical issues, we can say that knowledge is an immense collection of fragments of information, each fragment being interrelated with every other so as to form a pattern. The laminar neuropil provides the dense grid of neurons that can store and express the massive information and the connectivity required for each to share it with others in the field. These operations constitute the exercise of intelligence, which is defined as the ability to acquire and apply knowledge and skills. It is noteworthy that the laminar neuropil is most fully developed in the brains of the most intelligent animals of three phyla: Vertebrata (mammals), Arthropoda (bees), and Mollusca (cuttlefish). Intelligent life has emerged and evolved independently three times in the earth's geological history, each branch with very different brain architectures but similar neuropil and neural dynamics. Clearly, the neuropil is an electrochemical system made of the same atoms as all matter. It is also a thermodynamic system that uses metabolic energy to construct knowledge from information. By study of cortical temporospatial activity patterns, their neural mechanisms of construction and transmission prior to termination, we might aspire to better understand human mechanisms of intelligence and brain disorders and, perhaps, even construct intelligent machines that, in some useful sense, know what they are doing.

We find it profitable to look for concepts and tools in physics, mathematics, and engineering that we can use to design our experiments and simulate our observations of the properties of laminar neuropil. We rely most heavily on techniques for digital signal processing, by which we decompose EEG and ECoG time series into frequency bands and components (Chap. 2). We analyze frequency modulation by using time-frequency analysis (Chap. 3) and by using wavelets (Chap. 4). We adapt the filters to single-trial, single-channel evoked potentials in order to avoid ensemble averaging and reveal how trial-by-trial changes correlate with different brain processes (Chap. 5).

We also rely heavily on techniques from systems control theory (Chap. 6) and from digital imaging (Chap. 7). We show that, during normal cognitive operations, the cortical neuropil holds itself in a range we can characterize as linear, Gaussian, and time-invariant. Having done so, we can then simulate the major dynamic operations of the neuropil by using matrices of linear differential equations in piecewise linear approximations. The solutions of the equations give a family of linear basis functions –exponentials, sines, cosines, ramps, etc. – with which to measure the evoked potentials and the waves of spontaneous and induced cortical activity. Then we can use changes in the parameters and coefficients of the equations to represent the changes in cortical dynamics caused by intrinsic nonlinearities as well as the time-varying state changes that underlie arousal, learning, and the exercise of experience. We illustrate the categorization of spatiotemporal images with respect to behavior in the primitive allocortex in the olfactory system (Chap. 8), the more complex sensory neocortices (Chap. 9), and higher cognitive functions correlated with patterns in the ECoG and EEG (Chap. 10). We finish with a synthesis of our data in the context of the brain viewed as an open thermodynamic system operating far from equilibrium (Chap. 11), which uses the cortex to extract relevant sensory information and condense it into knowledge stored in widespread synaptic modifications that are retrieved for use, as revealed in macroscopic patterns and microscopic firing of multiple types of category cells (Chap. 6).

The greatest value of piecewise linear analysis is in the application of feedback control theory to the calculation of the strengths of functional synaptic connectivity. We define the various types of synaptic interaction strengths as the forward and feedback gains of the multiple neural loops formed by populations of excitatory and inhibitory neurons. The calculations of gain values are based on models of the topology of the types of connections in a hierarchy, called Katchalsky sets (K-sets, Freeman 1975; Kozma and Freeman 2001; Freeman and Erwin 2008). We evaluate the gains from measurements of the frequencies and rates of increase or decrease in the envelopes of oscillatory components of the EEG, ECoG, and evoked potentials. We use the gains to define the stable states of cortical neuropil, each with its attractor, the boundaries of each basin of attraction, and the state transitions that enable the temporospatial pattern of each attractor to emerge, flourish, and dissolve.

The mathematical details of the digital signal processing and systems control theory we use have been described in many textbooks and monographs. In our book, we present an overview of the main features and dynamics of spatiotemporal patterns, with only a minimum of the mathematics on which our analyses and conclusions rest. We cite suitable references for readers from clinical and biological domains to go beyond our qualitative descriptions. There is also an extensive literature on the theory of electric potentials and their applications to the study of the brain electrophysiological signals, serving particularly to locate the sources and sinks of evoked potentials and epileptic spikes in EEGs and ECoGs. In order to maintain focus and ensure brevity, we introduce only a bit of this theory in describing the limits of the spatial resolution of the high-density arrays of electrodes we use to reveal the textures of EEG and ECoG. For readers from mathematics, physics, and engineering, we also describe briefly the main physiology principles involved in the generation and interpretation of EEG and ECoG signals. For more details, we recommend introductory texts on neurobiology. We refer readers who propose to replicate our experimental results to our original reports for technical details. We have in mind also readers from psychology, psychiatry, cognitive science, and philosophy, and we hope that we have made a judicious selection of arcane details needed by anyone who seriously addresses the mind-body problem.

There is still more extensive literature on the interactions of the cortex with subcortical structures: the thalamus, striatum, cerebellum, and the modulatory aminergic and peptidergic nuclei in the brain stem. We refer to these mechanisms whenever we find it necessary to do so, but our focus is on the intracortical mechanisms that form and maintain spatiotemporal images, particularly those with frequencies in the theta, beta, and gamma ranges. We emphasize that the greater part of our understanding of cortical dynamics comes from sampling the fields with electrode arrays of unprecedented high density, spaced at intervals one tenth those of conventional clinical arrays (Chap. 7).

Exploration of the properties of high-resolution spatiotemporal images related to cognition opens enticing new avenues for the development of new brain theory by experts in physics and for the devising of new forms of machine intelligence by experts in robotics. Examples of exploration of the dynamics of very large systems are already emerging in other areas of knowledge that are being applied to brain imaging, include models of neuropil implemented in VLSI analog hardware (Principe et al. 2001); the use of random graph theory (Kozma 2007; Freeman et al. 2009); the use of many-body physics and quantum field theory (Vitiello 2001; Freeman and Vitiello 2010); and the use of nonequilibrium thermodynamics (Freeman et al. 2012), which is especially attractive for the possibility of combining EEG/ECoG/LFP imaging with recordings from concept cells and with the several measures estimating the oxidative metabolism of brains (Logothetis 2008; Freeman et al. 2009) into a unified science (Chap. 11). However enticing these new avenues may be, the business at hand is to describe the existence and detailed properties of macroscopic neural electrical activity patterns now known in order to acquire new data at even higher resolutions in the spatial, temporal, and spectral dimensions.

Berkeley, CA, USA — Walter J. Freeman
Leicester, UK — Rodrigo Quian Quiroga

References

Baars BJ, Banks WP, Newman JN (eds) (2003) Essential sources in the scientific study of consciousness. MIT Press, Cambridge, MA

Basar E (ed) (1998) Brain function and oscillations. Vol 1: Principles and approaches. Vol II: Integrative brain function. Neurophysiology and cognitive processes. Springer Series in Synergetics, Berlin

Bressler SL, Coppola R, Nakamura R (1993) Episodic multiregional cortical coherence at multiple frequencies during visual task performance. Nature 366:153–156

Bressler SL, Kelso JAS (2001) Cortical coordination dynamics and cognition. Trends Cogn Sci 5:2–36

Cooley JW, Tukey JW (1965) An algorithm for the machine calculation of complex Fourier series. Math Comput 19:297–301

Dawson GD (1954) A summation technique for the detection of small evoked potentials. Electroenceph Clin Neurophysiol 6:153–154

DeMott DW (1970) Toposcopic studies of learning. CC Thomas, Springfield

Eichele T, Specht K, Moosmann M, Jongsma M, Quian Quiroga R, Nordby H, Hugdahl K (2005) Assessing the spatio-temporal evolution of neuronal activation with single-trial ERP-fMRI. Proc Natl Acad Sci USA 102:17798–17803

Eichele T, Calhoun VD, Moosmann M, Specht K, Jongsma MLA, Quian Quiroga R, Nordby H, Hugdahl K (2008) Unmixing concurrent EEG-fMRI with parallel independent component analysis. Int J Psychophysiol 67:222–234

Freeman WJ (1975) Mass action in the nervous system. Academic Press, New York

Freeman WJ (2008) A pseudo-equilibrium thermodynamic model of information processing in nonlinear brain dynamics. Neural Networks 21:257–265. http://repositories.cdlib.org/postprints/2781

Freeman WJ, Ahlfors SM, Menon V (2009) Combining EEG, MEG and fMRI signals to characterize mesoscopic patterns of brain activity related to cognition. Special issue (Lorig TS, ed) Int J Psychophysiol 73(1):43–52

Freeman WJ, Baird B (1989) Effects of applied electric current fields on cortical neural activity. In: Schwartz E (ed) Computational neuroscience. Plenum Press, New York, pp 274–287

Freeman WJ, Erwin H (2008) Freeman K-set. Scholarpedia 3(2):3238. http://www.scholarpedia.org/article/Freeman_K-set

Freeman WJ, Kozma R, Bollobás B, Riordan O (2009) Chapter 7. Scale-free cortical planar network. In: Bollobás B, Kozma R, Miklös D (eds) Handbook of large-scale random networks. Bolyai mathematical studies, vol 18. Springer, New York, pp 277–324. http://www.springer.com/math/numbers/book/978-3-540-69394-9

Freeman WJ, Livi R, Opinata M, Vitiello G (2012) Cortical phase transitions, non-equilibrium thermodynamics and the time-dependent Ginzburg-Landau equation. Int J Mod Phys B 26(6):1250035. doi:10.1142/S021797921250035X

Freeman WJ, Rogers LJ (2002) Fine temporal resolution of analytic phase reveals episodic synchronization by state transitions in gamma EEGs. J Neurophysiol 87:937–945

Freeman WJ, Schneider W (1982) Changes in spatial patterns of rabbit olfactory EEG with conditioning to odors. Psychophysiology 19:44–56

Freeman WJ, Vitiello G (2010) Vortices in brain waves. Int J Mod Phys B 24(17):3269–3295. http://dx.doi.org/10.1142/S0217979210056025

Jordan JS (2008) Wild-Agency: nested intentionalities in neuroscience and archeology. Phil Trans Roy Soc B Biol Sci 363:1981–1991

Kelso JAS, Tognoli E (2006) Metastability in the brain. Neural Networks IJCNN'06: 363–368. doi:10.1109/IJCNN.2006.246704

Koch C, Tononi G (2008) The neural correlates of consciousness: an update. Ann New York Acad Sci 1124:239–261

Kozma R (2007) Neuropercolation. Scholarpedia 2(8):1360Kozma R, Freeman WJ (2001) Chaotic resonance: methods and applications for robust classification of noisy and variable patterns. Int J Bifurcat Chaos 10:2307–2322

Lehmann D (1971) Multichannel topography of human alpha EEG fields. EEG Clin Neurophysiol 31:439–49

Lilly JC (1954) Instantaneous relations between the activities of closely spaced zones on the cerebral cortex – electrical figures during responses and spontaneous activity. Am J Physiol 176:493–504

Livanov MN (1977) Spatial organization of cerebral processes. Wiley, New York

Logothetis NK (2008) What we can do and what we cannot do with fMRI. Nature 453:869–878. doi:10.1038/nature06976

Lopes da Silva F (1993) EEG analysis: theory and practice. In: Niedermeyer E, Lopes da Silva F (eds) Electroencephalography: basic principles, clinical applications and related fields. Williams and Wilkins, Baltimore, pp 1097–1123

Majumdar NS, Pribram KH, Barrett TW (2006) Time frequency characterization of evoked brain activity in multiple electrode recordings. IEEE Trans Biomed Eng 53(12):1–9

Maurer K (ed) (1989) Topographic brain mapping of EEG and evoked potentials. Springer, Berlin

Orsucci F (1998) The complex matters of the mind. World Sci Books, Singapore

Orsucci F (2009) Mind force. On human attractions. World Sci Books, Singapore

Pribram KH (1999) Brain and the composition of conscious experience. J Conscious Stud 6(5):19–42

Principe JC, Tavares VG, Harris JG, Freeman WJ (2001) Design and implementation of a biologically realistic olfactory cortex in analog VLSI. Proc IEEE 89:1030–1051

Quian Quiroga R, Sakowicz O, Basar E, Schürmann M (2001) Wavelet transform in the analysis of the frequency composition of evoked potentials. Brain Res Protocols 8:16–24

Quian Quiroga R, Atienza M, Jongsma M (2007) What can we learn from single-trial event-related potentials? Chaos Complexity Letters 2:345–65

Quian Quiroga R (2000) Obtaining single stimulus evoked potentials with wavelet denoising. Physica D 145:278–192

Quian Quiroga R (2012) Concept cells: The building blocks of declarative memory functions. Nature Reviews Neuroscience 13:587–597. doi: 10.1038/nrn3251

Quian Quiroga R, Garcia H (2003) Single-trial event-related potentials with wavelet denoising. Clin Neurophysiol 114:376–390

Seth AK (2009) Explanatory correlates of consciousness: theoretical and computational challenges. Cogn Comput 1:50–63

Tallon-Baudry C (2009) The roles of gamma-band oscillatory synchrony in human visual cognition. Front Biosci 14:321–332 (She assumes "the equivalence between power or phase-synchrony with local or long-distance oscillatory synchrony (p 322))

Tononi G (2008) Consciousness as Integrated Information: a provisional manifesto. Biol Bull 215(3):216–242

Tsuda I (2001) Towards an interpretation of dynamic neural activity in terms of chaotic dynamical systems. Behav Brain Sci 24:793–810

Vitiello G (2001) My double unveiled. John Benjamins, Amsterdam

Walter WG (1953) The living brain. WW Norton, New York

Yufik Y (1998) Virtual associative networks: a framework for cognitive modeling. Ch. 5. In: Pribram K (ed) Brain and values: is a biological science of values possible? Lawrence Erlbaum Associates, Mahway, pp 109–178

Acknowledgments

I was inspired by my father, Walter Freeman, to embark upon a career in brain science by applying my training in engineering and wartime electronics. My teachers at Yale included John Fulton in neurophysiology, inspired by his mentor Sir Charles Sherrington; Rafael Lorente de Nó in neuroanatomy, inspired by his mentor Santiago Ramón y Cajal; Warren McCulloch in neural networks; and Paul Yakovlev, Paul Maclean, and Karl Pribram in neuropsychiatry. At Berkeley I learned behaviorism from Frank Beach, linear control systems theory from Otto J M Smith, multivariate statistics from Lucien LeCam, probability theory from Michelle Loève, differential equations from Morris Hirsch, phenomenology from Hubert Dreyfus, and chaos theory from Leon Chua. The names of my students and colleagues who have taught me so much are listed as coauthors in the chapters. Prominent among them are Robert Kozma, who introduced me to random graph theory; Giuseppe Vitiello, who shared with me elements of quantum field theory; Terence Barrett, who engaged me with electromagnetic field theory; and Ichiro Tsuda, who showed me how far one can go with the mathematics of chaos theory.

Walter J. Freeman

Toward the end of my studies in physics, I had the fortune to meet Horacio Garcia, an Argentinean neurologist, who introduced me to the fascinating realm of brain function. Horacio was my first mentor, the one that passed me his enthusiasm in trying to understand how the brain works and spent endless hours explaining me the art of reading EEG recordings and evoked potentials. Peter Grassberger shaped my knowledge on dynamical systems, chaos theory, statistical mechanics, and signal processing, a baggage of extremely useful tools to study the brain. The big jump to neuroscience was during my time at Caltech thanks to my two mentors, Christof Koch and Richard Andersen, who gave me the foundations to start a career in neuroscience. I have also been very lucky to have excellent collaborators and students, too many to be named, who have made this quest to understand the brain truly a joy.

Rodrigo Quian Quiroga

Contents

Chapter 1
Electroencephalography

1.1 Introduction

Our knowledge about brain function increased dramatically in the last decades due to the development and refinement of several recording techniques. Such advances flourished at different levels, ranging from the study of synaptic activity at the microscopic level to the refinement of brain imaging techniques at a macroscopic level. Modern data acquisition systems and new electrode designs enabled the simultaneous recording from dozens of neurons at a larger scale, and powerful computers allowed more complex simulations and data analysis, thus giving rise to the field of computational neuroscience. A somewhat less spectacular but also remarkable and steady progress has been made at an intermediate mesoscopic level (Freeman 1975, 1999) in the analysis of electroencephalograms (EEGs).

The EEG measures the electrical activity of the brain at different sites of the head, typically using electrodes placed on the scalp. Its main advantages over other recording techniques are its high temporal resolution and the fact that it can be recorded noninvasively (i.e., without the need of a surgery). Due to their relatively low cost, EEG recordings are widely used both in clinical settings and research laboratories. This makes the EEG a very accessible and useful tool, which is particularly interesting for the analysis of high-level brain processes that arise from the group activity of large cell populations. Such processes can be well localized in time or they can be correlated to time varying patterns, like brain oscillations, which are beyond the time resolution of imaging techniques as functional magnetic resonance imaging (fMRI). The caveat of noninvasive EEGs is the fact that they reflect the average activity of a large number of sources far from the recording sites and, therefore, they do not have an optimal spatial resolution.

Although the way of recording EEG signals did not change as much as in the case of microscopic and macroscopic recordings (though in later chapters, we will describe basic guidelines for electrode designs that improve the spatial analysis of the EEGs), there have been significant advances in the methodology for analyzing

W.J. Freeman and R. Quian Quiroga, *Imaging Brain Function With EEG: Advanced Temporal and Spatial Analysis of Electroencephalographic Signals*,
DOI 10.1007/978-1-4614-4984-3_1,

EEG data. In fact, EEG recordings have been an ultimate challenge for most methods of signal processing due to their high complexity, low signal to noise ratio, nonlinearity, and nonstationarity. As we will describe in this book, the development and implementation of new algorithms that are specifically designed for complex signals such as the EEGs will allow us to get much more information than has been accessible with previous methods and the conventional visual inspection of the recordings, as done by trained electroencephalographers. These methods open a new window to the study of high-level cognitive processes in humans with noninvasive techniques and at no great expense.

1.2 Brief History of EEG

The history of human EEG recordings goes back to Hans Berger (1873–1941), a professor of psychiatry at the University of Jena, Germany. Following the work of Richard Caton (1842–1926), a surgeon from Liverpool who successfully recorded the electrical activity of exposed cerebral hemispheres from monkeys and rabbits in 1875, Hans Berger was the first one able to record electrical activity from the human scalp in 1924. After 5 years collecting data and reexamining his results, he finally published in 1929 "Über das Elektroenkephalogramm des Menschen." In this seminal work, Berger already reported the presence of brain oscillations of about 10 cycles per second, what he called alpha waves, seen with the subject in a relaxed state with eyes closed. When opening the eyes, these waves disappeared (alpha blocking) and oscillations of higher frequencies (beta waves) were observed (Fig. 1.1). A similar type of beta oscillations was also observed with eyes closed when the subjects performed mental arithmetic tasks.

The importance of Berger's work was not recognized until 1934 when Lord Edgar Adrian (1889–1977), at Cambridge, confirmed his results. From then on, the EEG technique triggered a revolution in the way to study normal and pathological brain processes (Fig. 1.2). Just to mention some of the major achievements, in the 30s Grey Walter, first in London and then at the Burden Neurological Institute in Bristol, reported slow oscillations (delta waves) over hemispheric brain tumors and introduced the concept of EEG topography to localize brain lesions. Immediately after, EEG research spread to the USA. At Harvard, Hallowell Davis, Frederic Gibbs, Erna Gibbs, and William Lennox started to study paroxysmal EEG patterns related to epilepsy. These abnormal patterns, such as spikes or spike-waves, are still used to help the diagnosis of epilepsy. The 1940s saw the beginning of sleep studies. At the end of this decade, the first human intracranial recordings were performed. In our days, these types of recordings are mainly used in patients that are candidates to epilepsy surgery in order to determine the origin of the seizures. In the 1950s, Wilder Penfield and Herbert Jasper, at McGill University in Montreal, used electrical stimulation, with open brain surgeries under local anesthesia, to localize areas involved in different brain processes. In the same decade, a major advance in the field was

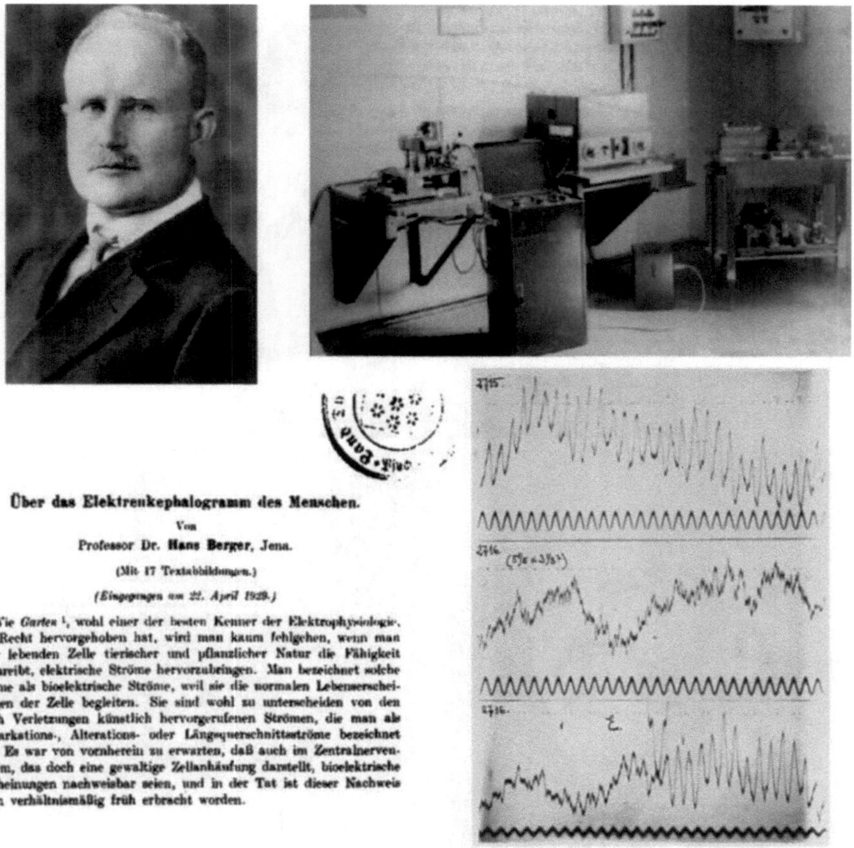

Über das Elektrenkephalogramm des Menschen.

Von

Professor Dr. **Hans Berger**, Jena.

(Mit 17 Textabbildungen.)

(Eingegangen am 22. April 1929.)

Wie *Garten* [1], wohl einer der besten Kenner der Elektrophysiologie, mit Recht hervorgehoben hat, wird man kaum fehlgehen, wenn man jeder lebenden Zelle tierischer und pflanzlicher Natur die Fähigkeit zuschreibt, elektrische Ströme hervorzubringen. Man bezeichnet solche Ströme als bioelektrische Ströme, weil sie die normalen Lebenserscheinungen der Zelle begleiten. Sie sind wohl zu unterscheiden von den durch Verletzungen künstlich hervorgerufenen Strömen, die man als Demarkations-, Alterations- oder Längsquerschnittsströme bezeichnet hat. Es war von vornherein zu erwarten, daß auch im Zentralnervensystem, das doch eine gewaltige Zellanhäufung darstellt, bioelektrische Erscheinungen nachweisbar seien, und in der Tat ist dieser Nachweis schon verhältnismäßig früh erbracht worden.

Fig. 1.1 Hans Berger and his laboratory at the University of Jena (*top*). Berger's seminal paper describing the EEG for the first time, and one of his recordings of alpha and beta oscillations (*upper traces*) and the appearance of alpha oscillations when closing the eyes (*lower trace*)

introduced by George Dawson, in London, who developed a summation technique to visualize average EEG responses to stimuli. Later on, a major breakthrough was the introduction of computers in the analysis of EEG signals, especially with the use of the fast Fourier transform developed by Cooley and Tukey (1965).

A significant slowdown in EEG research resulted as a consequence of the introduction of other methodologies for measuring brain activity, such as single neuron recordings in the 1950s and especially the emergence of imaging techniques and magnetoencephalography in the 1980s. In our days, EEG recordings are generally used for clinical diagnoses, like head injuries, brain tumors, and epilepsy. Neuroscientists also study different types of EEG activity during controlled behavior in human subjects and animals.

Fig. 1.2 Grey Walter (*top*), Lord Edgar Adrian and a picture of his laboratory in Cambridge (*bottom*)

What is the future of electroencephalography given the advances of the new recording techniques? There are three main advantages of EEG over other methods: (1) it is noninvasive and it is therefore possible to do experiments with normal human subjects; (2) it has very high time resolution – of the order of milliseconds – which permits to follow up the temporal dynamics of brain processes; and (3) it is relatively inexpensive. On top of that, there have been significant advances in the development of methods to study complex signals, and most of them are only starting to be used in EEG recordings. As we will see in later chapters, some of these methods offer a new perspective to study EEGs and brain processes in general.

1.3 Recording of EEG Signals

1.3.1 Scalp EEG Recordings

Scalp EEG recordings are performed using high conductance electrodes (i.e., with an impedance of less than 5 kW) placed on top of the head. Electrodes are distributed at specific locations, typically using the so-called 10–20 system, where 16–20 electrodes are separated by 10–20% the total distance around the circumference of the head (although it is becoming more common to use 32, 64,128, or 256 electrodes; see Sect. 10.5). The electrodes are placed on the head with gel in order to increase the conductivity with the skull. Electrodes used to be placed manually one-by-one, but now, they typically come already positioned in a cap, which can be easily fit around the subject's head. This is more practical and less time consuming, considering the possibilities of modern equipment, which allow the simultaneous recording of more than 200 channels.

The EEG can be recorded with reference to a common passive electrode – monopolar (referential) recordings – or it can be recorded differentially between pairs of contiguous electrodes – bipolar recordings. In the later case, there are several ways of choosing the electrode pairs according to montages designed to visualize the propagation of activity in different directions. Some particular montages may be very useful for visualizing the sources of different EEG patterns. This is the case of spikes whose localization, usually given by a polarity inversion of the signal, may help on the study of epileptic patients. It should be noted, however, that the utility of the different montages nowadays is limited to an on-line visualization of the data (which used to be stored in paper), since different derivations can be calculated off-line with the use of computers.

EEG signals are recorded with a sampling frequency of 100 Hz or higher. Modern acquisition systems can easily deal with high sampling rates, and it is now usual to record EEGs with a sampling frequency of 500 Hz or more to enable the study of high-frequency oscillations or fast transitions between the different electrodes.

Figure 1.3 shows the 10–20 electrode distribution (left side) and a typical monopolar recording of a normal subject with eyes open (right side). The reference is the common activity of a pair of linked electrodes placed at the earlobes (A_1 and A_2). The capital letters denote the different electrode locations, F for frontal, C for central, P for parietal, T for temporal, and O for occipital. Odd numbers correspond to left sites and even numbers to right sites, with z denoting the midline. Overall, the EEG recording has a peak-to-peak amplitude of less than 100 mV, which is relatively small in comparison to other type of physiological recordings. In the posterior sites (occipital electrodes, at the bottom of the plot), we observe oscillations of about 10 Hz, which constitute the alpha rhythm. Brain oscillations at different frequencies and localizations have been correlated with several functions, stages and pathologies of the brain and are one of the main building blocks in the analysis of brain activity at the EEG mesoscopic level. In the following sections, we will give more details of their analysis and interpretation. But the analysis of EEG signals is

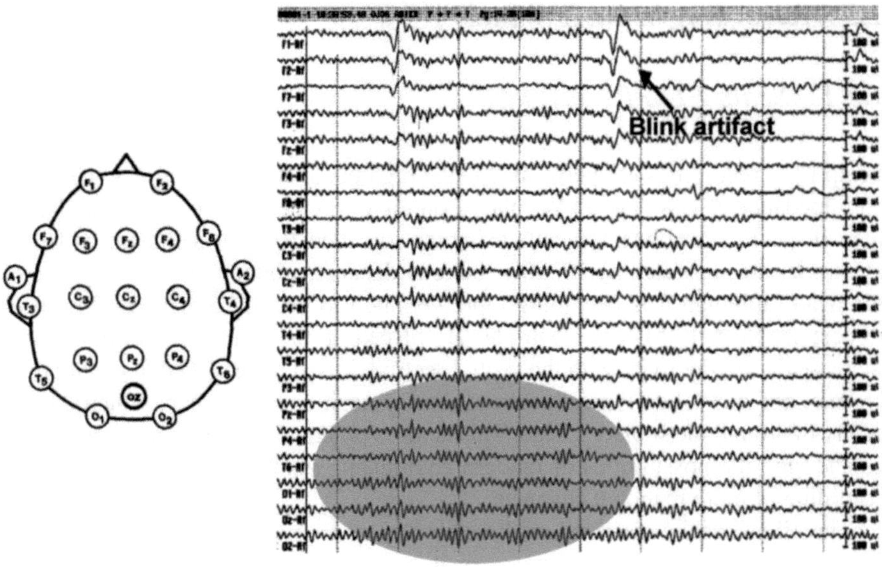

Fig. 1.3 An example of the placement of electrodes according to the 10–20 system (*left*) and a 10 s EEG recording at these locations. Note the presence of a blink artifact in the anterior locations (*top*) and alpha oscillations in the posterior ones (*bottom*), marked with the *grey area*

not limited to brain oscillations. In fact, stereotyped patterns in the EEGs have been also widely studied. Typical examples are the appearance of different type of spikes in the EEG recordings of epileptic patients, or patterns that are characteristic of different sleep stages.

1.3.2 Artifacts

Due to their very low amplitude, EEG signals are easily contaminated by external sources. These "artifacts," inherent of scalp EEG recordings, are produced by head movements, blinking, electrocardiogram, muscle activity, etc. In Fig. 1.3, we see an artifact produced by blinking. To the naïve eye, this may look like real brain activity, but an expert EEG researcher will easily recognize it as a blink artifact due to its morphology and spatial localization in the frontal sites. Eye blinks can be better identified by placing electrodes close to the eyes to measure *electrooculograms*. Other types of artifacts are given by muscle activity. Muscle artifacts can, for example, be generated by a tense posture, which usually correlates with high-frequency activity. These artifacts can in principle be eliminated by using standard digital filters, but unfortunately, in some cases, this is not possible because they overlap with the frequencies of interest. Head movements are correlated with low frequency activity, and in this case, the EEG typically shows a fluctuating baseline.

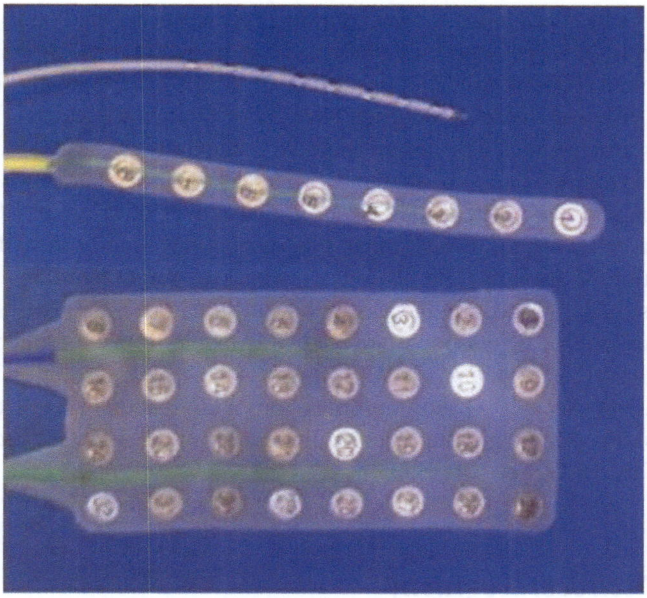

Fig. 1.4 Electrodes for intracranial recordings. From *top* to *bottom*: a depth electrode, a strip and a subdural grid

This activity can be eliminated with a high pass filter (usually set at 1 Hz), but again, such filtering is only adequate when low frequencies are of no interest.

One important point is that artifacts usually limit the length of EEG recordings that can be considered as stationary (i.e., segments in which the main characteristics of the signal, such as its mean, variance, and power spectrum do not change; see Sect. 6.4). Indeed, the relatively short duration of stationary EEG recordings is one of the major challenges for their analysis.

1.3.3 Intracranial Recordings

Scalp EEGs can be recorded noninvasively at a relative low cost and have become a standard diagnostic tool in clinical practice. In very particular cases, electrodes are placed inside the head to perform intracranial recordings. This is done, for example, in patients suffering from epilepsy refractory to medication that are candidates to epilepsy surgery. The goal of intracranial recordings in epileptic patients is to localize precisely the area initiating the epileptic seizures in order to evaluate an eventual surgical resection of the focus. According to the type of seizure and other clinical aspects, two main types of electrodes are used (see Fig. 1.4): (1) deep electrodes, which are used for recording in deep structures such as the hippocampus (an area that is in many cases involved in the generation of the seizures). They are needle-shaped and have usually seven to ten annular recording contacts distributed on its surface; (2) subdural grids, or strips, which are placed on top of the brain and are more suitable for an accurate spatial localization in the cortex. Strips have typically four to eight contacts distributed linearly, and grids have 32 or 64 contacts equally spaced in a rectangular arrangement. The design of grids, particularly in terms of aperture size and interelectrode distances for spatial digitizing, is described in Sect. 7.2.

Since intracranial electrodes are closer to the sources of origin of EEG activity, they have a better spatial resolution, and moreover, they significantly diminish the contamination of artifacts. The obvious drawback of such recordings is that they involve a surgical procedure and are therefore limited to very particular clinical cases. Recordings of ECoGs are more readily available from experimental animals but with nonverbal cognitive skills (Sect. 10.2). The high resolution ECoG in a surgical patient (Sect. 10.4) shows promise for yielding neural correlates of higher cognitive skills, but the greatest value will come from the informed recording and analysis of the EEG (Sect. 10.5).

1.3.4 Electrocorticography

In the previous section, we saw two types of electrodes for intracranial recordings. The depth electrodes have the advantage that they can be introduced through relatively small holes drilled in the skull, thus diminishing the risk of infection. The implantation of subdural grids requires a craniotomy, that is, exposing the surface of the brain on which the grid is placed. This is clearly a more invasive procedure that, however, allows the recording from large cortical structures and it is known as electrocorticography (ECoG). ECoG recordings can be done during surgery or chronically, in which case the epileptic patients have a grid implanted for several days (typically between 1 and 2 weeks) in order to record and study the onset and spread of spontaneous seizures. When recordings are done during surgery, it is possible to move the location of the grid to map epileptogenic activity in different areas. Moreover, in patients under local anesthesia, it is possible to use electrical stimulation to map the function of sensory, motor, and speech areas, in conjunction with the ECoG recordings. This procedure, championed by Penfield and Jasper (1954), has an important clinical relevance since it is therefore possible to determine the precise location of the epileptogenic activity and how it overlaps (or not) with different brain areas, something that it is critical to evaluate the consequences of a surgical resection of the focus. The spatial analysis of the ECoGs necessarily requires placement of arrays through openings large enough to accommodate the aperture of observation. The analysis of the spatial spectrum of the human ECoG shows that an optimal sample is provided by a 1×1 cm array that can be fitted onto a single gyrus (Sect. 7.2). Similar spatial analyses can be also obtained with EEG recordings, but in this case, the spatial patterns may be distorted and degraded by spread through the skull and scalp. In order to know what features to look for in spatial

EEG patterns, it is desirable to first look for these patterns in ECoG recordings. In fact, a major thrust of this book is the description and interpretation of spatial patterns in the ECoG and to extend the analysis and interpretation to the EEG. We show that the realization of the full value of the EEG as a clinical tool hinges on maximizing the resolution of measurements of the EEG in the spatial, temporal, and spectral domains.

1.3.5 Local Field Potentials

We should finally mention the local field potential (LFP) recordings. In animals, microwires are implanted to record the firing of neurons close to the electrode tip. The firing of the neurons is recorded extracellularly using a high sampling frequency (typically more than 15 KHz) and low pass filtering of the data. Due to storage constrains, the recording systems use to store just the time of firing of the neurons (and the shape of the action potentials, to distinguish the activity from different neurons), discarding the rest of the data. However, researchers studying single cell recordings have increasingly recognized the importance of the activity at lower frequencies (which constitute the LFP), and more recent acquisition systems allow the recording and storage of the whole broadband continuous data in order to obtain simultaneous readings of the spiking and LFP activity. We mention the LFPs because they are quite similar in nature to the EEGs and are therefore suitable for the type of analysis to be described in the rest of the book.

1.4 Evoked Potentials

In many scientific fields, especially in Physics and Engineering, one very useful way to learn about a system is by studying its reactions to perturbations. In brain research, it is also a common strategy to see how single neurons or large neuronal assemblies, as measured by the EEG, react to different types of stimuli. Evoked potentials (EPs) are the changes in the ongoing EEG activity due to stimulation. They are also used as well-defined inputs in ECoG studies (Sects. 6.5 and 8.3). They are time locked to the stimulus and have a characteristic pattern of response that is more or less reproducible under similar experimental conditions. They are characterized by their polarity and latency, for example, P100 meaning a positive deflection (P for positive) occurring 100 ms after stimulation. The recording of evoked potentials is done in the same way as the EEGs, with the stimulus delivery system sending triggers to the acquisition system in order to identify the stimuli onsets and offsets.

Evoked potentials can be classified as exogenous and endogenous. Exogenous are the ones elicited by the physical characteristics of the external stimulus, which is typically visual, auditory, or somatosensory. Endogenous EPs are elicited by

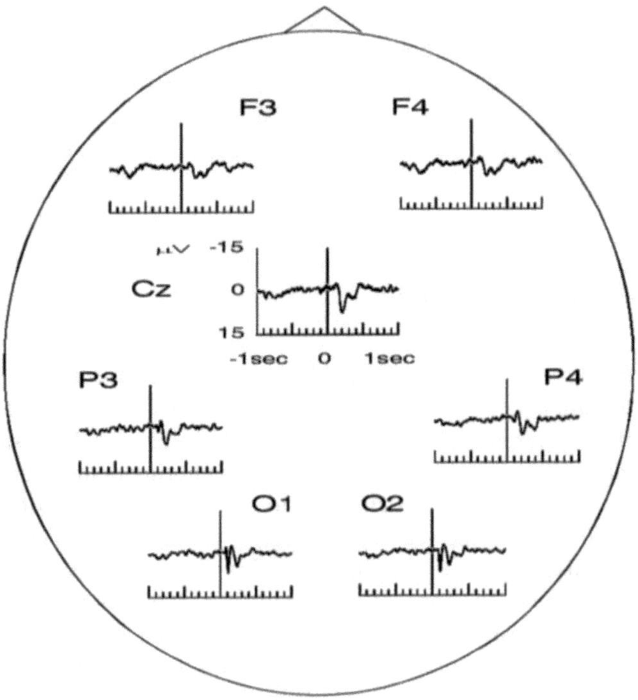

Fig. 1.5 Visual evoked potentials at seven scalp locations. Note a positive deflection (*positive going downwards*) at about 100 ms (the P100), followed by a negative potential at about 200 ms (N200). This P100-N200 response is best defined in the posterior locations

internal brain processes and respond to the significance of the stimulus. Endogenous EPs can be used to study cognitive processes as discussed in the next section.

1.4.1 Visual Evoked Potentials

Visual evoked potentials are usually evoked by light flashes or visual patterns such as a checkerboard. Figure 1.5 shows the grand-average visual evoked potential of ten subjects. Scalp electrodes were placed according to the 10–20 system, with a linked earlobes reference. The stimuli were a color reversal of the (black/white) checks in a checkerboard pattern. There is a positive deflection at about 100 ms after stimulus presentation (P100) followed by a negative rebound at 200 ms (N200). These peaks are best defined at the occipital electrodes, which are the closest to the primary visual area. The P100 is also observed in the central and frontal electrodes but not so well defined and appearing later than in the posterior sites. Visual EPs can be used in clinical practice to identify lesions in the visual pathway, such as the ones caused by optic neuritis and multiple sclerosis (Regan 1989; Celesia 1993).

1.4.2 Auditory Evoked Potentials

Auditory evoked potentials are usually elicited by tones or clicks. According to their latency, they are further subdivided into early, middle, and late latency EPs. Early EPs comprise (a) the electrocochleogram, which reflect responses in the first 2.5 ms from the cochlea and the auditory nerve, and (b) brain stem auditory evoked potentials (BSAEP), which reflect responses from the brain stem in the first 12 ms after stimulation and are recorded from the vertex. BSAEP are seen at the scalp due to volume conduction. Early auditory EPs are mainly used clinically to study the integrity of the auditory pathway (Celesia and Grigg 1993; Picton 1990). They are also useful for detecting hearing impairments in children and in subjects that cannot cooperate in behavioral audiometric studies. Moreover, the presence of early auditory EPs may be a sign of recovery from coma.

Middle latency auditory EPs are a series of positive and negative waves occurring between 12 and 50 ms after stimulation. Clinical applications of these EPs are very limited due to the fact that the location of their sources is still controversial (Picton 1990; Celesia and Grigg 1993). Late auditory EPs occur between 50 and 250 ms after stimulation and consist of four main peaks labeled P50, N100, P150, and N200 according to their polarity and latency. They are of cortical origin and have a maximum amplitude at vertex locations. Auditory stimulation can also elicit potentials with latencies of more than 200 ms. These are, however, responses to the context of the stimulus rather than to its physical characteristics and will be further described in the next section.

1.4.3 Somatosensory Evoked Potentials

Somatosensory EPs are obtained by applying short lasting currents to sensory and motor peripheral nerves and are mainly used to identify lesions in the somatosensory pathway (Erwin et al. 1993). In particular, they are used for the diagnosis of diseases affecting the white matter, like multiple sclerosis, for noninvasive studies of spinal cord traumas and for peripheral nerve disorders. They are also used for monitoring the spinal cord during surgery, giving an early warning of a potential neurological damage in anesthetized patients (Erwin et al. 1993).

1.5 Evoked Potentials and Cognition

Typically, the term evoked potentials refers to EEG responses to sensory stimulation. Sequences of stimuli can be organized in paradigms and subjects can be asked to perform different tasks. Event-related potentials (ERPs) constitute a broader category of responses that are elicited by "events," such as the recognition of a "target" stimulus or the lack of a stimulus in a sequence.

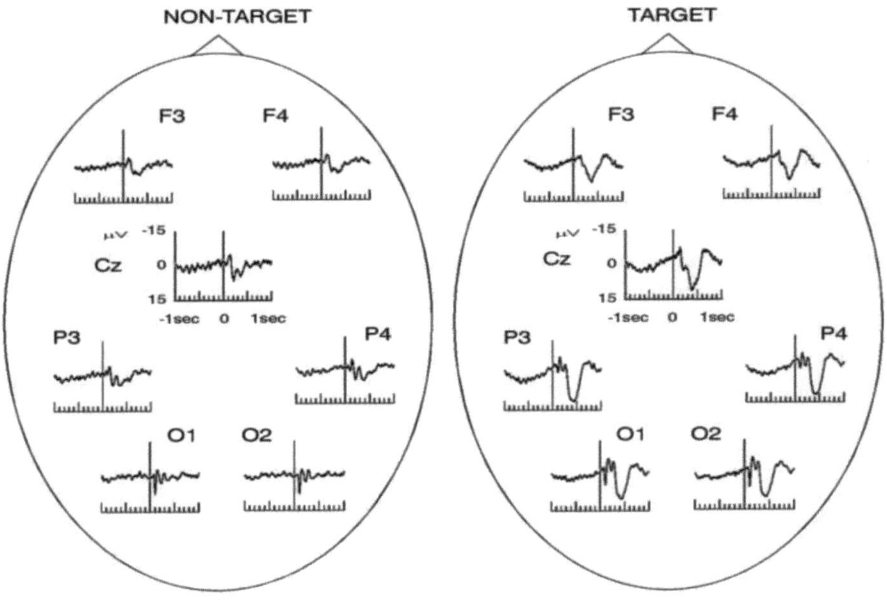

Fig. 1.6 Responses to target and nontarget stimuli using an oddball paradigm. For the target stimuli note the appearance of a large positive response (*positive plotted downwards*) at about 300–500 ms (P300)

1.5.1 Oddball Paradigm and the P300

The most common method to elicit ERPs is by using the oddball paradigm. Two different stimuli are distributed pseudo-randomly in a sequence: one of them appearing frequently (standard stimulus) and the other one being a target stimulus appearing less often and unexpectedly. Standard and target stimuli can be tones of different frequencies, or figures of different colors, different shapes, etc. Subjects are usually asked to count the number of target appearances in a session or to press a button whenever a target stimulus appears.

Figure 1.6 shows grand-average (ten subjects) visual evoked potentials elicited with an oddball paradigm. The left panel shows the average responses to the frequent (nontarget) stimuli and the right one to the targets. The experiment was the same as the one described in Fig. 1.5, but in this case, target stimuli were pseudo-randomly distributed within the frequent ones. Frequent stimuli (75%) were color reversals of the checks, as in the previous experiment, and target stimuli (25%) were also color reversals but with a small displacement of the checkerboard pattern (see Quian Quiroga and Schürmann 1999 for details). The subjects had to pay attention to the appearance of the target stimuli.

The response to the nontarget stimulus is qualitatively similar to the response to visual EPs (where there was no task) shown in Fig. 1.5. As in the case of pattern visual EPs, the P100-N200 complex can be observed both upon nontarget and target stimulation. These peaks are mainly related to primary sensory processing due to the fact that they do not depend on the task, they have a relatively short latency (100 ms), and they are best defined in the primary visual area (occipital lobe). Target stimulation led to a marked positive component, the P300, appearing between 300 and 500 ms and most marked in the central and posterior locations.

While the localization of the P300 in the scalp is well known, the localization of the sources of the P300 in the brain is still controversial (for a review see Molnar 1994). Since the P300 is task dependent and since it has a relatively long latency, it is traditionally related to cognitive processes such as signal matching, recognition, decision making, attention, and memory updating (Picton 1992). There have been many works using the P300 to study cognitive processes (for reviews, see Pritchard 1981; Picton 1992). Abnormal P300 responses can reflect pathologies where cognition is impaired, as it has been shown in depression, schizophrenia, dementia, and others (Picton 1992; Polich 1991).

The P300 can also be elicited using a passive oddball paradigm (i.e., an oddball sequence without any task). In this case, a P300-like response appears upon target stimulation, reflecting the novelty of the stimulus rather than the execution of a certain task. This response has been named P3a. It is earlier than the classic P300 (also named P3b), it is largest in frontal and central areas, and it habituates quickly (Polich 2002).

1.5.2 Mismatch Negativity (MMN)

Mismatch negativity is a negative potential elicited by auditory stimulation. It appears along with any change in some repetitive pattern and peaks between 100 and 200 ms after stimulation (Naatanen et al. 2001). It is generally elicited by the passive (i.e., no task) auditory oddball paradigm, and it is visualized by subtracting the frequent stimuli from the deviant one. MMN is generated in the auditory cortex. It is known to reflect auditory memory (i.e., the memory trace of preceding stimuli) and can be elicited even in the absence of attention (Naatanen 2003). It provides an index of sound discrimination and has therefore being used to study dyslexia (Naatanen 2003). Moreover, it has been proposed as an index for coma prognosis (Kane et al. 1993; Fischer et al. 1999).

1.5.3 Omitted Evoked Potentials

Omitted evoked potentials (OEPs) are similar in nature to the P300 and MMN, but they are evoked by the omission of a stimulus in a sequence (Simson et al. 1976; Ruchkin et al. 1981; Bullock et al. 1994). The nice feature of these potentials is that they are elicited without external stimulation, thus being purely endogenous components. Omitted evoked potentials mainly reflect expectancy (Jongsma et al. 2005) and are modulated by attention (Bullock et al. 1994). The main problem in recording OEPs is the lack of a stimulus trigger. This results in large latency variations from trial to trial, and therefore, OEPs may be difficult to visualize after ensemble averaging.

1.5.4 Contingent Negative Variation (CNV)

The CNV is a slowly rising negative shift appearing before stimulus onset during periods of expectancy and response preparation (Walter et al. 1964). It is usually elicited in tasks resembling conditioned learning experiments. A first stimulus gives a preparatory signal for a motor response to be carried out at the time of a second stimulus. The CNV reflects the contingency or association between the two stimuli. It has been useful for the study of aging and different psychopathologies, such as depression and schizophrenia (for reviews, see Birbaumer et al. 1990; Tecce and Cattanach 1993). Similar in nature to the CNVs are the "Bereitschaft" or "readiness" potentials (Kornhuber and Deeke 1965), which are negative potential shifts preceding voluntary movements (for a review, see Birbaumer et al. 1990).

1.5.5 N400

Of particular interest are ERPs showing signs of language processing. Kutas and Hillyard (1980) described a negative deflection between 300 and 500 ms after stimulation (N400), correlated with the appearance of semantically anomalous words in otherwise meaningful sentences. It reflects "semantic memory," that is, the predictability of a word based on the semantic content of the preceding sentence (Hillyard and Kutas 1983).

1.5.6 Error-related Negativity (ERN)

The ERN is a negative component that appears after negative feedback (Holroyd and Coles 2002; Nieuwenhuis et al. 2004). It can be elicited by a wide variety of reaction time tasks, and it peaks within 100 ms of an error response. It reaches its maximum over frontal and central areas, and convergent evidence from source localization analyses and imaging studies point toward a generation in the anterior cingulate cortex (Holroyd and Coles 2002).

1.6 Basic Analysis of Evoked Potentials

Figure 1.7 shows 16 single-trial visual ERPs from the left occipital electrode of a typical subject. These are responses to target stimuli using the oddball paradigm described in the previous section. Note that it is very difficult to distinguish the single-trial ERPs due to their low amplitude and their similarity to spontaneous fluctuations in the EEG. The usual way to improve the visualization of the ERPs is

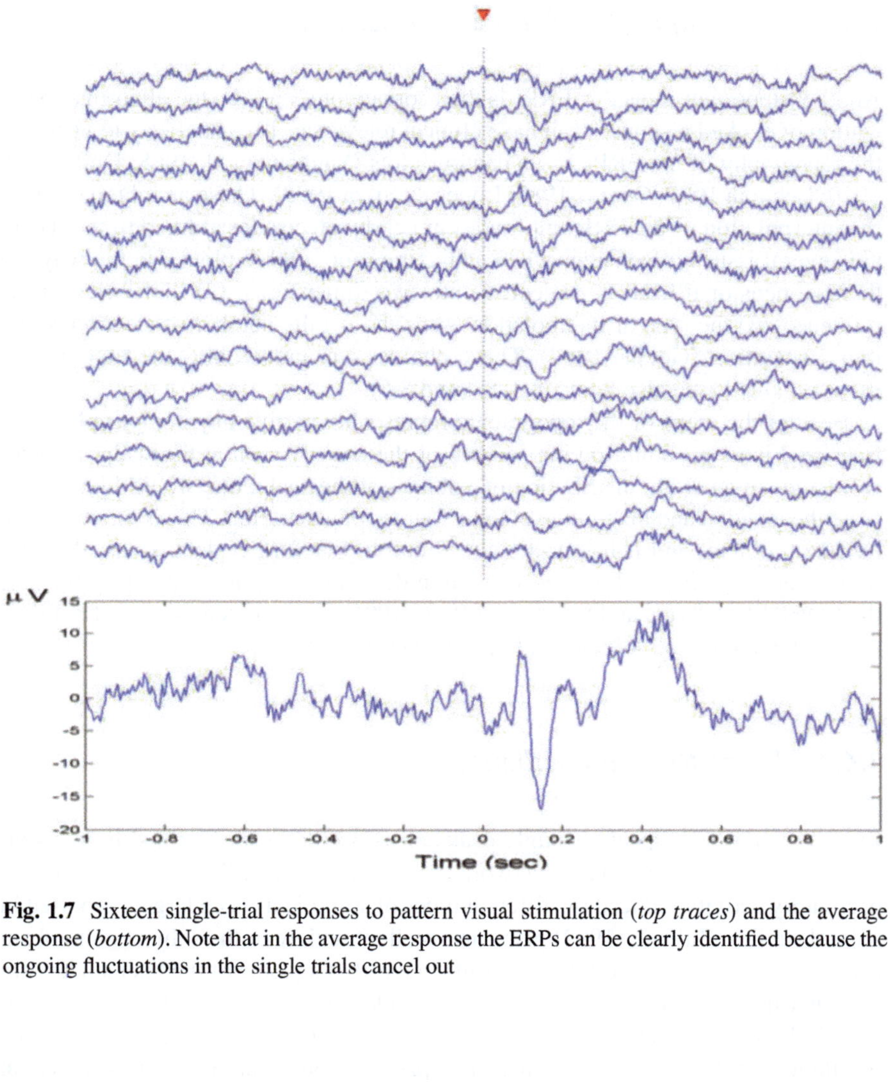

Fig. 1.7 Sixteen single-trial responses to pattern visual stimulation (*top traces*) and the average response (*bottom*). Note that in the average response the ERPs can be clearly identified because the ongoing fluctuations in the single trials cancel out

by averaging the responses of several trials. Since evoked potentials are locked to the stimulus onset (but of course, with some latency variability), their contribution will add, whereas the one of the ongoing EEG will cancel. The bottom plot in Fig. 1.7 shows the average evoked potential. Here, it is possible to identify the P100, N200, and P300 responses described in the previous section.

The main quantification of the average ERPs is by means of the peak amplitudes and latencies. Most research using ERPs compare statistically the distribution of peak amplitudes and latencies between groups of subjects, tasks, or conditions. Such comparisons can be also used clinically, and in general, pathological cases show peaks with long latencies and small amplitudes (Niedermeyer and Lopes da Silva 1993; Regan 1989).

1.6.1 Topography and Source Localization

Another important aspect of ERPs is their topography. In fact, the abnormal localization of evoked responses can have clinical relevance. The usual way to visualize the topography of the ERPs is via contour plots (Vaughan et al. 1968; Duffy et al. 1979; Lehman 1987; Gevins 1987; Lopes da Silva 1993). These are obtained from the interpolation of the ERP amplitudes in nearby electrodes at fixed times. There are several issues to consider when analyzing topographic plots: (1) the way the three-dimensional head is projected onto two dimensions, (2) the choice of the reference, (3) the type of interpolation used, and (4) the number of electrodes and their separation (Gevins 1987). These choices can indeed bias the topographic maps obtained (see also Sects. 8.3, 8.4 and 8.5).

Besides the merit of the topographic representation given by contour plots, the final goal is to get a hint on the sources of the activity seen at the scalp. In other words, given a certain distribution of voltages at the scalp, one would like to estimate the location and magnitude of their sources of generation. This is known as the inverse problem, and it has no unique solution. For further details, see Pascual-Marqui et al. (2002) and references therein describing the use and applications of the LORETA software and Scherg and Berg (1996) and an extensive list of publications using the BESA software at http://www.besa.de).

1.6.2 Event-Related Oscillations

Evoked responses appear as single peaks or as oscillations generated by the synchronous activation of a large network (see Sects. 9.2 and 9.3). The presence of oscillatory activity induced by different types of stimuli has been largely reported in animal studies. Bullock (1992) gives an excellent review of the subject going from earlier studies by Adrian (1942) to results in the 1990s (some of the later studies are included in Basar and Bullock 1992). Examples are event-related oscillations of 15–25 Hz in the retina of fish in response to flashes (Bullock et al. 1991), gamma oscillations in the olfactory bulb after odor presentation in cats and rabbits (Freeman 1975; Freeman and Skarda 1981), and beta oscillations in the olfactory system of insects (Laurent and Naragui 2003; Laurent et al. 1996). Moreover, it has been proposed that these brain oscillations play a role in information processing (Freeman 1975). This idea became very popular after the report of gamma activity correlated to the binding of perceptual information in anesthetized cats (Gray et al. 1989) and humans (Rodriguez et al. 1999).

Intracranial event-related oscillations in animals are quite robust and in many cases visible by naked eye. In humans, this activity is noisier and localized in time. Consequently, more sophisticated time-frequency representations – as the one given by the wavelet transform – are needed in order to precisely localize event-related oscillations both in time and frequency. We finish this section with a cautionary note about event-related oscillations that is particularly important for human studies.

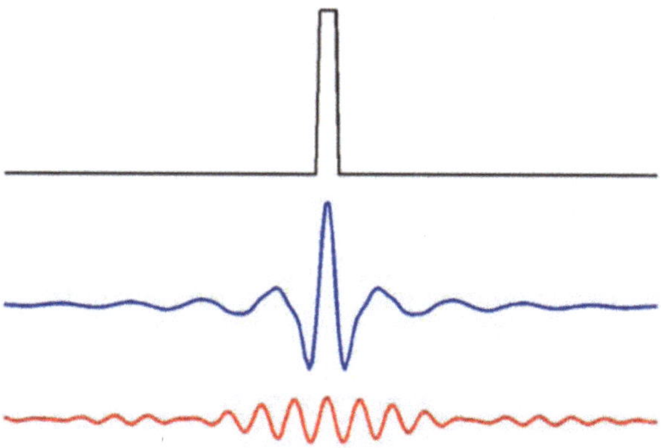

Fig. 1.8 Effect of bandpass filtering a single pulse. Filtering introduces ringing effects, making the pulse look like an oscillation

Since oscillations are usually not clear in the raw data, digital filters are used in order to visualize them. However, one should be aware that digital filters can introduce "ringing effects" and single peaks in the original signal can look like oscillations after filtering. In Fig. 1.8, we exemplify this effect by showing a delta function (upper plot) filtered with a broad-and a narrow-band elliptic filter (middle and lower plot, respectively). Note that the original delta function can be mistaken for an oscillation after filtering, especially with the narrow-band filter (see also Bullock 1992).

References

Adrian ED (1942) Olfactory reactions in the brain of the hedgehog. J Physiol 100:459–473

Basar E, Bullock T (eds) (1992) Induced rhythms in the brain. Birkhauser, Boston

Birbaumer N, Elbert T, Canavan AGM, Rockstroh B (1990) Slow potentials of the cerebral cortex and behavior. Physiol Rev 70:1–41

Bullock TH (1992) Introduction to induced rhythms: a widespread, heterogeneous class of oscillations. In: Basar E, Bullock T (eds) Induced rhythms in the brain. Birkhauser, Boston

Bullock TH, Hofmann MH, New JG, Nahm FK (1991) Dynamics properties of visual evoked potentials in the tectum of cartilaginous and bony fishes, with neuroethological implications. J Exp Zool Suppl 5:142–155

Bullock TH, Karamursel S, Achimowics JZ, McClune MC, Başar-Eroglu C (1994) Dynamic properties of human visual evoked and omitted stimulus potentials. Electroencephalogr Clin Neurophysiol 91:42–53

Celesia GG (1993) Visual evoked potentials and electroretinograms. In: Niedermeyer E, Lopes da Silva F (eds) Electroencephalography: basic principles, clinical applications and related fields. Williams and Wilkins, Baltimore

Celesia GG, Grigg MM (1993) Auditory evoked potentials. In: Niedermeyer E, Lopes da Silva F (eds) Electroencephalography: basic principles, clinical applications and related fields. Williams and Wilkins, Baltimore

Cooley JW, Tukey JW (1965) An algorithm for the machine calculation of complex Fourier series. Math Comput 19:297–301

Duffy FH, Burchfiel JL, Lombroso CT (1979) Brain electrical activity mapping (BEAM): a method for extending the clinical utility of EEG and evoked potential data. Ann Neurol 5:309–321

Erwin CW, Rozear MP, Radtke RA, Erwin AC (1993) Somatosensory evoked potentials. In: Niedermeyer E, Lopes da Silva F (eds) Electroencephalography: basic principles, clinical applications and related fields. Williams and Wilkins, Baltimore

Fischer C, Morlet D, Bouchet P, Luante J, Jourdan C, Salford F (1999) Mismatch negativity and late auditory evoked potentials in comatose patients. Clin Neurophysiol 11:1601–1610

Freeman WJ (1975) Mass action in the nervous system. Academic, New York

Freeman WJ (1999) How the brains make up their minds. Weidenfeld & Nicholson, London

Freeman WJ, Skarda CA (1981) Spatial EEG-patterns, non-linear dynamics and perception: the neo-Sherringtonian view. Brain Res Rev 10:147–175

Gevins AS (1987) Overview of computer analysis. In: Gevins AS, Remond A (eds) Methods of analysis of brain electrical and magnetic signals. Elsevier, Amsterdam

Gray CM, Koenig P, Engel AK, Singer W (1989) Oscillatory responses in cat visual cortex exhibit inter-columnar synchronization which reflects global stimulus properties. Nature 338:334–337

Hillyard SA, Kutas M (1983) Electrophysiology of cognitive processing. Annu Rev Psychol 34:33–61

Holroyd CB, Coles GH (2002) The neural basis of human error processing: reinforcement learning, dopamine, and the error-related negativity. Psychol Rev 109:679–709

Jongsma MLA, Eichele T, Quian Quiroga R, Jenks KM, Desain P, Honing H, VanRijn CM (2005) The effect of expectancy on omission evoked potentials (OEPs) in musicians and non-musicians. Psychophysiology 42:191–2001

Kane NM, Curry SH, Butler SR, Gummins BH (1993) Electrophysiological indicator of awakening from coma. Lancet 341:688

Kornhuber HH, Deeke L (1965) Hirnpotentialanderungen bei Willkurbewegungen und passiven Bewegungen des Menschen. Bereitschaftspotential und reafferente Potentiale. Pfluegers Arch 248:1–17

Kutas M, Hillyard SA (1980) Reading senseless sentences: brain potentials reflect semantic incongruity. Science 207:203–205

Laurent G, Naraghi M (2003) Odorant-induced oscillations in the mushroom bodies of the locust. J Neurosci 14:2993–3004

Laurent G, Wehr M, Davidowitz H (1996) Odour encoding by temporal sequences of firing in oscillating neural assemblies. J Neurosci 16:3837–3847

Lehmann D (1987) Principles of spatial analysis. In: Gevins AS, Remond A (eds) Methods of analysis of brain electrical and magnetic signals. Elsevier, Amsterdam

Lopes da Silva F (1993) EEG analysis: theory and practice. In: Niedermeyer E, Lopes da Silva F (eds) Electroencephalography: basic principles, clinical applications and related fields. Williams and Wilkins, Baltimore

Molnar M (1994) On the origin of the P3 event-related potential component. Int J Psychophysiol 17:129–144

Naatanen R (2003) Mismatch negativity: clinical research and possible applications. Int J Psychophysiol 48:179–188

Naatanen R, Tervaniemi M, Sussman E, Paavilainen P, Winkler I (2001) 'Primitive intelligence' in the auditory cortex. Trends Neurosci 24:283–288

Niedermeyer E, Lopes da Silva F (eds) (1993) Electroencephalography: basic principles, clinical applications and related fields. Williams and Wilkins, Baltimore, pp 1097–1123

Nieuwenhuis S, Holroyd CB, Mol N, Coles MGH (2004) Reinforcement-related brain potentials from medial frontal cortex: Origins and functional significance. Neurosci Biobehav Rev 28:441–448

Pascual-Marqui RD, Esslen M, Kochi K, Lehmann D (2002) Functional imaging with low resolution brain electromagnetic tomography (LORETA): a review. Methods Find Exp Clin Pharmacol 24:91–95

Penfield W, Jasper HH (1954) Epilepsy and the functional anatomy of the human brain. Little, Brown, Boston

Picton TW (1990) Auditory evoked potentials. In: Daly DD, Pedley TA (eds) Current practice of clinical electroencephalography. Raven, New York

Picton TW (1992) The P300 wave of the human event-related potential. J Clin Neurophysiol 9:456–479

Polich J (1991) P300 in clinical applications: meaning, method and measurement. Am J EEG Technol 31:201–231

Polich J (2002) Neuropsychology of P3a and P3b: a theoretical overview. In: Arikan K, Moore N (eds) Advances in electrophysiology in clinical practice and research. Kjellberg, Inc, Wheaton

Pritchard WS (1981) Psychophysiology of P300. Psychol Bull 89:506–540

Quian Quiroga R, Schürmann M (1999) Functions and sources of event-related EEG alpha oscillations studied with the Wavelet Transform. Clin Neurophysiol 110:643–655

Regan D (1989) Human brain electrophysiology. Evoked potentials and evoked magnetic fields in science and medicine. Elsevier, Amsterdam

Rodriguez E, George N, Lachaux JP, Martinerie J, Renault B, Varela FJ (1999) Perception's shadow: Long-distance synchronization of human brain activity. Nature 397:430–433

Ruchkin DS, Sutton S, Munson R, Silver K, Macar F (1981) P300 and feedback provided by absence of the stimulus. Psychophysiology 18:271–282

Scherg M, Berg P (1996) New concepts of brain source imaging and localization. Electroencephalogr Clin Neurophysiol Suppl 46:127–137

Simson R, Vaughan HG Jr, Ritter W (1976) The scalp topography of potentials associated with missing visual or auditory stimuli. Electroencephalogr Clin Neurophysiol 40:33–42

Tecce JJ, Cattanach L (1993) Contingent negative variation (CNV). In: Niedermeyer E, Lopes da Silva F (eds) Electroencephalography: basic principles, clinical applications and related fields. Williams and Wilkins, Baltimore, pp 1097–1123

Vaughan HG Jr, Costa D, Ritter W (1968) Topography of the human motor potential. Electroencephalogr Clin Neurophysiol Suppl 27:61–70

Walter WG, Cooper R, Aldridge VJ, McCallum WC, Winter AL (1964) Contingent negative variation. An electric sign of sensorimotor association and expectancy in the human brain. Nature 203:380–384

Chapter 2
Frequency Analysis

2.1 Introduction

Jean Baptiste Joseph Fourier (1768–1830), a brilliant French mathematician, had the grace (or disgrace) to live at the time of Napoleon's conquest of the civilized world. He joined Napoleon's expedition to Egypt as scientific advisor, later becoming an Egyptologist and administrator for Napoleon's government. It was during his time as prefect in Grenoble when he did his major work on heat conduction. It took him, however, nearly two decades to publish this work, mainly due to the proposal of a novel – and at the time controversial – way to decompose periodic signals into weighted sums of sine and cosine functions. This decomposition, in our days known as Fourier series, has been his major contribution to science, largely transcending its original application to heat conduction.

Following Fourier ideas, signals as the ones recorded from scalp EEG surface electrodes can be represented in the time domain or alternatively in terms of their decomposition into sines and cosines in the frequency domain. Take for example the oscillatory signal of Fig. 2.1a and suppose you want to transmit it to somebody else. You could in principle dictate all the time points of the sinusoid one by one, or alternatively you can just say that it is a sinusoid with a frequency of 10 Hz (i.e., a cycle repeating itself every 100 ms), as represented in the frequency plot of the lower left panel. These two views seem analogous, though you may also say that the frequency representation appears to be more compact and simple. Take now the example of Fig. 2.1b on the right hand side. It is quite hard to get an understanding of this signal from the time representation in the upper plot. However, the frequency representation in the lower plot gives a good grip of its nature: it is just the superposition of three sinusoids of different frequencies. This simple example illustrates the idea of why we use frequency representations. Basically, we try to get a simpler picture of some of the basic characteristics of the signal, which are usually not obvious from noisy and complex time representations.

The frequency representation of a signal is given by its Fourier Transform, which has innumerable applications in different scientific disciplines. In the specific case

W.J. Freeman and R. Quian Quiroga, *Imaging Brain Function With EEG: Advanced Temporal and Spatial Analysis of Electroencephalographic Signals*, DOI 10.1007/978-1-4614-4984-3_2, © Springer Science+Business Media New York 2013

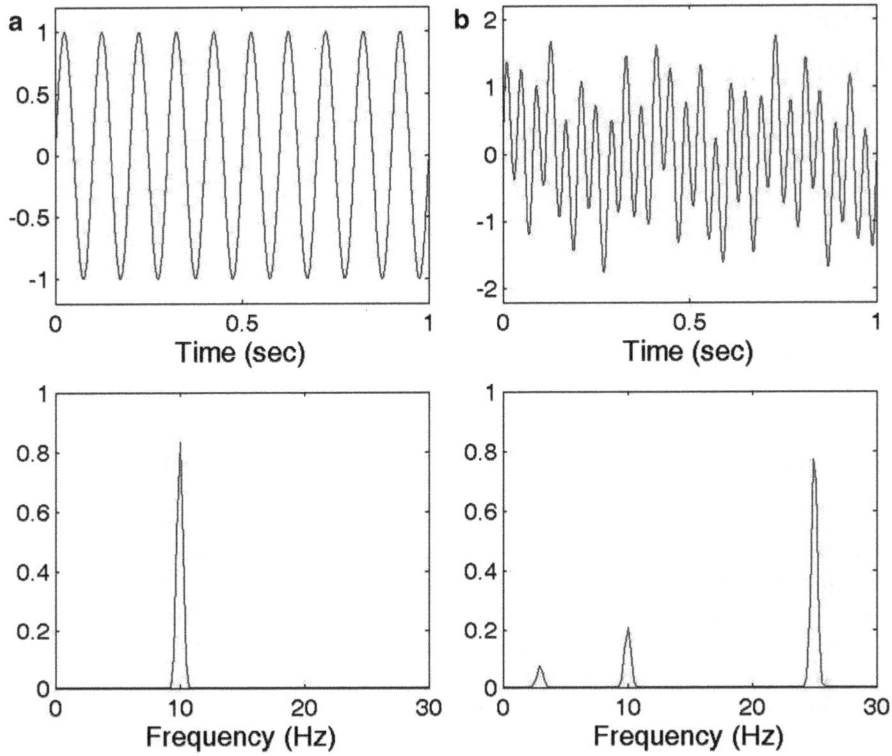

Fig. 2.1 (**a**) A sinusoidal signal in the time (*upper plot*) and frequency (*bottom plot*) domains. (**b**) A quasi-periodic signal. In this case, the Fourier Transform gives a simpler representation

of EEG signals, it is by far the most used quantitative tool, especially after the introduction of a very efficient and fast algorithm to calculate it, the Fast Fourier Transform (FFT; Cooley and Tukey 1965). In the next sections we will describe the basic ideas of the Fourier Transform and its implementation to the analysis of EEG signals together with some applications.

2.2 The Continuous Fourier Transform

There are four different types of Fourier Transforms, depending on whether the signal is continuous or discrete and on whether it is periodic or not. The derivations of these four transforms can be found in mathematical textbooks (see e.g., Oppenheim and Schaffer 1999). Here we focus on the general case of non-periodic signals, starting in this section with the continuous Fourier Transform and its basic properties.

The *continuous* Fourier Transform of a function *x(t)* is defined as[1]:

$$X(\omega) = \int_{-\infty}^{+\infty} x(t) e^{-j\omega t} dt \tag{2.1}$$

Where $e^{-j\omega t} = \cos \omega t - j \sin \omega t$ are complex exponentials and ω is the angular frequency related to the linear frequency f by $\omega = 2\pi \cdot f$. Equation 2.1 quantifies the amount of activity at each frequency ω of the original signal. The inverse Fourier Transform is defined as:

$$x(t) = \frac{1}{2\pi} \int_{-\infty}^{+\infty} X(\omega) e^{j\omega t} d\omega \tag{2.2}$$

and it gives back the original signal *x(t)* expressed as a sum (or an integral to be precise) of sine and cosine functions of different frequencies, weighted by the Fourier coefficients $X(\omega)$. Note the symmetry of Eqs. 2.1 and 2.2, in the sense that one can exchange *x(t)* by $X(\omega)$ just by changing the sign of the complex exponential and adding a normalization factor.

The Fourier Transform can be seen as the correlation between the signal *x(t)* and the complex sinusoidal functions $e^{-j\omega t}$:

$$X(t) = \left\langle x(t), e^{-j\omega t} \right\rangle \tag{2.3}$$

This gives a very intuitive idea of the Fourier Transform. Indeed it is just the 'matching' between the original signal *x(t)* and complex exponentials (or sine and cosine functions, if you prefer) of different frequencies.

2.3 The Discrete Fourier Transform

Digital signals have a finite length and are sampled with a certain sampling frequency. This finite length and sampling introduces several problems that we will discuss in this and the following sections.

Let us consider a discrete signal $x[n]$ $n = 1, \ldots, N$, which has been derived from a continuous signal *x(t)* by sampling at equal time intervals Δt (i.e. with a sampling frequency $f_s = \frac{1}{\Delta t}$). Obviously, the length of the signal is $T = N * \Delta t$. Analogous to the continuous case (Eqs. 2.1 and 2.2), the *discrete* Fourier Transform is defined as:

$$X[k] = \sum_{n=0}^{N-1} x[n]\, e^{-j2\pi kn/N} \qquad k = 0, \ldots, N-1 \tag{2.4}$$

[1] To simplify convergences issues (see Mallat 1999; Strang and Nguyen 1997; Chui 1992 for details) let us just state that the Fourier transform exists for *absolutely integrable* functions; i.e. $\int |x(t)|\, dt < \infty$, or in mathematical notation $x(t) \in L^1(\mathrm{R})$. For discrete signals we require that $\sum |x[n]| < \infty$, which is always fulfilled by real signals since they have a finite length. We can recover the original signal using the inverse Fourier Transform (2.2) if $X(\omega) \in L^1(\mathrm{R})$.

and the signal *x[n]* can be reconstructed with the inverse discrete Fourier Transform:

$$x[n] = \frac{1}{N}\sum_{n=0}^{N-1} X[k]\, e^{j2\pi kn/N} \tag{2.5}$$

The Fourier coefficients $X[k]$ are complex numbers that can be represented in Cartesian or polar forms, as:

$$X[k] = X_R[k] + jX_I[k] = |X[k]|\, e^{j\phi}, \tag{2.6}$$

where X_R and X_I denote the real and imaginary parts in the Cartesian representation, and $|X[k]|$ and ϕ denote the amplitude and phase in the polar representation. If we consider only real sequences $x[n]$, it can be easily shown that $X[k] = X^*[N-k]$ (where * denotes complex conjugation). Then, the Fourier Transform gives a total of *N/2* independent complex coefficients; that means *N* independent values. Since we can reconstruct a signal of *N* data points from the same number of independent Fourier values, the Fourier Transform is non-redundant.

From the time series $x[n]$ the discrete Fourier Transform gives the activity at frequencies f_k, with

$$f_k = \frac{k}{N\Delta t} \tag{2.7}$$

Clearly, the frequency resolution will be given by:

$$\Delta f = \frac{1}{N\Delta t} = \frac{1}{T} \tag{2.8}$$

According to the *Shannon Sampling Theorem* (Mallat 1999), the *Nyquist frequency* is defined as the highest frequency that can be resolved with a sampling period Δt:

$$f_N = \frac{1}{2\Delta t} = \frac{f_s}{2} \tag{2.9}$$

and it corresponds to $k = N/2$ in Eq. 2.7.

Note from Eq. 2.8 that we can increase the frequency resolution by increasing the signal length *T*. For a given signal length, decreasing the sampling period Δt does not change the frequency resolution, but the Nyquist frequency.

2.4 Aliasing

Let us illustrate the idea of aliasing with the example of Fig. 2.2. Suppose we sample a continuous sinusoidal signal with a relatively large sampling period Δt. Since our sampling is too sparse, we will not be able to resolve the underlying sinusoidal

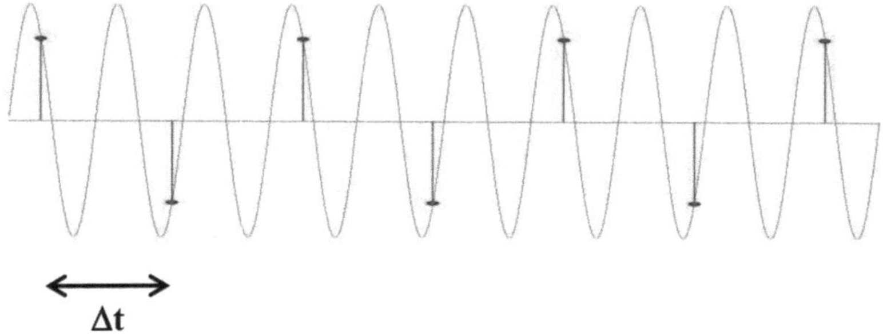

Fig. 2.2 Illustration of aliasing: An inadequate (*sparse*) sampling of the signal introduces spurious low frequency oscillations

signal and, even worse, we will see a slow oscillation that was not present in the original signal (just follow the markers of the digital samples). This effect is called aliasing: the introduction of spurious low frequencies due to an inadequate sampling of the signal.

Intuitively, to resolve a given oscillation we need at least two data points per period or in other words, the sampling frequency should be at least two times the frequency of the signal. This is just another way to state the Shannon Sampling Theorem and the Nyquist frequency of Eq. 2.9.

In real cases, such as for EEG recordings, we do not have a sinusoid of a given frequency for which we can set an appropriate sampling rate. Far from it, we have noisy signals with activity in different frequency bands and the sampling rate is set by our recording system. Then, in order to avoid aliasing, we have to guarantee that the Shannon Sampling Theorem is verified, namely, that the maximum frequency of the signal f_{max} fulfills $f_{\max} < \frac{f_s}{2} = \frac{1}{2\Delta t}$. This is achieved by using low pass 'anti-aliasing' filters. It has to be remarked that anti-aliasing filtering has to be performed by hardware before the digitization of the signal. Once the signal has been digitized, there is no way to get rid of aliasing!

2.5 Fast Fourier Transform

The calculation of the discrete Fourier Transform with Eq. 2.4 requires N^2 complex multiplications, because for each of the N discrete frequencies k we have to calculate a sum of N multiplications with complex exponentials. This may take too long for large N but, fortunately, it is possible to reduce dramatically the computation speed by using the Fast Fourier Transform algorithm (FFT; Cooley and Tukey 1965). The introduction of the FFT has revolutionized the analysis of digital signals and, in particular, it boosted the study of EEGs in the frequency domain. A detailed description of the FFT algorithm is outside the scope of this book and can be found in most signal processing textbooks (see e.g., Oppenheim and Schaffer 1999).

The basic idea is to avoid redundancies given that in Eq. 2.4 we end up calculating the same multiplications several times. In particular, the complex exponential of Eq. 2.4 is periodic and the permutation of n and k give the same result. So, it is possible to reduce the number of calculations to be done. If N is a power of 2 (e.g., 64, 128, 256, …), it can be shown that the original N-point discrete Fourier Transform can be expressed in terms of two $N/2$-point transforms. Since the computing time goes as $O(N^2)$ this results in a faster processing time. Even better, each of the two $N/2$-point transforms can also be expressed in terms of $N/4$-point transforms and so on until we are left with 2-point sequences. It can then be shown that the computing time in this case is of the order $N \log_2 N$, which is clearly faster than N^2. The difference in processing time becomes critical for large datasets. For example, for 64 data points the FFT is about ten times faster than the direct calculation of Eq. 2.4, and for a million data points (just about half an hour recording of one channel with a sampling frequency of 500 Hz) the FFT is over 50,000 times faster!

2.6 Power Spectrum

From the complex Fourier coefficients $X[k]$ of Eq. 2.4 we can define the *periodogram* as:

$$I_{xx}[k] = |X[k]|^2 = X[k] \cdot X^*[k] \tag{2.10}$$

Considering that the signal is a stationary stochastic process, the periodogram is a raw estimation of the *power spectral density* of the signal (the power spectrum).

In Sect. 2.3 we stressed that the Fourier Transform is non-redundant. This means that if we have a real signal with N data points, the Fourier Transform gives N independent values (or $N/2$ complex coefficients) from which we can get back the original signal. No information is gained or lost. This is true both for linear and nonlinear signals. However, it is well known that the Fourier Transform is only suited for linear signals and cannot characterize nonlinear patterns. How can this be?

Recall Eq. 2.6 where we showed that the Fourier coefficients can be written in polar form in terms of an amplitude and a phase. A stationary nonlinear signal, say a sequence of epileptic spikes, is represented in the Fourier domain as a sum of sinusoids, each of them added with a particular phase to reproduce the nonlinear spike shapes. But if we disregard the phases, we loose critical information that characterizes the nonlinear pattern of the original signal (i.e. the spikes). Now look again at Eq. 2.10. It is just the square of the amplitude of the Fourier coefficients defined in Eq. 2.6. The problem is that we usually look only at the power spectrum of the signal and we disregard the phase. This is the reason why we loose information about nonlinear structures with the Fourier Transform. But even if we keep the phase information, the representation of nonlinear patterns as sums of sinusoids at particular phases seems quite cumbersome. In practice, we use the Fourier Transform to extract the linear characteristics of the signals and we turn to other methods to study nonlinear processes.

2.7 Leakage and Windowing

We mentioned that the periodogram is a raw estimate of the power spectrum. Let us illustrate this with the example of Fig. 2.3. The sinusoid on the left has an exact number of cycles in the 0.5 s period of the signal and its periodogram gives a single peak at 6 Hz. The sinusoid on the right, on the other hand, has a non-integer number of cycles in the period considered and its periodogram gives an activity that is spread between 2 and 8 Hz. This smearing of the power spectrum estimation is called leakage.

To understand where leakage comes from we first need to realize that every real signal has a limited duration and that when we calculate the discrete Fourier Transform we make the implicit assumption that the signal repeats itself periodically outside the time range in which it has been recorded. It will take us too long to demonstrate this, but the basic idea is that discretizing the signal (as we do by sampling it) imposes that the Fourier Transform will be periodic, and discretizing

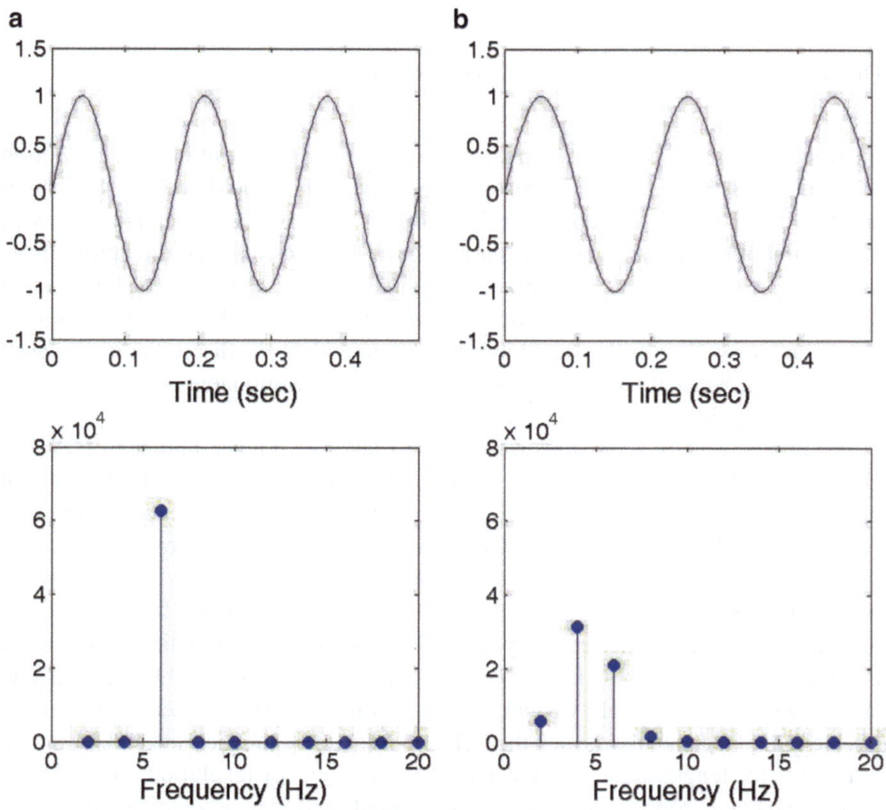

Fig. 2.3 Example of leakage. The sinusoid in (**a**) has an integer number of cycles and its power spectrum gives a single peak at 6 Hz. The sinusoid in (**b**) has a non-integer number of cycles and its power spectrum is smeared around 5 Hz

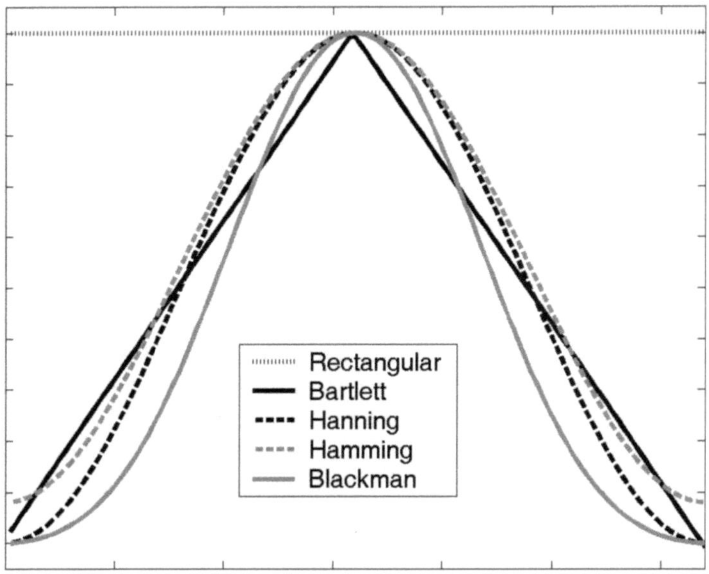

Fig. 2.4 Taper windows used to diminish leakage effects

the frequencies (as we also have to do, since we cannot get a continuous distribution of frequencies for real data) imposes periodicity in the time domain (Oppenheim and Schaffer 1999). If we repeat the signal in Fig. 2.3a over and over again, we will have a smooth sinusoid, since the starting point is exactly the continuation of the final one. On the contrary, if we repeat the signal on Fig. 2.3b we will be introducing discontinuities. These discontinuities are the ones causing leakage. In other words, if we want to synthesize the signal in Fig. 2.3b, including the discontinuities caused by repetition, we will have to use in principle all the components of the spectrum and especially those between 2 and 8 Hz.

A simple way to avoid this would be to take an integer number of cycles. However, real signals have activity at different frequencies and it is in general not possible to define a single periodicity. An alternative approach to avoid these discontinuities is by tapering the borders of the signal using an appropriate window function. This procedure is known as *windowing*. There is, however, a tradeoff when windowing because on the one hand, it diminishes leakage effects but on the other hand it also decreases the frequency resolution. A precise mathematical formulation is outside the scope of this book (see Oppenheim and Schaffer 1999; Jenkins and Watts 1968), but intuitively we can see that a strong tapering decreases the 'effective' length of the signal in which the different frequencies are defined and, as shown in Eq. 2.8, the length of the signal determines its frequency resolution. Several windows have been proposed to optimize this tradeoff and their advantages and disadvantages depend on the application. Among these, the most popular windows are the Bartlett, Hanning, Hamming and Blackman, as shown in Fig. 2.4. For a comprehensive review of the properties of these windows we refer to Oppenheim and Schaffer (1999) and Jenkins and Watts (1968).

2.8 Variance of the Power Spectrum: Periodogram Averaging

It can be shown that, besides the problem of leakage, the periodogram is not a statistically consistent estimate of the power spectrum because its variance does not approach zero with increasing data length (Oppenheim and Schaffer 1999; Dumermuth and Molinari 1987). Furthermore, for large data sets the periodogram tends to vary rapidly with frequency. These variations arise from the estimation process per se and result in a 'random looking' power spectrum. To get a smoother estimate, Bartlett proposed to average several periodograms (Bartlett 1953). This can be done by dividing the dataset into a number of segments, calculating the periodogram for each segment and then averaging the periodograms. This method was further developed by Welch who showed that better estimates are obtained by using half-overlapping windows (Welch 1967). Periodogram averaging, known as the Bartlett or Welch method, makes the power spectrum smother and reduces its variability (which now tends to zero for large data). It can also be shown (Oppenheim and Schaffer 1999) that this averaging procedure is equivalent to a smoothing of the original periodogram with a spectral window. Another advantage is that, due to the averaging that is involved, it is possible to estimate error bars and confidence intervals.

Periodogram averaging copes with the problem of variability of the power spectrum estimation but, as usual, there is no free lunch! Again we face a tradeoff. On the one hand, the more segments we use for averaging the smaller the variability and the smoother the power spectrum will look. But on the other hand, the more segments we use, the less number of data points per segment and, consequently, the lower the spectral resolution (see Eq. 2.8).

2.9 Practical Remarks for Estimating the Power Spectrum of EEG Signals

From the discussions of the previous sections we may conclude that the design of an optimal frequency analysis of EEG signals is more a sort of art than a standardized procedure. Indeed there are many tradeoffs and limitations we should be aware of. Having said this, there are many common situations we face over and over again when doing a frequency analysis of EEG signals. Therefore, we can set some general guidelines for its implementation. Let us deal with each problem one at a time:

- *Sampling rate*: It should be at least two times the maximum frequency of interest in the EEG signal. In our days it is relatively common (and not expensive) to have recording systems with sampling frequencies of 500 Hz or higher. With 500 Hz it is possible to study frequencies of up to 250 Hz, which should be enough for most applications.[2]

[2] For cases like the study of fast propagating seizures, the analysis of early evoked potentials, or the analysis of fast ripples, a sampling of 1,000 Hz is the minimum admissible given that it is crucial to have a resolution of the order of a millisecond or less.

- *Aliasing*: Ideally it should be dealt with by the acquisition system with a low-pass 'anti-aliasing' filter set in the hardware. If this is not possible, the signal should be oversampled when recording (e.g., with 1,000 Hz), then low pass filtered and decimated.
- *Power spectrum estimation*: In order to reduce the variability of the power spectrum and wash out a 'noisy-looking' appearance, we can use the periodogram averaging method described in Sect. 2.8. This approach also helps to cope with other problems such as stationarity and artifacts as we will see in the next points. The number of segments to be used is determined by the length of the dataset and it should be in principle more than 10 (ideally 30 or more, depending on the application).
- *Segment length*: The length of the segments used for the average periodogram determines the frequency resolution. The larger the segments the better the frequency resolution. However, we want to use segments that are not too long, so they do not include artifacts, and the signal can be considered stationary to a first approximation. In practice, segments of 2 s seems appropriate, thus giving a frequency resolution of 0.5 Hz (see Eq. 2.8). Due to details of the implementation of the Fast Fourier Transform algorithm, to increase computational speed it is also desirable that the length of the segments is a power of 2 (e.g., 64, 128, 256, etc.). In search for spatial images of amplitude and phase in the EEG and ECoG (part II of this book) the optimal window duration may be reduced to the range of 0.1 s, which gives poor frequency resolution but excellent temporal resolution for the calculation of the amplitude and phase. The coarse graining of frequency turns out to be an advantage, because it facilitates tracking of EEG and ECoG images with frequency modulation (FM).
- *Leakage*: To diminish leakage we can taper each of the segments used for periodogram averaging with e.g., a Hanning window.
- *Stationarity*: One of the main assumptions we make to estimate the power spectrum of EEGs is that they can be treated as stationary stochastic signals. If they are not stationary, then the spectrum may be meaningless (see Figure 3.1). For periodogram averaging we must then assume that all segments correspond to the same stochastic process. This can be actually checked by observing the variability of the periodograms and any particular trend. An obvious pitfall would be to include segments corresponding to different brain states, such as mixing periods of normal and epileptic EEG activity or periods of awake and sleep EEG. Furthermore, we should also check that the signal can be considered stationary within each segment, which imposes a limitation to the segment length.
- *Artifacts*: Due to its large amplitude, artifacts can seriously contaminate the power spectrum. For estimating the power spectrum using the Welch method (i.e. averaging the periodograms of different segments of the signal), it is therefore advisable to select segments that are artifact free. Artifacts can be checked either visually or with advanced artifact detection methods, such as Independent Component Analysis (Jung et al. 2000).

2.10 Applications of EEG Frequency Analysis

2.10.1 EEG Frequency Bands

In the first report of human EEG recordings, Hans Berger already noted the presence of different brain oscillations (Berger 1929). In particular, he reported rhythmic activity of around ten cycles per second, most pronounced in the occipital electrodes with eyes closed. These oscillations, which he named *alpha rhythms*, were dramatically decreased by the influx of light with eyes opening. This effect is what in our days we call *alpha blocking* and it is one of the most dramatic and simplest demonstrations of how the EEG reflects brain processes. We define *reactivity* as the ratio of alpha activity with eyes closed and eyes open. The degree of reactivity varies from subject to subject, but it is generally accepted that a lack of reactivity is an abnormal finding (Niedermeyer 1993). Berger also described oscillatory activity of higher frequencies, which he called the *beta rhythms*. They appeared with eyes open and to some degree also with eyes closed when the subjects performed mental calculations. Following Berger's seminal work, different EEG oscillations and their correlation to different brain states, functions and pathologies had been thoroughly studied, especially after the introduction of digital recordings and the Fast Fourier Transform (Cooley and Tukey 1965). Based mainly on their function and localization, EEG oscillations have been grouped into frequency bands. Here we just give a brief summary of them and we refer to the excellent review of Niedermeyer (1993) for more details.

Figure 2.5 shows an EEG recording of 20 s and its corresponding power spectrum. The vertical lines mark the limits of the standard EEG frequency bands:

- Alpha rhythms (7.5–12.5 Hz): they appear spontaneously in normal adults during wakefulness, under relaxation and mental inactivity conditions. They are best seen with eyes closed, most pronounced in the occipital locations.
- Beta rhythms (12.5–30 Hz): they are best defined in central and frontal locations, with less amplitude than alpha waves. They are enhanced upon mental calculations, expectancy or tension over the entire surface of the scalp (Fig. 2.6 and Fig.10.9.
- Theta rhythms (3.5–7.5 Hz): They are typical during deep sleep. They play an important role in infancy and childhood. In the awake adult, high theta activity is considered abnormal and related to brain disorders, such as epilepsy.
- Delta rhythms (0.5–3.5 Hz): They are also characteristic of deep sleep stages. Depending on their morphology, localization and rhythmicity, delta oscillations can be normal as in slow wave sleep or pathological as in brain tumors.
- Low Gamma rhythms (30–60 Hz in human EEG, 30–80 Hz in animal ECoG): Of minor interest until the 1990s, gamma oscillations became very popular after they have been proposed to play a major role in linking stimulus features into a single perception (binding theory; Gray et al. 1989). Although the validity of the

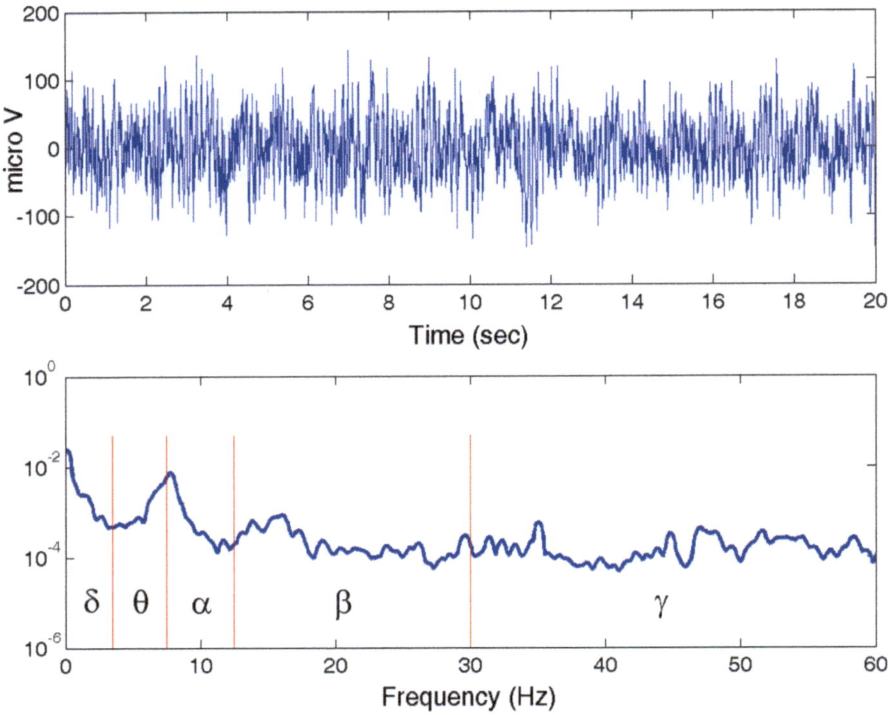

Fig. 2.5 EEG frequency bands

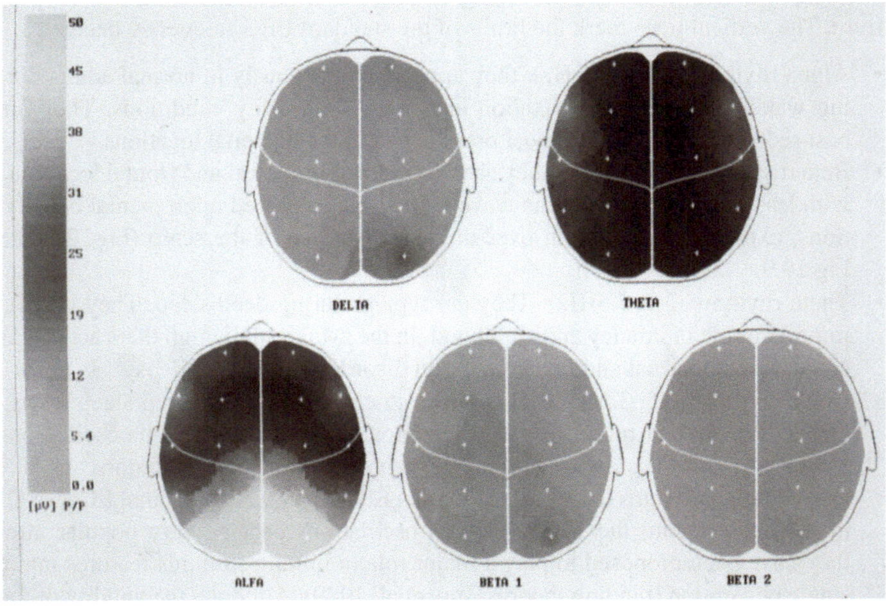

Fig. 2.6 Topographical mapping of the different EEG frequency bands from a normal EEG recording taken with eyes closed

binding theory is still under dispute, several follow up works have shown correlations of gamma activity with different sensory and cognitive processes, notably during visual, auditory, somatic and olfactory perception (see Chaps. 8 and 9) as well as with attention.

- High gamma rhythms (variously defined between 80–120+) also called epsilon rhythms have been found in human and animal ECoG in association with chattering action potentials (Ray and Maunsell 2011)

It should be emphasized that not all EEG oscillations of the same frequency have the same function. For example, delta oscillations are normal during slow wave sleep but are a clear sign of abnormality in awake states, given that 3 Hz spike and wave discharges are a characteristic sign of absence seizures. Mu rhythms have a frequency similar to that of the occipital alpha rhythm, but they are observed in central locations and are related to motor functions. Spindle oscillations have also a periodicity of about 10 Hz, but they are characteristic of sleep stage 2.

Since EEG patterns are quite variable and complex, visual inspection is still one of the preferred ways to analyze EEG recordings by expert electroencephalographers. This is more an art than an exact science and it requires years of training. Clearly, visual inspection is very subjective and a quantitative approach should in principle be preferred. However, it must be said that some training in EEG visual inspection is particularly useful before embarking in automatic and quantitative EEG analysis, at least to have a feeling of what type of patterns and characteristics of the raw EEG we are trying to quantify. Most of quantitative electroencephalography deals with the analysis of the EEG frequency bands described above. Several parameters have been defined to quantify them, such as the relative power between bands (being the most used the alpha/theta ratio), reactivity (ratio between eyes closed/eyes open alpha activity), asymmetry index (the difference between the left and right power), etc. (Nuwer et al. 1994). Moreover, statistical techniques can be used in order to establish normal ranges and their deviations with several pathologies (John et al. 1987).

2.10.2 Topographic Analysis

The information from the different electrodes can be arranged in topographic maps (Gotman 1990; Gevins 1987; Lehmann 1987; Lopes da Silva 1993a). These algorithms usually use linear or quadratic interpolations between the 3 or 4 nearest recording sites. One critical point is the election of the reference, since the use of single electrode references can distort the maps near the reference site (Lehmann 1971, 1987) (see Fig. 10.9). Several suggestions were proposed in order to avoid this distortion, among them the use of averaged references and the use of the average of the derivatives of the EEG signals (Lehmann 1987; Lopes da Silva 1993b). Another important issue to be considered is how to project a three dimensional head into a two dimensional map.

The use of topographic plots started more than 30 years ago (Walter et al. 1966; Lehmann 1971), but it was after the introduction of color topographic maps by Duffy et al. (1979) that they became widely accepted and started to be used in several medical centers. With these plots it is easy to visualize asymmetries and to localize the activity of the different frequency bands. Furthermore, the topographic maps complement the quantitative parameters described in the previous sections for the characterization of normal EEG patterns and the study and diagnosis of several pathologies (Duffy 1986; Maurer and Dierks 1991; Pfurtscheller and Lopes da Silva 1988).

Figure 2.6 shows the topographic map of the EEG recording of a normal subject with eyes closed. Five different frequency bands are plotted. As expected, the power is homogeneously distributed in all the frequency bands, except in alpha, where there is a symmetrical increase in the posterior locations. This increase reflects the presence of the normal spontaneous alpha activity described above.

Before leaving this section we stress that topographic maps give a static picture of the brain activity. Later we will describe a radically different approach to study how EEG activity propagates not only in time but also in space. These are the EEG images of the title of our book. In our chapters we will demonstrate the methods by which we succeed in extracting finely textured patterns of amplitude from the blandly uniform distributions of potential shown in Fig. 2.6 (see Sect. 10.5). The textured images are formed by amplitude and phase modulation (AM and PM) of spatially coherent carrier waves in the beta range. They recur at rates in the theta range as brief epochs that resemble cinematic frames. They contain cognitive information, because they are classifiable with respect to sensory stimuli that human subjects are perceiving as the patterns fly past. Part II of this book is directed to describe how to find these patterns, explain how they are generated, and interpret what they contribute to our understanding of human cognitive neurodynamics.

2.11 Summary

In this chapter we reviewed the basic background of the Fourier Transform and its use in the analysis of EEG signals. One important application is the comparison between the power at different frequency bands and their topological distribution. Normative values have been established and large deviations from them can reflect pathological cases. Moreover, deviations from background values in subjects who are engaged in cognitive tasks may direct us to discover EEG and ECoG correlates of cognition. This analysis is already adapted to many commercial systems and it is used in several medical centers. Although quantitative parameters are very useful and can be easily extracted from the EEG in an automated way, the visual inspection of the recordings should not be left aside, in order to avoid misinterpretations due to non-stationarity, artifacts, etc. In fact, topographic mapping and quantitative values should be considered as a complement and not as a replacement of visual inspection of the EEG.

The frequency analysis allowed by the Fourier Transform has been by far the most useful quantification of EEG activity. It has, however, three main limitations:

1. The Fourier Transform requires stationarity of the signal. For the purpose of estimating the power spectrum, the EEGs can be regarded as quasi-stationary only on the order of a few seconds (Blanco et al. 1995). Obviously, the Fourier Transform is also not well suited for the analysis of transient responses as in the case of Evoked potentials.
2. The Fourier Transform is very accurate at characterizing the frequency composition of a signal, but it gives no time information. In other words, we can very well define the activity at a particular frequency but we can not tell when exactly this frequency occurs and how it evolves in time. This is of course related to the issue of stationarity. It justifies the use of time varying methods, like the short-time Fourier Transform, Wavelets or the Hilbert transform to be described in the next chapters.
3. The Fourier Transform is not optimal to characterize non-linear signals. As we described in Sect. 2.6, non-linear patterns, for example epileptic spikes, are represented in the Fourier domain as complex combinations of different frequencies with precise phase relationships. Since we usually only look at the power spectrum and disregard the phase information, the non-linear nature of the signal is lost. But even if we decide to keep the phases, describing a spike as a sum of sinusoids with a certain phase relationship is very cumbersome and other methods such as wavelets (see Chap. 4) may be preferred.

References

Bartlett MS (1953) An introduction to stochastic processes with special reference to methods and applications. Cambridge University Press, Cambridge, MA

Berger H (1929) Uber das Elektrenkephalogramm des Menschen. Arch Psychiat Nervenkr 87:527–570

Blanco S, García H, Quian Quiroga R, Romanelli L, Rosso OA (1995) Stationarity of the EEG series. IEEE Eng Med Biol 14:395–399

Cooley WJ, Tukey JW (1965) An algorithm for the machine calculation of complex Fourier series. Math Comput 19:297–301

Duffy FH, Burchfiel JL, Lombroso CT (1979) Brain electrical activity mapping (BEAM): a method for extending the clinical utility of EEG and evoked potential data. Ann Neurol 5:309–321

Duffy FH (1986) Topographic mapping of brain electrical activity. Butterworths, London

Dumermuth G, Molinari L (1987) Spectral analysis of EEG background activity. In: Gevins A, Rémond A (eds) Handbook of electroencephalography and clinical neurophysiology, Vol. I: Methods of analysis of brain electrical and magnetic signals. Elsevier, Amsterdam, pp 85–130

Gevins AS (1987) Overview of computer analysis. In: Gevins A, Rémond A (eds) Handbook of electroencephalography and clinical neurophysiology, Vol. I: Methods of analysis of brain electrical and magnetic signals. Elsevier, Amsterdam, pp 31–83

Gotman J (1990) The use of computers in analysis and display of EEG and evoked potentials. In: Daly DD, Pedley TA (eds) Current practice of clinical electroencephalography, 2nd edn. Raven Press, New York, pp 51–83

Gray C, König P, Engel A, Singer W (1989) Oscillatory responses in cat visual cortex exhibit inter-columnar synchronization which reflects global stimulus properties. Nature 338:335–337

Jenkins GM, Watts DG (1968) Spectral analysis and its applications. Holden-Day, San Francisco

John ER, Harmony T, Valdes-Sosa P (1987) The use of statistics in electrophysiology. In: Gevins A, Rémond A (eds) Handbook of electroencephalography and clinical neurophysiology, Vol. I: Methods of analysis of brain electrical and magnetic signals. Elsevier, Amsterdam, pp 497–540

Lehmann D (1971) Multichannel topography of human alpha EEG fields. Electroencephalogr Clin Neurophysiol 31:439–449

Lehmann D (1987) Principles of spatial analysis. In: Gevins A, Rémond A (eds) Handbook of electroencephalography and clinical neurophysiology, Vol. I: Methods of analysis of brain electrical and magnetic signals. Elsevier, Amsterdam, pp 309–354

Lopes da Silva FH (1993a) EEG analysis: theory and practice. In: Niedermeyer E, Lopes da Silva FH (eds) Electroencephalography: basic principles, clinical applications and related fields, 3rd edn. Williams and Wilkins, Baltimore, pp 1097–1123

Lopes da Silva FH (1993b) Computer-assisted EEG diagnosis: pattern recognition and brain mapping. In: Niedermeyer E, Lopes da Silva FH (eds) Electroencephalography: basic principles, clinical applications and related fields, 3rd edn. Williams and Wilkins, Baltimore, pp 1063–1086

Jung T-P, Makeig S, Humphries C, Lee TW, McKeown MJ, Iragui V, Sejnowski TJ (2000) Removing electroencephalographic artifacts by blind source separation. Psychophysiology 37:163–78

Maurer K, Dierks T (1991) Atlas of brain mapping. Springer, Berlin

Niedermeyer E (1993) The normal EEG of the waking adult. In: Niedermeyer E, Lopes da Silva FH (eds) Electroencephalography: basic principles, clinical applications and related fields, 3rd edn. Williams and Wilkins, Baltimore, pp 131–152

Nuwer MR, Lehmann D, Lopes da Silva FH, Matsuoka S, Sutherling W, Vivert JF (1994) IFCN guidelines for topographic and frequency analysis of EEGs and EPs. Report of and IFCN committee. Electroencephalogr Clin Neurophysiol 91:1–5

Oppenheim A, Schaffer R (1999) Discrete-time signal processing. Prentice Hall, London

Pfurtscheller G, Lopes da Silva FH (eds) (1988) Functional brain imaging. Huber, Toronto

Ray S, Maunsell JHR (2011) Different origins of gamma rhythm and high-gamma activity in macaque visual cortex. PLoS Biol 9:e1000610

Walter D, Rhodes J, Brown D, Adey W (1966) Comprehensive spectral analysis of human EEG generators in posterior cerebral regions. Electroencephalogr Clin Neurophysiol 20:224–237

Welch PD (1967) The use of Fast Fourier Transform for the estimation of power spectra: a method based on time-averaging over short modified periodograms. IEEE Trans Audio-Electroacoust 15:70–73

Chapter 3
Time-Frequency Analysis

3.1 Introduction

In the previous chapter we mentioned that one of the main limitations of the Fourier transform is that it does not have time resolution. For calculating the Fourier transform, we assume that the signal is stationary and, consequently, that the activity at different frequencies is constant throughout the whole signal. In many occasions, however, signals have time-varying features that cannot be resolved with the Fourier transform. This is the case of music, speech, animal sounds, radar data, and many other signals (see examples in Cohen 1995). For EEG signals, this limitation is critical when we analyze processes that change in time, such as the response to a particular stimulus or the development of an epileptic seizure.

To illustrate this, let us consider the linear chirp – i.e., a signal with a steadily rising frequency – shown in Fig. 3.1a. Of course, it is not possible to track the varying frequency of the chirp with the Fourier transform. In fact, the Fourier transform integrates the frequency activity along the whole signal, and it gives a broad power spectrum (Fig. 3.1b). This representation is quite misleading because the chirp signal gives the same power spectrum as the one of a broadband random signal.

Intuitively, we can overcome the lack of time resolution of the Fourier transform by chopping the data into pieces and then calculating the power spectrum for each piece or, even better, by using a time-evolving window to focus at different segments of data. In order to avoid leakage (see Sect. 2.7), we can also taper the windowed data with an appropriate function (Cohen 1995; Chui 1992). This procedure is called the *short-time Fourier transform* (STFT) or windowed Fourier transform. If the window used is a Gaussian, it is called Gabor transform, in honor to Denis Gabor, a Hungarian physicist that first developed these ideas to analyze the frequency variations of sounds (Gabor 1946).[1] With the STFT it is possible to track the

[1] Besides this seminal contribution, Gabor is in our days most recognized for being the father of holography, an achievement for which he received the Nobel Prize in Physics in 1971.

W.J. Freeman and R. Quian Quiroga, *Imaging Brain Function With EEG: Advanced Temporal and Spatial Analysis of Electroencephalographic Signals*,
DOI 10.1007/978-1-4614-4984-3_3,

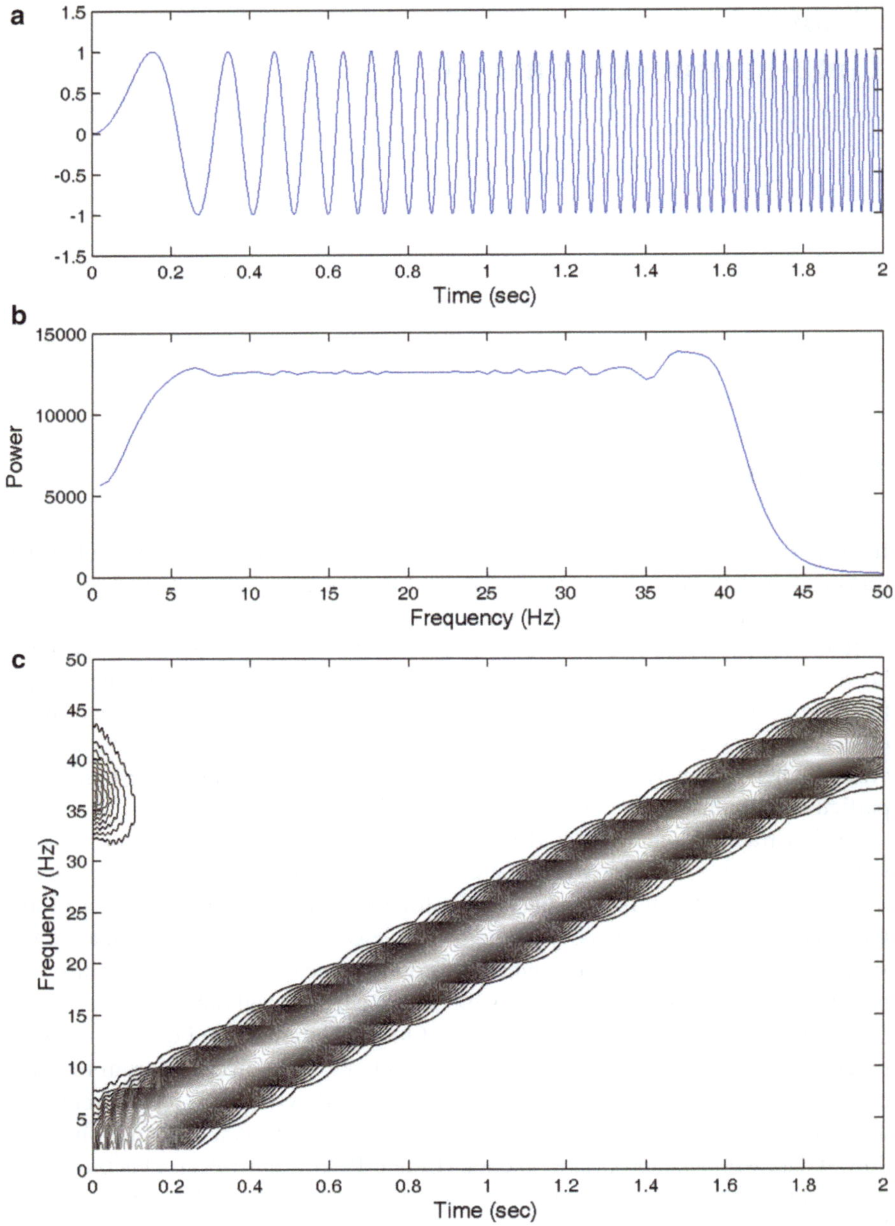

Fig. 3.1 A linear chirp (**a**), its Fourier transform (**b**), and its time-frequency representation obtained with the short-time Fourier transform (**c**)

time evolution of the different frequencies and the stationarity requirement is satisfied by considering that the signals are quasi-stationary within each window (Lopes da Silva 1993; Blanco et al. 1995). Figure 3.1c shows the time-frequency spectrum of the chirp signal obtained with the STFT. In this case, we used a Gaussian window of 0.5s length. As expected, the time-frequency plot shows how the frequency of the signal increases linearly with time.

In the next sections we will give the basic theoretical background of the STFT, and we will illustrate its use in the analysis of scalp EEG recordings during tonic-clonic seizures.

3.2 Short-Time Fourier Transform

Let us now formalize the concepts introduced in the previous section. The STFT of a signal *x(t)* is defined as

$$G_D(\omega,t) = \int_{-\infty}^{+\infty} x(t')g_D^*(t'-t)\, e^{-j\omega t'} dt' \tag{3.1}$$

where * denotes complex conjugation. Note that $G_D(f,t)$ is the same as the Fourier transform (Eq. 2.1) but with the introduction of the window $g_D^*(t'-t)$ of width *D* and center in *t*. The STFT quantifies the activity of the signal around time *t* and frequency ω. As in the case of the Fourier transform, it can also be shown that the signal *x(t)* can be reconstructed from the coefficients $G_D(\omega,t)$ (Mallat 1999). With respect to the particular window g_D to be used, Gabor (1946) proposed to use a Gaussian function:

$$g_\alpha(t) = \left(\frac{\alpha}{\pi}\right)^{1/4} e^{-\frac{\alpha}{2}t^2}, \tag{3.2}$$

given that its Fourier transform is also a Gaussian, thus giving a simultaneous localization in time and frequency. The constant α determines the width of the Gaussian, and it is the main parameter that sets the effective size of the window. Since in the following we will only consider Gaussian windows, we will refer indistinctly to the length of the window *D* (used in Eq. 3.1) or to the width of the Gaussian α. Note that Gaussian functions do not have a compact support; i.e., they extend to plus and minus infinity. However, since they approach asymptotically zero, they can be easily truncated without introducing any major distortion.

The STFT can be expressed as the inner product between the signal *x(t)* and the complex sinusoidal functions $e^{-j\omega t'}$ modulated by the Gaussian window g_α:

$$G_\alpha(\omega,t) = \left\langle x(t'), g_\alpha(t'-t)e^{-j\omega t'} \right\rangle \tag{3.3}$$

Then, it can be seen as the "matching" between the original signal *x(t)* and the oscillatory functions $W_\alpha(\omega,t) = g_\alpha(t'-t)e^{-j\omega t'}$. Figure 3.2 shows the real and imaginary part of the function $W_\alpha(\omega,t)$, called the Gabor or Morlet function.

As in the case of the Fourier transform (see Eq. 2.10) it is possible to define an estimation of the time-varying power spectrum, the *spectrogram*, as

$$I(\omega,t) = |G(\omega,t)|^2 = G^*(\omega,t) \cdot G(\omega,t), \tag{3.4}$$

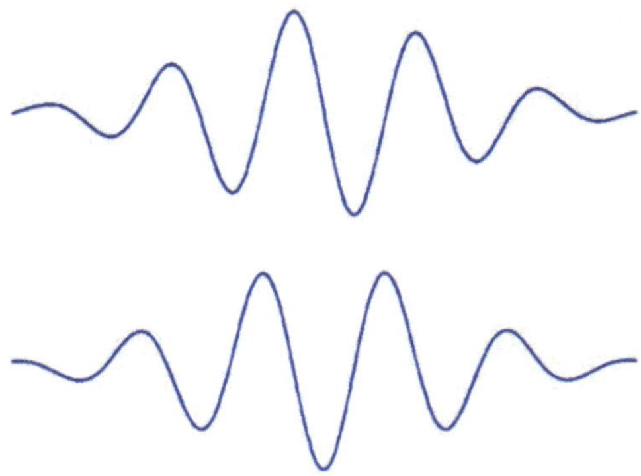

Fig. 3.2 Real (*top*) and imaginary (*bottom*) parts of the function $W_\alpha(\omega,t)$

and we can represent the time-resolved frequencies with contour plots as shown in Fig. 3.1c.

Let us now consider a discrete signal *x[n] (n = 1,…,N)*. The *discrete* STFT is defined as

$$G[k,n]=\sum_{m=0}^{N-1}x[m]g[m-n]e^{\frac{-j2\pi km}{N}}, \tag{3.5}$$

where n and k denote the discrete time and frequency localizations, respectively. The discrete STFT is very redundant because it gives a frequency representation for every time point. Indeed, from a discrete signal with N values, we get a time-frequency representation with a total of *NxD* values (with D the number of data points in the window). In order to decrease this redundancy, a *sampled* STFT can be defined by considering only a subset of all the possible time and frequency values. This decreases the redundancy and saves computation time. However, the price to pay is that the reconstruction of the original signal from the STFT is no longer straightforward (Qian and Chen 1996).

3.3 Uncertainty Principle

We just saw how to get time resolution from the Fourier transform by windowing the data. There is still a critical point to be discussed for the implementation of the STFT. We have to decide how to choose the size of the window D, or more specifically, the rate of decay of the Gaussian function α. The bad news is that this is not so straightforward and, again, a compromise has to be taken. If the window is too narrow, it will give a good time resolution but frequencies will not be well resolved (remember from Eq. 2.8 that the frequency resolution is inversely proportional to the data length). On the contrary, if the window is too large we will have a good frequency resolution, but the time localization will be lost. There is a trade-off between frequency and time resolution. In analogy to Heisenberg's uncertainty

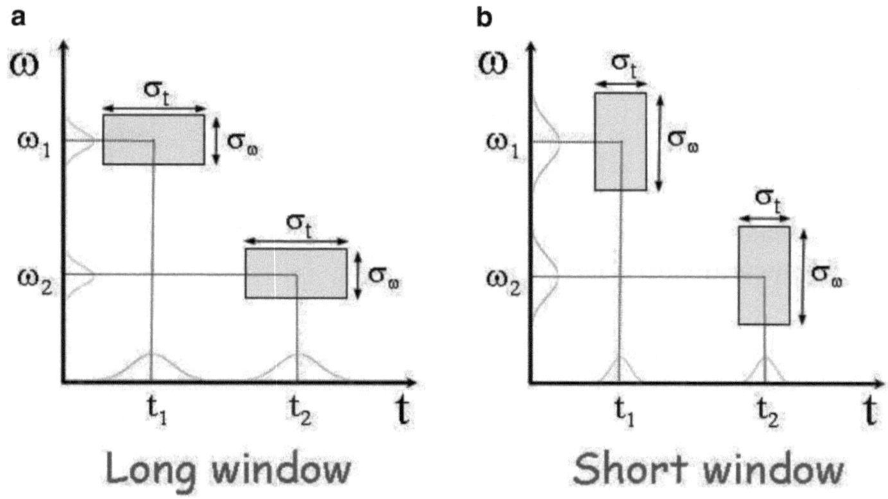

Fig. 3.3 Heisenberg uncertainty boxes with a long (**a**) and a short (**b**) window. A long window has good frequency (but not time) localization, and a short window has good time (but not frequency) localization. The area of the uncertainty boxes has a minimum value given by Eq. 3.6

principle in quantum mechanics,[2] this is called the uncertainty principle of signal analysis: frequency and time resolution cannot be made arbitrarily small at the same time. In other words, sharp localization in time and frequency are mutually exclusive because we need several data points to define a frequency. If we denote by σ_t the time uncertainty and by σ_ω the frequency uncertainty, the uncertainty principle can be mathematically expressed as (Cohen 1995; Chui 1992; Mallat 1999)

$$\sigma_t \, \sigma_\omega \geq \frac{1}{2} \tag{3.6}$$

This limitation becomes important when the signal has transient components localized in time, as in the case of some EEG activity and evoked potentials. As mentioned above, Gabor (1946) suggested the use of a Gaussian window due to its good localization both in time and frequency. In fact, a Gaussian function gives the best possible time-frequency localization, and Eq. 3.6 becomes an equality.

It is standard to represent the uncertainty principle graphically using *Heisenberg boxes*. Figure 3.3a shows the time and frequency resolutions for 2 time-frequency pairs (t_1, ω_1) and (t_2, ω_2) using a relatively large window. In this case, we have a small frequency uncertainty σ_ω, but the time uncertainty σ_t is too large. Equation 3.6 means that the areas of the Heisenberg boxes have a minimum value. Note also that the boxes have the same shape for the two time-frequency pairs. This is just show-

[2] Heisenberg's uncertainty principle states that it is not possible to determine with arbitrary accuracy the position and velocity of a given particle at the same time.

ing the fact that the window size, and therefore the compromise between time and frequency resolution, is the same for all frequencies. Figure 3.3b shows the time-frequency resolution with a short window. In this case, the time resolution is increased but at the cost of a lower frequency resolution.

3.4 Measures Derived from the Spectrograms: Spectral Entropies

Although the spectrograms give an elegant visual representation of the time evolution of the different frequencies (e.g., Fig. 3.1c and Fig. 3.4b), this information is still qualitative. To quantify the frequency distribution at a given time and particularly to see its evolution, we can calculate the entropy of the power spectrum. Entropy is a measure of randomness or, in other words, the information content of a signal. Random signals are unpredictable and every new data point gives new information. On the contrary, with ordered signals new data points can be predicted from the previous values and therefore carry less information.

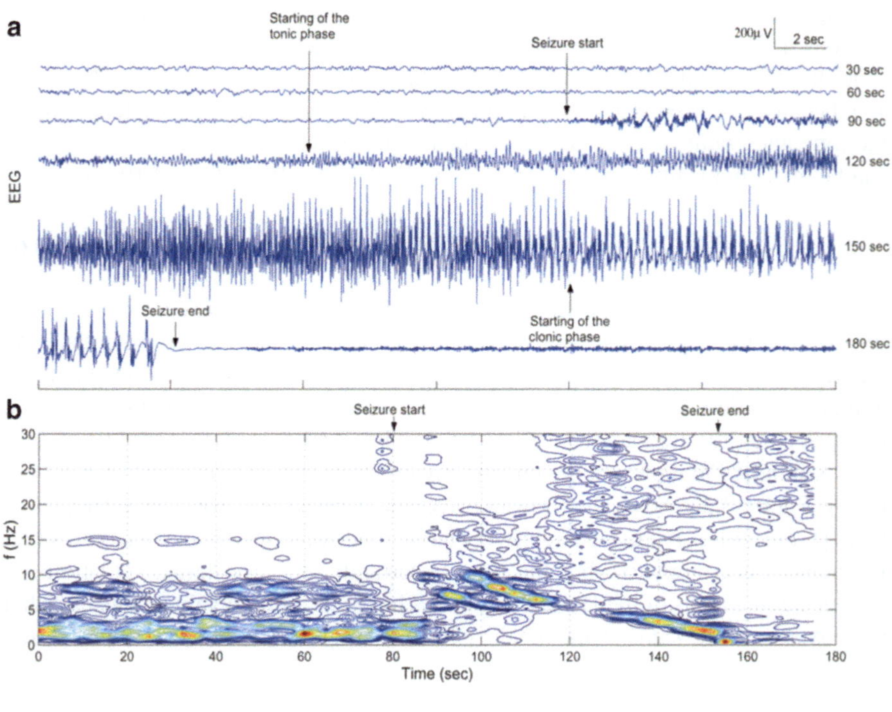

Fig. 3.4 (**a**) Three-minute EEG recording with a grand mal (tonic-clonic) seizure and (**b**) its spectrogram. The seizure is characterized by oscillatory activity localized at around 10 Hz, which slows downs as the seizure progresses

Let us consider a discrete variable $x[n]$ $(n = 1,...,N)$ with K possible outcomes x_k $(k = 1,...K)$, each one with a normalized probability p_k.[3] The *Shannon entropy* of the distribution of outcomes is defined as

$$H = -\sum_k p_k \log_2 p_k, \tag{3.7}$$

which is measured in *bits* if the logarithm is taken with base 2. For example, imagine we generate a series with the outcomes of 100 throws of a dice. Then N will be 100 and the dice has $M = 6$ possible outcomes. For a fair dice, each outcome has an equal probability $p_k = 1/6$, and Eq. 3.7 gives a maximum value of $H = 2.58$ bits. This means, each dice throw is unpredictable, and once we know the outcome, we get a maximum of new information. For a heavily loaded dice, we get always the same number which has a probability $p_k = 1$, and all other numbers have a probability equal 0. In this case, we get $H = 0$ bits,[4] which in other words means that we already know the outcome beforehand, and each throw of the dice does not give any new information.

To quantify the entropy of the power spectrum, in Eq. 3.7 we take p_k as the normalized spectral density I for a given frequency k, at a time t:

$$p_k(t) = \frac{I(f_k,t)}{\sum_k I(f_k,t)} \tag{3.8}$$

The Shannon entropy is equal to 0 for a delta distribution (i.e., a sinusoidal signal) and positive otherwise, reaching a maximum of $H = \log_2 N$ for a “flat” distribution. In terms of the power spectrum, low entropy means that the signal is concentrated in a few frequency bands, and high entropy means that the frequency spectrum looks like broadband noise.

Let us now assume we have two probability distributions (e.g., two different power spectra) $p = \{p_k\}$ and $q = \{q_k\}$. We can define the *Kulback-Leibler* or relative entropy as

$$K(p \mid q) = \sum_k p_k \log_2 \frac{p_k}{q_k} \tag{3.9}$$

The Kullback-Leibler entropy is positive and equals 0 if $p_k \equiv q_k$. It measures the degree of similarity between the two probability distributions p and q: the more dissimilar the distributions, the larger it gets. This is very useful to compare the power spectrum in different states or to analyze how the power spectrum changes with time with respect to a reference spectrum q_k, as we will see in Fig. 3.6.

---

[3] The probabilities p_k satisfy $p_k \geq 0$ and $\sum_{k=1}^{K} p_k = 1$

[4] Where $0 \cdot \log_2 0 \equiv 0$

3.5 Time-Frequency Analysis of Grand Mal Seizures

About 20% of epileptic patients do not respond to antiepileptic drugs, and, depending on the type of epilepsy and clinical considerations, they may be candidates for curative surgery. As part of a comprehensive evaluation, these patients stay in hospital for about a week or two, where they are continuously monitored with video cameras and scalp EEG recordings. The goal of this procedure is to record spontaneous seizures with these video-EEG recordings to localize the focus of the seizures and evaluate the possibility of surgical resection. When the information from noninvasive techniques (i.e., scalp EEG, neuropsychological tests, fMRI, and clinical manifestations) is not conclusive, these patients may undergo the implantation of intracranial electrodes.

Figure 3.4a shows a tonic-clonic (grand mal) seizure recorded from a right central (C4) scalp electrode. From the video recording, it was possible to establish that the seizure started at second 80 with oral automatisms, followed 20 s later by a generalized tonic contraction. In the EEG recording, the starting of the seizure correlates with a burst of slow waves with high frequency activity of lower amplitude superposed to it. Afterward, the seizure developed further, and the analysis of the EEG becomes more complicated due to muscle artifacts. The clonic phase started about 60 s after the beginning of the seizure, and the seizure ended at second 155 where there is an abrupt decay of the signal.

Figure 3.4b shows the spectrogram of the seizure recording. The pre-ictal activity was dominated by delta frequencies, and shortly after the seizure onset, there was a dramatic change in the spectrogram with the appearance of oscillatory activity localized at 10 Hz. This frequency gradually decayed as the seizure progressed, and it went down to 1 Hz just before the seizure end. Muscle artifacts that contaminated the EEG recording are also identifiable from the spectrogram, as a widespread pattern going up to frequencies larger than 30 Hz. However, they did not obscure the time-frequency dynamics seen in the spectrogram. The starting of the clonic phase was correlated with a localized activity at 3 Hz, which was due to a slowing of the 10-Hz activity that appeared at the beginning of the seizure. Consequently, it seems that clonic contractions were a response to brain activity that could only be established when brain oscillations were slow enough to be followed by the muscles.

In summary, from the spectrogram we can postulate that tonic-clonic seizures are a single process, with frequencies initially localized at about 10 Hz slowing down to about 1 Hz at the seizure end (this was the case for over 70% of the seizures studied in Quian Quiroga et al. 1997). The tonic and clonic phases, and even the tremors that can sometimes be seen between these two phases, are then muscular responses determined by such frequency evolution (Quian Quiroga et al. 1997; Quian Quiroga et al. 2002). This pattern was not clear from the scalp EEG recordings due to muscle contamination. The analysis of intracranial recordings without such muscle contamination showed a similar frequency dynamics (Quian Quiroga et al. 2000).

The spectrogram of Fig. 3.4b was calculated with half-overlapped windows of 5-s length. In Fig. 3.5, we show the spectrograms obtained using half-overlapped

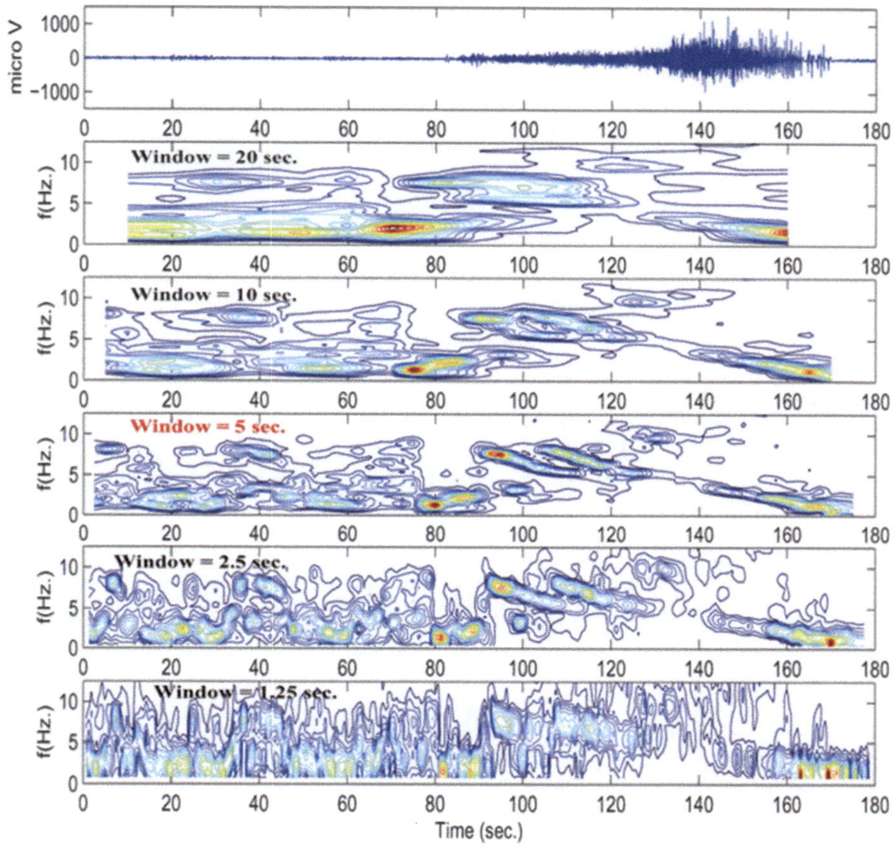

Fig. 3.5 The tonic-clonic seizure of Fig. 3.3 and its spectrograms using different window sizes

windows of 20, 10, 5, 2.5, and 1.25 s, respectively. As discussed in Sect. 3.3, for large windows (e.g., 20 s) there is a good frequency resolution, but the patterns of the spectrogram are not localized in time. On the contrary, for short windows (e.g., 1.25 s) there is a better time resolution, but the frequencies are less localized. Note that the window of 5 s gives an optimal compromise between time and frequency resolution (time resolution: 2.5 s; frequency resolution: 0.2 Hz).

Figure 3.6 shows the same tonic-clonic seizure in a compressed form, the relative intensity ratio (RIR) of the different frequency bands, the Shannon entropy (Eq. 3.7), and the Kullback-Leibler entropy (Eq. 3.9). For the latter one, we took as a reference spectrum the one of the first time window, corresponding to pre-ictal activity. In agreement with the time-frequency patterns described in Sect. 3.4, the analysis of the power in the different frequency bands (RIR) shows that pre-seizure activity was characterized by a high power in the delta band, which decayed abruptly during the seizure given the dominance of theta and alpha frequencies (Quian Quiroga et al. 1997). The Shannon entropy showed an increase during the seizure,

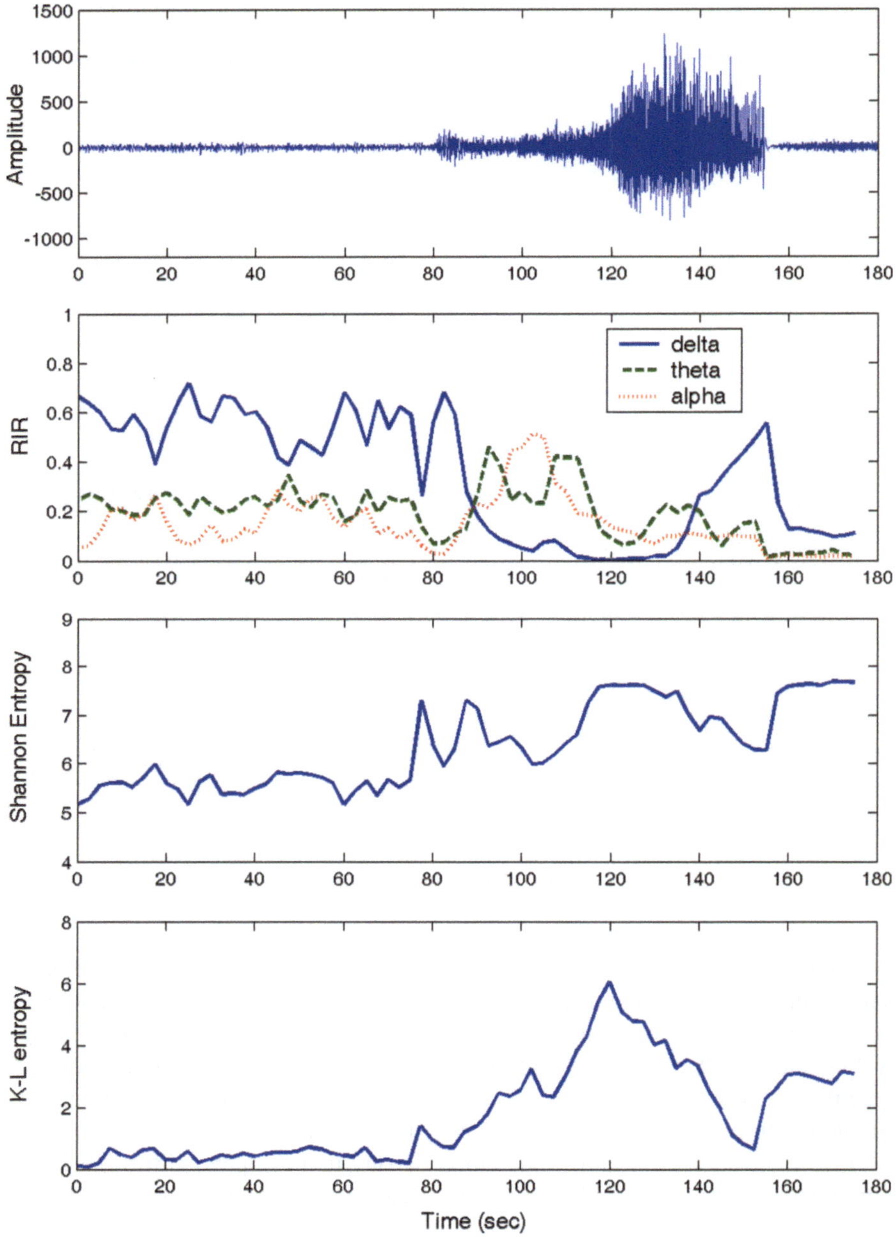

Fig. 3.6 From *top* to *bottom*, the tonic-clonic seizure of Fig. 3.3, the band relative intensity ratio, Shannon entropy and Kullback-Leibler entropy

but this change was more clearly seen with the Kullback-Leibler entropy, due to the difference of the power spectrum in this state compared to the one in pre-seizure epoch. As shown in Fig. 3.7, a similar pattern was also observed with intracranial recordings, with the largest entropy change appearing in the electrode closest to the

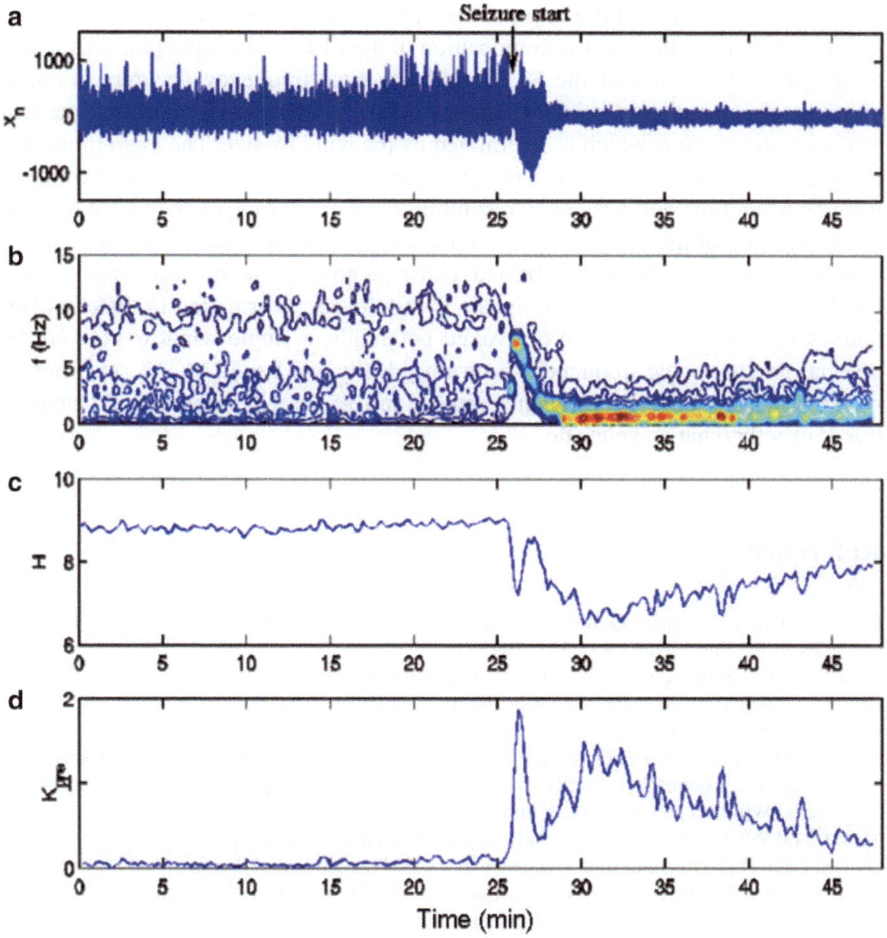

Fig. 3.7 (**a**) Intracranial EEG recording from the electrode contact within the seizure-generating area, (**b**) its corresponding power spectrum, (**c**) Shannon entropy, and (**d**) Kullback-Leibler entropy taking a pre-seizure reference window

seizure focus (Quian Quiroga et al. 2000). Then, the Kullback-Leibler entropy may be a potentially useful tool to help to localize the source of the epileptic seizures.

3.5.1 Summary

In this chapter we showed that with the STFT it is possible to localize in time the activity of the different EEG frequencies, and we used this method to track the frequency dynamics during tonic-clonic seizures. We also showed that it is possible to define spectral entropies from the time-varying power obtained with the STFT, which showed changes in the spectral composition of the EEG during epileptic seizures.

A critical limitation of the STFT, and any time-frequency decomposition, is given by the uncertainty principle. Indeed, there is a trade-off between time and frequency resolutions which is determined by the window size. The main limitation of the STFT is that the window length is fixed and it may give an optimal compromise between time and frequency resolution for a given frequency but not for others. In the case of the tonic-clonic seizures, most of the interesting activity changed slowly and occurred below 10 Hz, and, consequently, a window of 5 s was optimal for their analysis. However, for signals with relevant information in different frequency ranges, as in the case of evoked potentials, a single window may not be optimal for the whole frequency spectrum. Ideally, we would like to set different window sizes for different frequency ranges, and, as we will see in the next chapter, this is exactly what wavelets do.

References

Blanco S, García H, Quian Quiroga R, Romanelli L, Rosso OA (1995) Stationarity of the EEG series. IEEE Eng Med Biol 14:395–399

Chui C (1992) An introduction to wavelets. Academic, San Diego

Cohen L (1995) Time-frequency analysis. Prentice-Hall, New Jersey

Gabor D (1946) Theory of communication. J Inst Elec Eng 93:429–459

Lopes da Silva FH (1993) EEG analysis: theory and practice. In: Niedermeyer E, Lopes da Silva FH (eds) Electroencephalography: Basic Principles, Clinical Applications and Related Fields, 3rd edn. Williams and Wilkins, Baltimore

Mallat S (1999) A wavelet tour of signal analysis. Academic, San Diego

Qian S, Chen D (1996) Joint time-frequency analysis. Prentice Hall, London

Quian Quiroga R, Blanco S, Rosso OA, Garcia H, Rabinowicz A (1997) Searching for Hidden Information with Gabor Transform in Generalized Tonic-Clonic Seizures. Electroencephalogr Clin Neurophysiol 103:434–439

Quian Quiroga R, Arnhold J, Lehnertz K, Grassberger P (2000) Kullback-Leibler and Renormalized Entropy: Applications to EEGs of Epilepsy Patients. Phys Rev E 62:8380–8386

Quian Quiroga R, Garcia H, Rabinowicz A (2002) Frequency evolution during tonic-clonic seizures. Electromyogr Clin Neurophysiol 42:323–331

Chapter 4
Wavelets

4.1 Introduction: Brief History

In the previous chapter, we showed that a key issue with the short-time Fourier transform (STFT) is the choice of the window length, given the basic limitation imposed by the uncertainty principle of signal analysis. Short windows give good time (but bad frequency) resolution, and conversely, long windows give good frequency (but bad time) resolution (see Sect. 3.3). In the late 1970s, Jean Morlet, a geophysicist working for a French oil company, realized that the STFT was not suitable for the study of his seismic data. He observed that a good compromise between time and frequency resolution was not possible because high-frequency patterns had a shorter duration compared to the low-frequency ones. So, a single window for all frequencies would not do. His solution was quite straightforward: he just took different window sizes for different frequencies, or more accurately, he took a cosine function tapered with a Gaussian (a Gabor function, see Sect. 3.2) and compressed it or stretched it in time to get the different frequencies (see Fig. 4.1). Then, instead of always having the same window size, he had the same wave shape at different scales, that is, with a variable size. With this simple trick, he created the basis of wavelets!

In our days, wavelets are recognized as a very powerful signal decomposition tool with a large number of applications in different fields, including the analysis of brain signals. The development of wavelets significantly profited from an interaction between typically disparate disciplines. In particular, mathematicians working on wavelets theory profited from the knowledge on filter banks and its applications from the engineering world (for image processing, data compression, denoising, etc.), and conversely, engineers in this area profited from the development of a very solid theoretical foundation of wavelets, based on harmonic analysis in mathematics. The story of such interactions started in a few cases due to fortuitous events (for a beautiful account told by one of the main researchers involved in the development of wavelets, see Daubechies 1996). Sensing that his idea for processing the seismic data had wider applications, Morlet contacted Alexander Grossmann, a theoretical

W.J. Freeman and R. Quian Quiroga, *Imaging Brain Function With EEG: Advanced Temporal and Spatial Analysis of Electroencephalographic Signals*, DOI 10.1007/978-1-4614-4984-3_4, © Springer Science+Business Media New York 2013

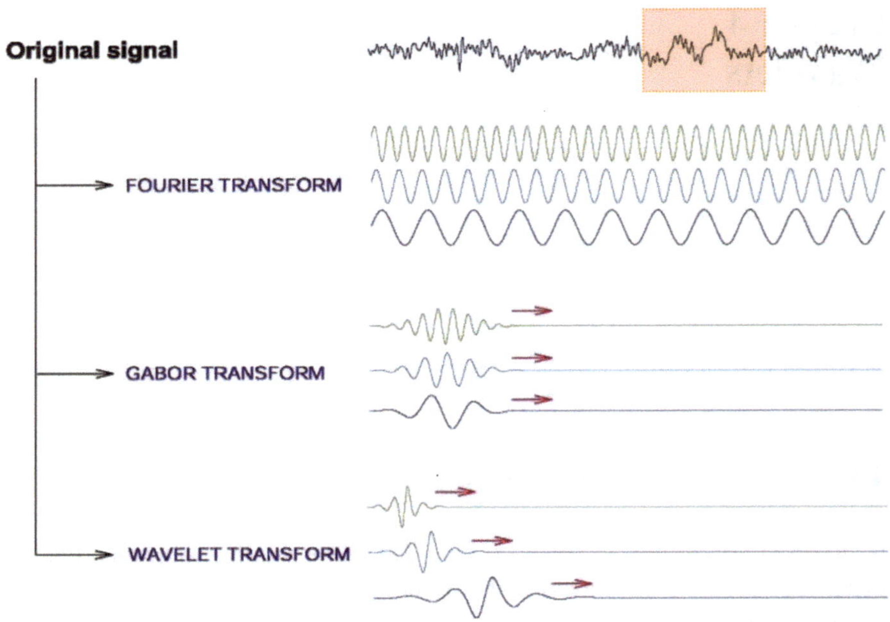

Fig. 4.1 Basic idea of wavelets. The Fourier transform is the inner product of the signal and sinusoids of different frequencies. To get time resolution, the Gabor (or short-time Fourier) transform uses *windowed* sinusoids tapered with a Gaussian function. With wavelets, by stretching or compressing a wavelet function, the size of the window is variable and we therefore obtain an optimal compromise between time and frequency resolution for all frequencies

physicist working on similar formalisms in quantum mechanics, who constructed an exact inverse formula to reconstruct the original signal. A pure mathematician, Yves Meyer, heard by chance about Morlet and Grossmann's ideas while waiting in line to make some photocopies and ended up playing a key role in putting the recently born wavelets into the context of harmonic analysis. Another interesting encounter occurred a few years later, when Stefane Mallat, a student working on computer vision and image analysis, met an old friend in a French beach who happened to be a student of Meyer. This encounter was the spark that started Mallat's work linking research on filter banks in the engineering community with harmonic analysis and wavelets in mathematics. Mallat's contribution was not limited to this but also to the development of a very powerful and fast algorithm, the multiresolution decomposition, to calculate the wavelet transform.

4.2 Basic Idea

There are a few excellent books dedicated exclusively to the description of wavelets and their very solid theoretical background (Chui 1992; Mallat 1998; Strang and Nguyen 1996), but the basic idea behind wavelets is very simple and hardly requires the use of any formula to grasp it (see 4.7 for a more formal description). For this, let us revisit the Fourier transform. As shown in Eq. 2.3, the Fourier transform can be seen at the matching between a signal, as the one shown on top of Fig. 4.1, and complex exponentials – sine and cosine functions – of different frequencies. The main problem of the Fourier transform is the assumption of stationarity and the consequent lack of time resolution. For example, in the signal of Fig. 4.1, we observe a transient oscillation marked with a light-red box. This gives a peak in the Fourier power spectrum at this particular frequency, but from such decomposition we can't tell when this oscillation happened or whether it was localized in time at all. The solution to this problem was presented in the previous chapter: we just chop the data into consecutive pieces and calculate the respective Fourier transforms, or even better, we calculate successive Fourier transforms using time-localized sinusoids (by tapering them with a Gaussian function) that slide through the data. This short-time Fourier (or Gabor) transform can be seen as the matching between the original signal and the time-localized Gabor functions (Eq. 3.3). But as described in the previous chapter, the key issue is the choice of the window length due to the uncertainty principle (Sect. 3.3). Looking at the high-frequency sinusoid of the Gabor transform in Fig. 4.1, one may say that there are too many oscillations within the window – that is, the frequency resolution is very good because there are several cycles of the sinusoid that can be matched to the signal – and it may be better to gain more time resolution by shortening the window. However, this would be a very bad idea when considering the low-frequency sinusoid, because in this case there would be hardly one oscillation inside the window, and therefore, the frequency resolution would be too low. The key idea of wavelets is to take different window sizes for different frequencies, which is done by compressing or stretching the same function, called the *mother wavelet*. Note also that the function to be used does not necessarily need to be a tapered sinusoid. This gives a second advantage: there is a *dictionary* of wavelet functions to choose from according to their properties and the application in mind. Since we now don't necessarily have sinusoids of different frequencies but functions at different scales, we say that by doing the inner product (i.e., the matching) between the signal and these wavelet functions at different times and scales, we get a *time-scale* (instead of time-frequency) decomposition.

4.3 Two Common Misconceptions

A common mistake is to think that wavelets give a better time-frequency resolution compared to, for example, the Gabor transform. This is not the case. In fact, the best possible time-frequency resolution is already given by Gabor functions (Cohen 1995). In other words, with wavelets the areas of the Heisenberg uncertainty boxes of Fig. 3.3 do not get smaller (on the contrary, for most wavelet functions they get a bit larger). But the key point is that the shape of the Heisenberg boxes is variable and it is optimal for each frequency. As shown in Fig. 4.2, for high frequencies (ω_1), the wavelet has a small time window, thus increasing the time resolution at the cost

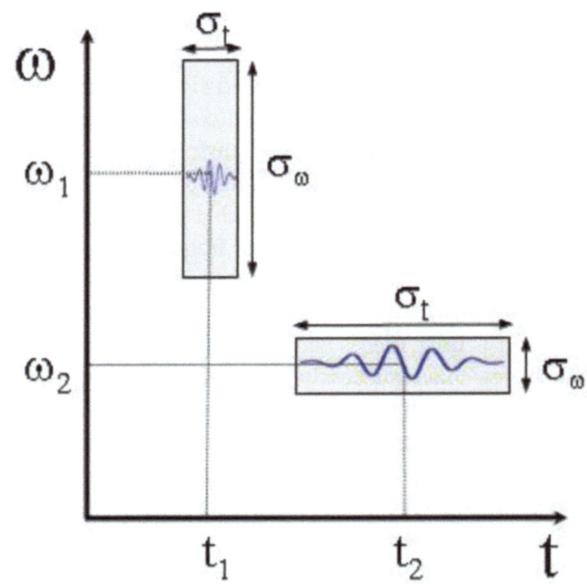

Fig. 4.2 Heisenberg time-frequency boxes for wavelets. Compared to the STFT (see Fig. 3.3), the main advantage of wavelets is that the shape of the box is variable, thus giving an optimal compromise between time and frequency resolution for all frequencies

of frequency resolution. Conversely, for low frequencies (ω_2), the size of the time window is large, thus having a better frequency resolution at the cost of a larger time uncertainty. As in many other disciplines where wavelets have proven to be extremely useful, this is exactly what we typically want for the analysis of EEG signals. For example, it is of little relevance whether a gamma oscillation is of 50 Hz or 52 Hz, as for these high-frequency oscillations we usually prefer to have a better time resolution (e.g., we may want to determine whether the gamma oscillation appears at 100 or 110 ms after a given stimulus). Conversely, the same error of 2 Hz but now confusing an oscillation of 3 Hz with one of 5 Hz would be too bad, as we would be mixing delta and theta oscillations, which have completely different functions (see Sect. 2.10). Establishing whether a delta oscillation started 100 or 110 ms after stimulation is clearly less important because such latency differences can be defined less accurately for slow oscillations.

Another misconception is to confuse the time-frequency localization of a given wavelet function with the time-frequency localization of the function applied to the signal. To illustrate this, let us consider the wavelet transform of a delta function. If we use a Morlet wavelet (a complex sinusoid tapered with a Gaussian function, see Fig. 4.3), the wavelet decomposition of the delta function will be spread across many scales, because the Fourier transform of a delta function has components all over the frequency spectrum. In other words, to generate a delta function, we need to add sinusoids of all frequencies. On the contrary, if we use a Haar wavelet – which is just a square function (see Fig. 4.3) – the decomposition will have a better time-frequency localization, in the sense that there will be fewer wavelet coefficients representing the signal. This is simply because the Haar wavelet matches more naturally the patterns of the delta function. Summarizing, when choosing a mother wavelet we should know that even if a certain wavelet function has an excellent time-frequency localization, it may not be optimal for our data. The key point is that the wavelet function should look similar to the type of patterns we want to localize, so that these are correlated with a few coefficients. Having only a few relevant coefficients simplifies the description of the data and any further analysis we may want to do, as it will be the case for denoising in the next chapter.

4.4 Choice of the Mother Wavelet

There are many different functions that can be used as wavelets,[1] each one having different characteristics whose importance typically depends on the application in mind. Indeed, the possibility of choosing a mother wavelet from a dictionary of possible functions is one of the main strengths of wavelets. Figure 4.3 displays four widely used wavelet functions.

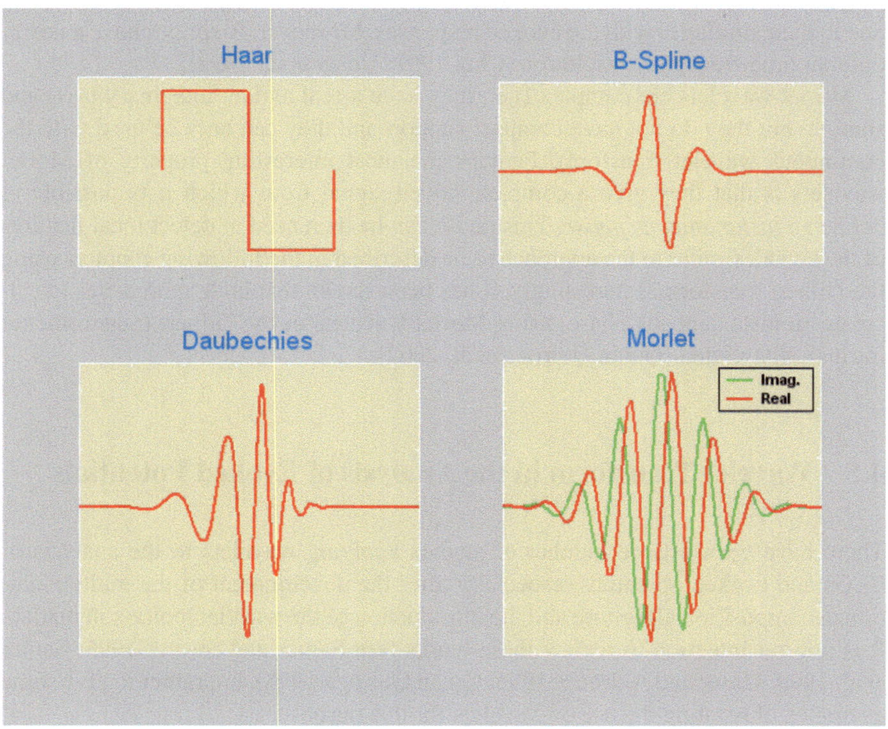

Fig. 4.3 Example of wavelet functions

[1] Simply put, a wavelet is a function with zero average, which implies that the function has to be oscillatory or a *wave* (for details see Mallat 1998; Chui 1992).

The Haar wavelet is simply a square function with some interesting properties: (1) it has compact support, or in other words, it is localized in time; (2) it is orthogonal (i.e., the inner product of two different Haar wavelets is zero), although this is seldom required for practical applications; (3) it is (anti) symmetric; and (4) it can be used both for the continuous and the dyadic wavelet transform. A drawback of the Haar wavelet is that it is discontinuous, which could be a disadvantage because it tends to give staircase-looking signals when used for filtering or denoising. Daubechies wavelets have been also used for several applications. They are smooth, orthogonal, and suitable both for the continuous and the dyadic transform (see Sect. 4.7.1), and they have compact support. Daubechies wavelets are largely non-symmetric, thus giving different results when analyzing the signal in the forward or backward direction. B-spline wavelets have compact support, they are suitable for both the continuous and the dyadic transforms, and they are smooth, (anti) symmetric, and not orthogonal.

Besides the mathematical properties of the wavelet to choose, a basic requirement is that it looks similar to the patterns to be localized in the signal (see Sect. 4.3). This allows a good localization of the structures of interest in the wavelet domain. For the analysis of evoked potentials, B-spline wavelets are in general a good choice due to their similarity with the evoked responses. Moreover, B-splines have a nearly optimal time-frequency resolution (Chui 1992; Unser et al. 1992).

Morlet wavelets are complex (i.e., they have a real and an imaginary part) and smooth, but they do not have compact support and they can only be used with the continuous wavelet transform. Perhaps the most interesting property of Morlet wavelets is that they give a complex-valued signal from which it is possible to define an *instantaneous phase*. This phase can be then used to detect local features of the signal, similar to the approach to be described in the following chapters using the Hilbert transform. Interestingly, it has been shown that both approaches to calculate an instantaneous phase, using Morlet wavelets or the Hilbert transform, are intrinsically related (Quian Quiroga et al. 2002).

4.5 Wavelet Transform in the Analysis of Evoked Potentials

There have been a large number of studies applying wavelets to the analysis of EEGs and evoked potentials, especially after the development of the multiresolution decomposition algorithm and the introduction of the wavelet toolbox in matlab. It is not our intention to review these works (see Samar and Swartz 1995; Samar et al. 1999; Unser and Aldroubi 1996; Quian Quiroga 1998), but rather to give some examples illustrating the use of wavelets for this purpose.

Before doing any calculation, perhaps the first question to be asked is whether wavelets are really the right tool for a particular data and research question in mind. We already showed in the last chapter that with the short-time Fourier transform it was possible to get a good characterization of the dynamics of grand mal seizures. Would wavelets give a better result for this data? Not really. Actually, the STFT

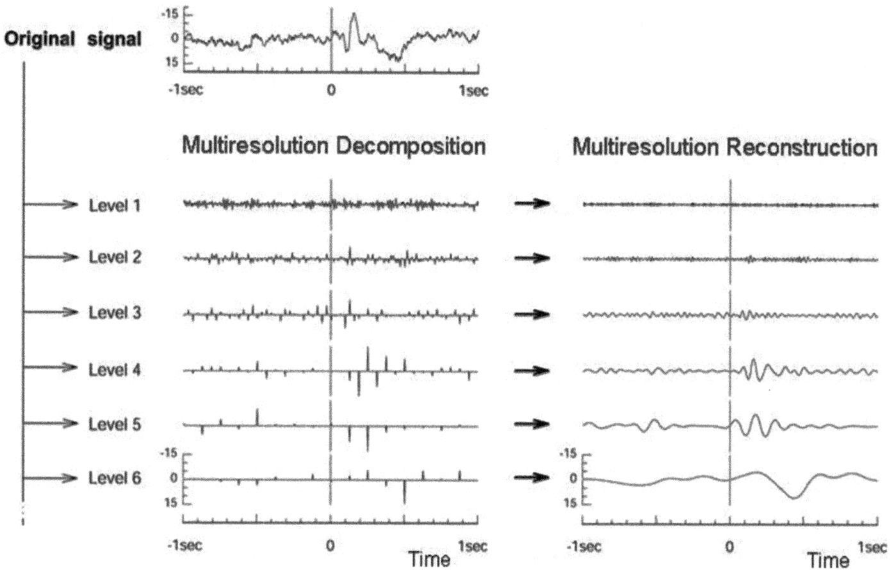

Fig. 4.4 Multiresolution decomposition and reconstruction of an average evoked potential

already gives the best possible time-frequency resolution since it uses Gabor functions. So, depending on the wavelet function we use, we can get a similar resolution or worse. The key point is that the interesting activity in the seizure period had a relatively short frequency span (from about 0 to 10 Hz) and it was therefore possible to find a single window length that was optimal within this range. A different case is the one of evoked potentials, where we can have frequencies of interest going from delta or theta (less than 10 Hz) to gamma (up to 80 Hz). In this case, we don't have a single window length that is suitable for all frequencies. Of course, we can try setting different window lengths by hand, but this was exactly the starting point of Morlet which led to wavelets more than 30 years ago. So, we would be just rediscovering the wheel! This type of data is indeed the one that cries for a method as wavelets.

Figure 4.4 shows a five-level multiresolution decomposition of an average evoked potential (same data as in Fig. 1.7). The left part of the figure shows the wavelet coefficients, and the right part shows the corresponding reconstructed waveforms for each scale. The sum of all the reconstructions gives back the original signal. Given that the sampling rate of the signal was 250 Hz, the frequency bands corresponding to each scale are the following (see Sect. 4.7.2): 62–125 Hz for D_1, 31–62 Hz (gamma) for D_2, 16–31 Hz (beta) for D_3, 8–16 Hz (alpha) for D_4, 4–8 Hz (theta) for D_5, and 0.5–4 Hz band (delta) for the last approximation A_5. Note that the frequency ranges of the scales approximately match the EEG frequency bands described in Sect. 2.10.

Let us now describe event-related oscillations in the alpha band for these evoked potentials, recalling that they were obtained with an oddball experiment with pattern visual stimulation (see Sect. 1.5). For this, we need to band-pass filter the

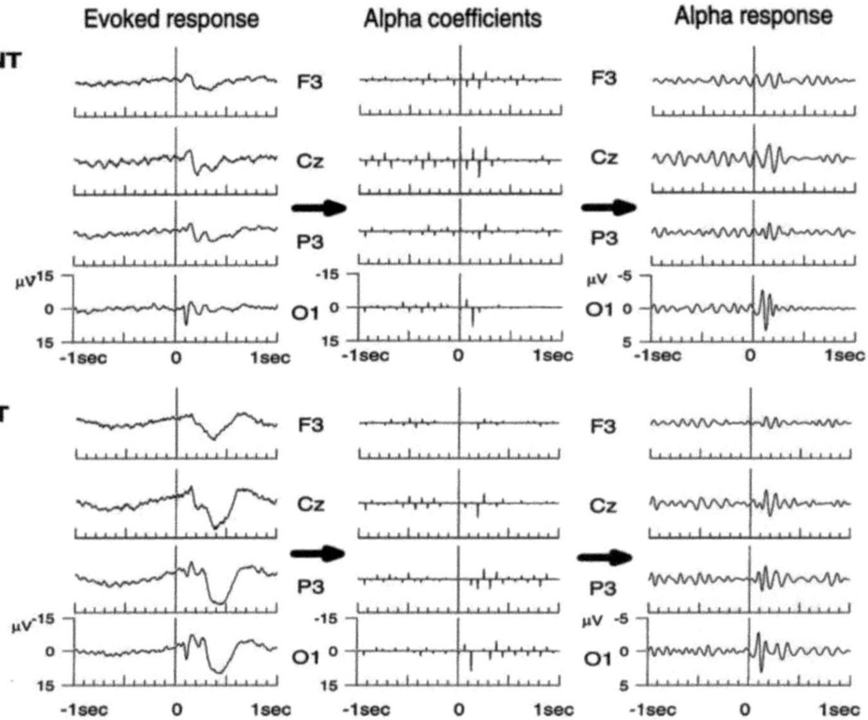

Fig. 4.5 Left panels: evoked responses in the F3, Cz, P3, and O1 electrodes (from *top* to *bottom*) for the nontarget (NT) and target (T) stimuli. Positive values are plotted downward. Note the P100 for both types of stimuli and the P300 only for the target stimuli. Middle panels: corresponding wavelet decompositions in the alpha band. Right panels: event-related alpha activity obtained by reconstructing the signals from the alpha coefficients

signal, which is done by keeping only the coefficients of one of the scales (level 4 in Fig. 4.4) and then doing the reconstruction. The grand average (across subjects) evoked potentials are shown in left side of Fig. 4.5. The upper plots show the responses to nontarget (NT) stimuli and the lower plots to target (T) stimuli. Only left electrodes and Cz are shown, given that the responses of the right electrodes were qualitatively similar. The P100-N200 complex is clearly visible in all modalities, and it is best defined in occipital locations. Target stimulation led to a marked P300 response that is largest in the parietal and occipital electrodes. The middle and right plots show the alpha band wavelet coefficients and the filtered evoked potentials obtained from these coefficients. Amplitude increases after stimulation are distributed over the entire scalp for both stimulus types and are most marked in the occipital electrode. Moreover, these responses appear first in the occipital electrode, with an increasing delay in the parietal, central, and frontal locations. The identification of these delays is important to establish the presence of a propagating activity rather than a single source appearing in all these electrodes due to

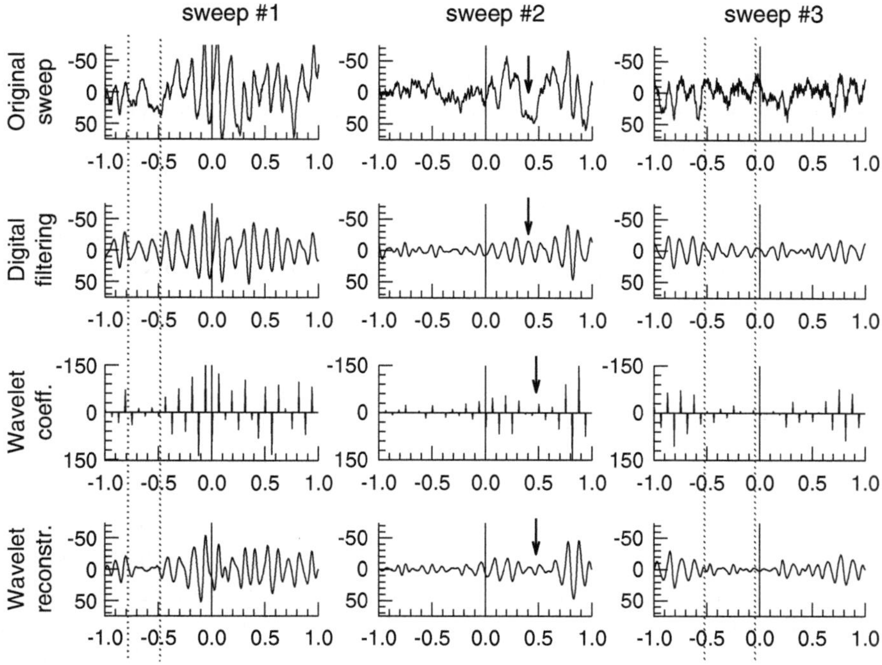

Fig. 4.6 Example of 3 sweeps showing the better resolution of wavelets compared to an "ideal" digital filter

volume conduction.[2] We should remark that the differences in the time onsets would have been smeared without the optimal resolution of wavelets. In this respect, it has been shown that a Fourier-based digital filter introduced ringing effects (sometimes giving event-related oscillations starting before stimulus onset) and was in general not suitable to accurately localize this type of activity (Quian Quiroga and Schürmann 1999; Quian Quiroga et al. 2001).

Figure 4.6 gives some examples of single-trial responses to illustrate the advantage of wavelets over a digital filter in the analysis of evoked potentials. The digital filter used was an "ideal filter," obtained by band-pass filtering the signal in the Fourier domain. The filter limits were the same as those obtained with the multiresolution decomposition for the alpha band. In between the dotted vertical lines in sweep #1, a transient with a frequency clearly lower than the range of alpha band is observed. The digital filtering does not resolve this transient, and it spuriously interpolates alpha oscillations in between the ones that precede and follow the transient. On the contrary, the wavelet coefficients and the reconstructed waveform from these coefficients correctly show a decrease in this time segment. At the time marked with an arrow in the second sweep, the digital filter (but not wavelets) gives spurious

[2] In the following chapters, we will describe how to analyze this type of propagating activity in more detail.

alpha oscillations not present in the original signal. In the signal of sweep #3, there is a theta oscillation of about 4–6 Hz between the dotted lines and the digital filter gives a spurious alpha oscillation not present in the original signal. On the contrary, the wavelet coefficients and the reconstructed signal correctly show a clear decrease in alpha for this time range.

One could in principle argue that it should be possible to design a better digital filter for the alpha band, but the specifications of this filter (not just the cutoff frequencies) should be optimized again if we want to study the activity in other frequencies. The advantage of wavelets is that we are already getting this optimization for free!

4.6 Summary: Cautionary Note

Several works have shown the utility of wavelets for the analysis of EEGs and evoked potentials. Wavelets are particularly useful to analyze time-localized patterns, especially if these span a wide frequency range that cannot be optimally covered with the short-time Fourier transform. In this respect, we described the use of wavelets for the analysis of evoked potentials, which typically have activity in a wide frequency range. Wavelets have been also used for the analysis of EEG signals, among others, for extracting features of seizure EEG recordings (Schiff et al. 1994a), for the automatic detection of spike complexes (Schiff et al. 1994b; Senhadji et al. 1995; Clark et al. 1995; Sartoretto and Ermani 1999; D'attellis et al. 1997), and for the automatic classification of different sleep states (Kiymik et al. 2004).

We finish this section with a cautionary note: wavelets are not a "magic bullet" that bypasses the limitations imposed by the uncertainty principle of signal analysis or the distortions that can be introduced by a digital filter. It is true that wavelets give an optimal time-frequency resolution and that they diminish filtering distortions (if a proper wavelet function is used). But it is important to remark that, as shown in Fig. 1.8, the correlation between a single pulse and a wavelet function will still typically look like an oscillation, especially if a high-order wavelet (a wavelet with several oscillations) is used. This goes back to the discussion of Sect. 1.6 of whether evoked responses should be considered as time-locked activity added to independent background EEG or as a reorganization of the ongoing EEG due to phase locking. In particular, it has been shown that most measures used to demonstrate the contribution of phase locking of ongoing oscillations to the generation of the evoked responses should be revisited and used with care, given that similar results were obtained with simulations in which a phasic response was added to independent background EEG activity (Yeung et al. 2004). The key point is that the latency of an added phasic response translates into the phase of an oscillation obtained after filtering. Whereas one prefers to base the analysis on the latencies of the original evoked responses or the phases of the filtered signals is a question of personal preference. The problem comes with the interpretation, given that an oscillatory-looking response after filtering could be just a filtering artifact.

4.7 Appendices

4.7.1 Continuous and Dyadic Wavelet Transforms

Having provided an intuitive introduction to wavelets and their main advantages over other decompositions, let us now formalize these ideas. The wavelet transform gives a time-frequency (or more accurately time-scale) representation that is defined as the correlation between the signal *x(t)* and the wavelet functions $\psi_{a,b}(t)$.

$$W_{\psi}X(a,b)=\langle x(t),\psi_{ab}\rangle \equiv |a|^{-1/2}\int_{-\infty}^{+\infty} x(t)\psi^{*}\left(\frac{t-b}{a}\right)dt \tag{4.1}$$

where * denotes complex conjugation and $\psi_{a,b}(t)$ are scaled and shifted versions of a unique mother wavelet $\psi(t)$ (see Fig. 4.7):

$$\psi_{a,b}(t)=|a|^{-\frac{1}{2}}\psi\left(\frac{t-b}{a}\right) \tag{4.2}$$

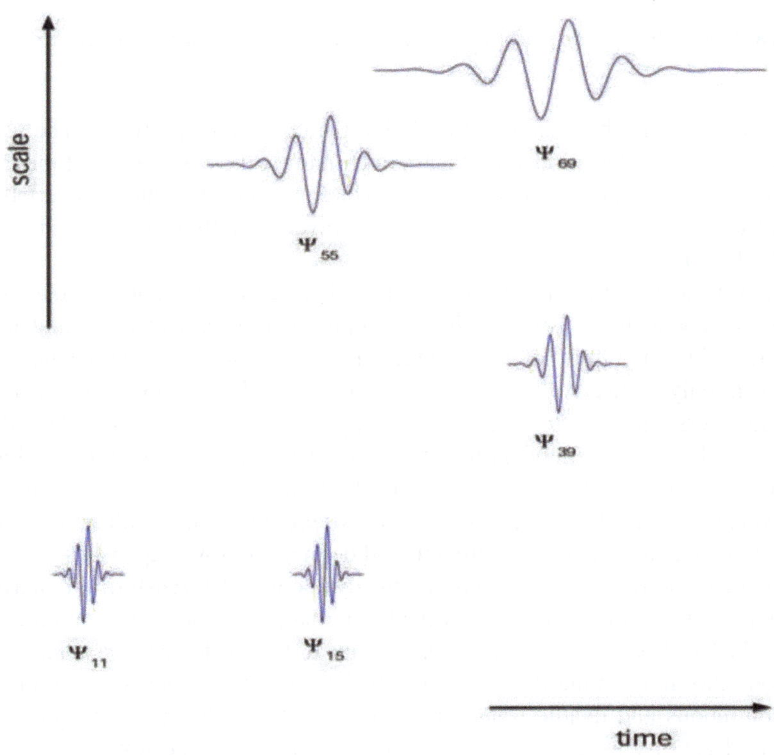

Fig. 4.7 A wavelet function $\psi_{a,b}(t)$ at different scales *a* and times *b*

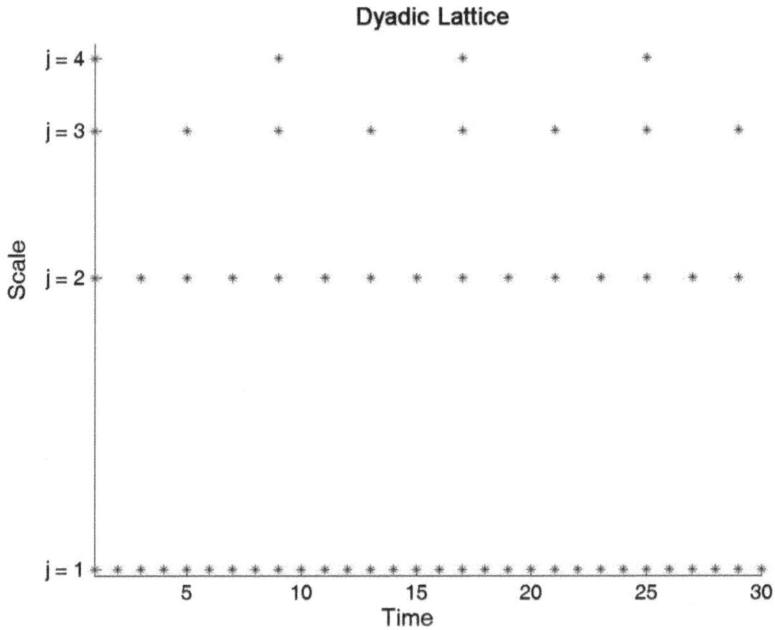

Fig. 4.8 Lattice showing the points at which the dyadic wavelet transform is calculated

where $a, b \in R$ are the scale and translation parameters, respectively. The wavelet transform gives a decomposition that is maximum at those scales and times where the wavelet best matches the signal *x(t)*. Moreover, Eq. 4.1 can be inverted, thus giving the reconstruction of *x(t)* from the wavelet coefficients (Grossmann and Morlet 1984).

The wavelet transform maps a signal of one independent variable *t* onto a function of two independent variables *a, b*. This representation is overly redundant, and without losing any information, it is sometimes more practical to define the wavelet transform only at discrete scales *a* and discrete times *b* by choosing the *dyadic* set of parameters $\{a_j = 2^j, b_{j,k} = 2^j k \quad j,k \in \mathbb{Z}\}$, as shown in Fig. 4.8. This dyadic sampling gives a nonredundant transform that has many samples for the high frequencies – where we actually want to have high time resolution-and less and more spaced samples for the lower frequencies – where high time resolution is not that crucial given that precise times are not well defined for low frequencies.

Figure 4.9 shows the continuous and the dyadic wavelet transform of an average evoked potential. In the average evoked potential, we observe two main components: the P100-N200 (a positive peak at about 100 ms followed by a negative peak at about 200 ms) and the P300 response (see Sect. 1.5 for details). Note that both the continuous and dyadic transforms give essentially the same information: an increased activity reflecting the P100-N200 complex in the low (i.e., high frequency) scales and an increased activity correlated with the P300 response in the higher (low frequency) scales. The continuous plot may look smoother but the main

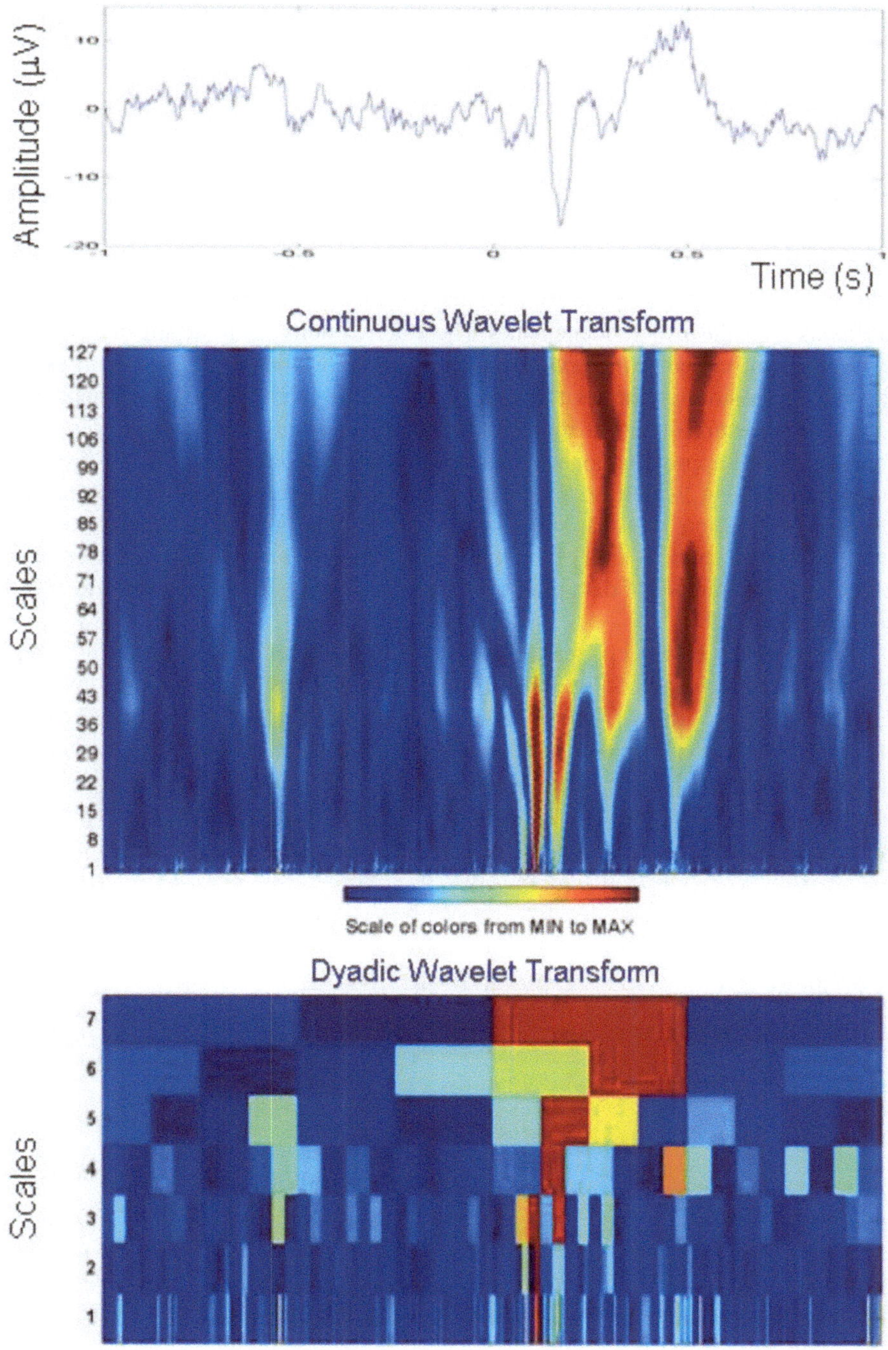

Fig. 4.9 The continuous and dyadic wavelet transform of the average evoked potential shown in Fig. 4.4

advantage of the dyadic transform is computational speed, especially considering that this transform can be implemented with a very fast algorithm, as we will see in the next section.

4.7.2 Multiresolution Decomposition

The correlation of the signal *x* with contracted versions of the dyadic wavelets of Eq. 4.2 gives the high-frequency components, and the correlation with the dilated versions gives the low-frequency ones. These correlations can be arranged in a recursive algorithm called *multiresolution decomposition* (Mallat 1989). The multiresolution decomposition separates the signal into *details* (high-frequency components) and *approximations* (coarser representations of the signal) at different scales. Each detail (D_j) and approximation (A_j) at a given scale *j* is obtained from the previous approximation (A_1) (see Fig. 4.10). This pyramidal scheme makes the multiresolution decomposition very fast, even faster than the fast Fourier transform: the time required for the computation of the multiresolution decomposition is of the order of *N* (with *N* the number of data points), whereas for the fast Fourier transform is $N * log\ N$ (Mallat 1989).

Let us now see the steps for the decomposition and reconstruction of the signal following the scheme of Fig. 4.10. First, the signal *x* is high-pass and low-pass filtered using the filters *G* and *H*, respectively. Both sets of coefficients obtained after filtering are decimated by two (one every two data points is deleted), thus giving the first level detail D_1, containing the activity in the upper half of the frequency spectrum (i.e., from $f_s/4$ to $f_s/2$),[3] and the first approximation A_1, containing the

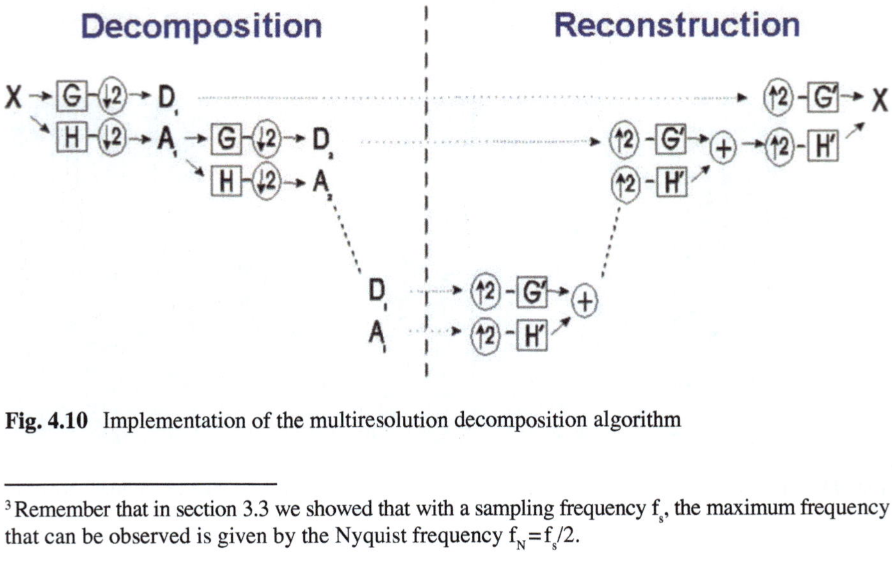

Fig. 4.10 Implementation of the multiresolution decomposition algorithm

[3] Remember that in section 3.3 we showed that with a sampling frequency f_s, the maximum frequency that can be observed is given by the Nyquist frequency $f_N = f_s/2$.

lower half (from 0 to $f_s/4$). After decimation, the number of data points of D_1 plus the ones of A_1 is equal to the number of data points of x, thus obtaining a nonredundant representation. Then, the approximation is further decomposed, and the whole procedure is repeated j times, where j is the number of chosen levels. As a result we obtain the signal x decomposed into D_1 to D_j details and one final approximation A_j. Note that the rescaling of the mother function is given by the decimation of the coefficients.

From this set of coefficients (the details and approximations), the reconstruction of the signal x is done in a similar way using the inverse filters G' and H' and upsampling the data (i.e., inserting zeros between samples), as shown in the right side of Fig. 4.10.

References

Chui C (1992) An introduction to wavelets. Academic, San Diego

Clark I, Biscay R, Echeverria M, Virues T (1995) Multiresolution decomposition of non-stationary EEG signals: a preliminary study. Compt Biol Med 25:373–382

Cohen L (1995) Time-frequency analysis. Prentice-Hall, New Jersey

D'attellis C, Isaacson S, Sirne R (1997) Detection of epileptic events in electroencephalograms using wavelet analysis. Ann Biomed Eng 25:286–293

Daubechies I (1996) Where do wavelets come from? – A personal point of view. Proc IEEE 84:510–513

Grossmann A, Morlet J (1984) Decomposition of Hardy Functions into square integrable wavelets of constant shape. SIAM J Math Anal 15:723–736

Kiymik MK, Akin M, Subasi A (2004) Automatic recognition of alertness level by using wavelet transform and artificial neural network. J Neurosci Methods 139:231–240

Mallat S (1989) A theory for multiresolution signal decomposition: the wavelet representation. IEEE Trans Pattern Analysis Machine Intell 2:674–693

Mallat S (1998) A wavelet tour of signal processing. Academic, San Diego

Quian Quiroga R. (1998). Quantitative analysis of EEG signals: Time-frequency methods and chaos theory. PhD thesis. University of Lubeck, Germany.

Quian Quiroga R, Schürmann M (1999) Functions and sources of event-related EEG alpha oscillations studied with the Wavelet Transform. Clin Neurophysiol 110:643–655

Quian Quiroga R, Sakowicz O, Basar E, Schurmann M (2001) Wavelet transform in the analysis of the frequency composition of evoked potentials. Brain Res Protoc 8:16–24

Quian Quiroga R, Kraskov A, Kreuz T, Grassberger P (2002) Performance of different synchronization measures in real data: A case study on electroencephalographic signals. Phys Rev E 65:041903

Samar VJ, Swartz KP (1995) Multiresolution analysis of event-related potentials by wavelet decomposition. Brain Cogn 27:398–438

Samar VJ, Bopardikar A, Rao R, Swartz K (1999) Wavelet analysis of neuroelectric waveforms: A conceptual tutorial. Brain Lang 66:7–60

Sartoretto F, Ermani M (1999) Automatic detection of epileptiform activity by single-level wavelet analysis. Clin Neurophysiol 110:239–249

Schiff S, Aldroubi A, Unser M, Sato S (1994a) Fast wavelet transformation of EEG. Electr Clin Neurophysiol 91:442–455

Schiff S, Milton J, Heller J, Weinstein S (1994b) Wavelet transforms and surrogate data for electroencephalographic spike and seizure localization. Optical Eng 33:2162–2169

Senhadji L, Dillenseger JL, Wendling F, Rocha C, Linie A (1995) Wavelet analysis of EEG for three-dimensional mapping of epileptic events. Annals of Biom Eng 23:543–552

Strang G, Nguyen T (1996) Wavelets and filter banks. Wellesley-Cambridge, Wellesley
Unser M, Aldroubi A (1996) A review of wavelets in biomedical applications. Proc IEEE 84:626–638
Unser M, Aldroubi A, Eden M (1992) On the asymptotic convergence of B-Spline wavelets to Gabor functions. IEEE Trans Inf Theory 38:864–872
Yeung N, Bogacz R, Holroyd CB, Cohen JC (2004) Detection of synchronized oscillations in the electroencephalogram: An evaluation of methods. Psychophysiology 41:822–832

Chapter 5
Single-Trial Evoked Potentials: Wavelet Denoising

5.1 Introduction: Single-Trial Evoked Potentials

Evoked potentials are typically very small in comparison with the ongoing electroencephalogram and, consequently, they are hardly visible in the individual trials. As discussed in Sect. 1.6, to improve the visualization of the evoked responses, it is a common practice to average several presentations of the same stimulus. Then, the ongoing EEG activity cancels out and the amplitude of the evoked potentials relative to the background EEG increases proportional to the square root of the number of trials. From the average responses, it is possible to identify evoked components, whose amplitudes, latencies, and topographies have been correlated with different sensory and cognitive functions (Regan 1989; Niedermeyer and Lopes da Silva 1993; Quian Quiroga 2006).

Although the averaging of individual responses improves the signal-to-noise ratio, it assumes that the evoked potentials are an invariant pattern time-locked to the stimulus, laying on an independent stationary and stochastic EEG signal. These assumptions are in strict sense not valid (see, e.g., Başar 1980). In particular, it has been shown that the spectral content of the background EEG at the time of stimulation does have a strong influence on the evoked waveforms (Brandt et al. 1991; Jongsma et al. 2000). But this is not the major problem. Even if there is some relationship between the ongoing EEG and the evoked responses, and even if the EEG cannot be strictly considered as additive noise, ensemble averaging improves the identification of the evoked potentials. In fact, ensemble averaging has been successfully used since the 1950s and there is no doubt about its clinical and scientific value. The major problem, illustrated in Fig. 5.1, is somehow more fundamental in nature. In the *bottom* plot of Fig. 5.1, we can clearly identify the evoked responses averaged over 16 trials (a P100, a N200, and a P300; see Sect. 1.5), but how do these responses change trial-by-trial during the experiment? Averaging implies a loss of information that is crucial to study the time course of dynamic cognitive processes. Moreover, these variations might affect the reliability of the average evoked potentials as a representation of the single-trial responses. For example, the wide

W.J. Freeman and R. Quian Quiroga, *Imaging Brain Function With EEG: Advanced Temporal and Spatial Analysis of Electroencephalographic Signals*, DOI 10.1007/978-1-4614-4984-3_5, © Springer Science+Business Media New York 2013

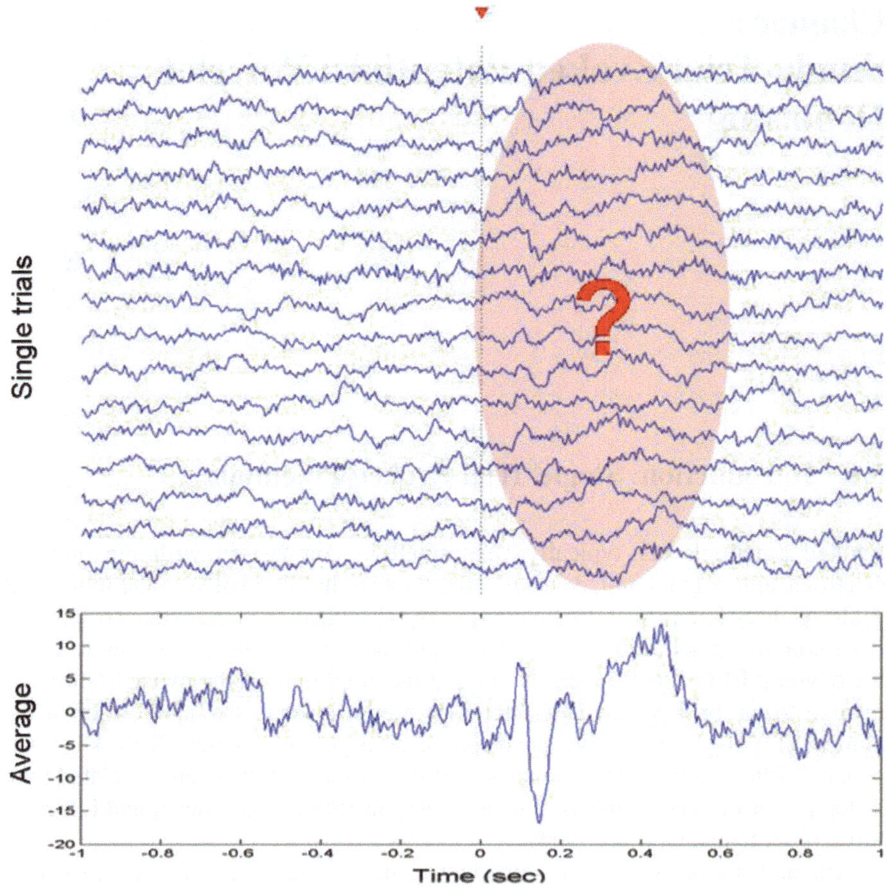

Fig. 5.1 16 single-trial (*above*) and average (*below*) evoked potentials upon target stimulation using pattern visual stimulation. In the average, we observe a P100, a N200, and a P300. But how do these responses change in the single trials?

P300 peak could have been generated by narrower single-trial responses with a variable latency.

Growing evidence has shown the important contributions of single-trial analyses in cognitive neuroscience (see, e.g., Quian Quiroga et al. 2007). From a physiological perspective, one might expect that neural responses are modified after several repetitions of the same stimulation pattern, or that they change during the emergence and consolidation of new brain representations, as it occurs during learning processes. Moreover, the identification of single-trial responses allows better averages, by eliminating trials with poor responses or by aligning peak latencies, as we will see in the next sections. This can have clinical applications because it can reduce the number of trials needed for obtaining robust average responses, for example, for pain evoked potentials – where the need to reduce the number of trials is obvious – or evoked potentials in children, who can typically be engaged in a task only for a few trials. But if we are able to study single-trial responses – as we will show in the next sections by using a denoising method based on the wavelet transform – perhaps the most important application is to do a radical paradigm shift in the way we design evoked potential experiments since more than 50 years! Following the "classic school," experimental paradigms typically tend to avoid single-trial variations to obtain reliable averages. The new type of paradigms we propose go in the opposite direction, given that we will actually try to elicit single-trial changes that could be correlated to cognitive processes.

5.2 Previous Approaches

With the classic evoked potential paradigms, a compromise should be taken when deciding the number of trials to be used. On the one hand, it is desirable to have several trials to get a good signal-to-noise ratio but, on the other hand, the number of trials should not be too large or the averages will be influenced by varying arousal levels, degrees of attention, etc. Moreover, in some cases the first few trials are discarded to avoid deviant responses due to, for example, habituation or sensitization effects (see Sect. 5.5). The problem of variability across an experiment can be partially solved by using sub-ensemble averages, which are consecutive averages of a few single trials. But this approach is limited, especially when there are few trials available or when the evoked responses change from one trial to the next.

Several methods have been proposed to filter the average evoked potentials (Lopes da Silva 1993). The success of these methods would imply the need of fewer trials, potentially leading to single-trial identification. There have also been attempts to directly filter the single-trial traces, in particular using techniques based on the Wiener formalism (Walter 1969; Doyle 1975). Wiener filters are constructed from the power spectrum of the average evoked potentials to filter the frequency activity not present in the average responses. However, these filters have the common drawback of considering the signal as a stationary process, and, given that the evoked potentials are compositions of transient responses with different time and frequency localizations, they do not give optimal results (Quian Quiroga and Garcia 2003). A natural step forward is then to implement time-varying filters using, for example, the wavelet transform.

The use of wavelets for filtering average evoked potentials or for visualizing single-trial responses was first reported in the early 1990s (Bartnik et al. 1992; Bertrand et al. 1994; Thakor et al. 1993). However, these works proposed denoising implementations based exclusively on the average responses, without considering latency variations in the single trials. To overcome this problem, other studies proposed to use latency corrections (Effern et al. 2000a) or an embedding of the single-trial responses in phase space (Effern et al. 2000b). But the caveat of these methods is that the former assumes that there is a single evoked response to be corrected (which is typically not the case) and the latter assumes that the shape of the evoked responses is similar in the different trials, even if appearing at different latencies.

This is also not true in general and furthermore, single-trial responses are typically not distinguishable from spontaneous EEG patterns.

The method we will present in the next section is more straightforward and it explicitly uses the knowledge of the time and frequency ranges in which the single-trial evoked responses are expected to occur (Quian Quiroga 2000; Quian Quiroga and Garcia 2003). The obvious disadvantage is that it requires heuristic adjustments, but once the wavelet coefficients are chosen, it does not need to be readjusted for different signal-to-noise ratios, number of trials, etc.

5.3 Wavelet Denoising

In the multiresolution decomposition of the average evoked potential of Fig. 5.2. (*gray traces*) the P100–N200 response is mainly correlated with the first post-stimulus coefficients in the levels 4 and 5 (details D_4–D_5), and the P300 is correlated with the coefficients at about 400–500 ms in the level 6 (A_5). This correspondence is easily identified because (1) the coefficients appear in the same time (and frequency) range of the evoked responses, (2) they are relatively larger than the rest due to phase-locking between trials (i.e., coefficients related with background oscillations cancel out in the average), and (3) the time-frequency composition of the evoked responses can be identified from the reconstructed waveforms of the right plots. In consequence, a straightforward way to filter the fluctuations related to the ongoing EEG is just by equaling to zero those coefficients not correlated with the evoked responses. However, the choice of these coefficients should not be solely based on the average evoked potential and it should also consider the time ranges in which the single-trial evoked potentials are expected to occur (i.e., some neighbor coefficients may be included to allow for latency jitters). In this respect, we can choose the coefficients correlated with the evoked responses from the average signal and then heuristically adjust this set of coefficients by comparing the outcomes of the denoised single-trial responses with the raw data (Quian Quiroga 2000; Quian Quiroga and Garcia 2003).

The coefficients used to denoise the evoked potentials are shown in *red* in Fig. 5.2. The two upper plots in this figure show the average evoked potentials before (*gray*) and after denoising (*red*). Note that in the denoised waveform the background EEG oscillations are filtered. This is usually difficult to achieve with a standard digital filter due to the different time and frequency localizations of the P100-N200 and the P300 responses, and also due to the overlapping frequency components of these peaks and the ongoing EEG. In particular, a bandpass filter to obtain the P100–N200 would have filtered the P300 and conversely, a low-pass filter to extract the P300 would have filtered the P100–N200.

Once the coefficients of interest are identified from the average evoked potentials, we can apply the same procedure to the single trials, as shown in Fig. 5.3. Note that after denoising we can distinguish the P100–N200 and the P300 in most of the trials. Note also that these responses are not easily identified in the original signal due to their similarity with the ongoing EEG. For an easier visualization, in Fig. 5.4 we display the single-trial evoked potentials (with and without denoising) using

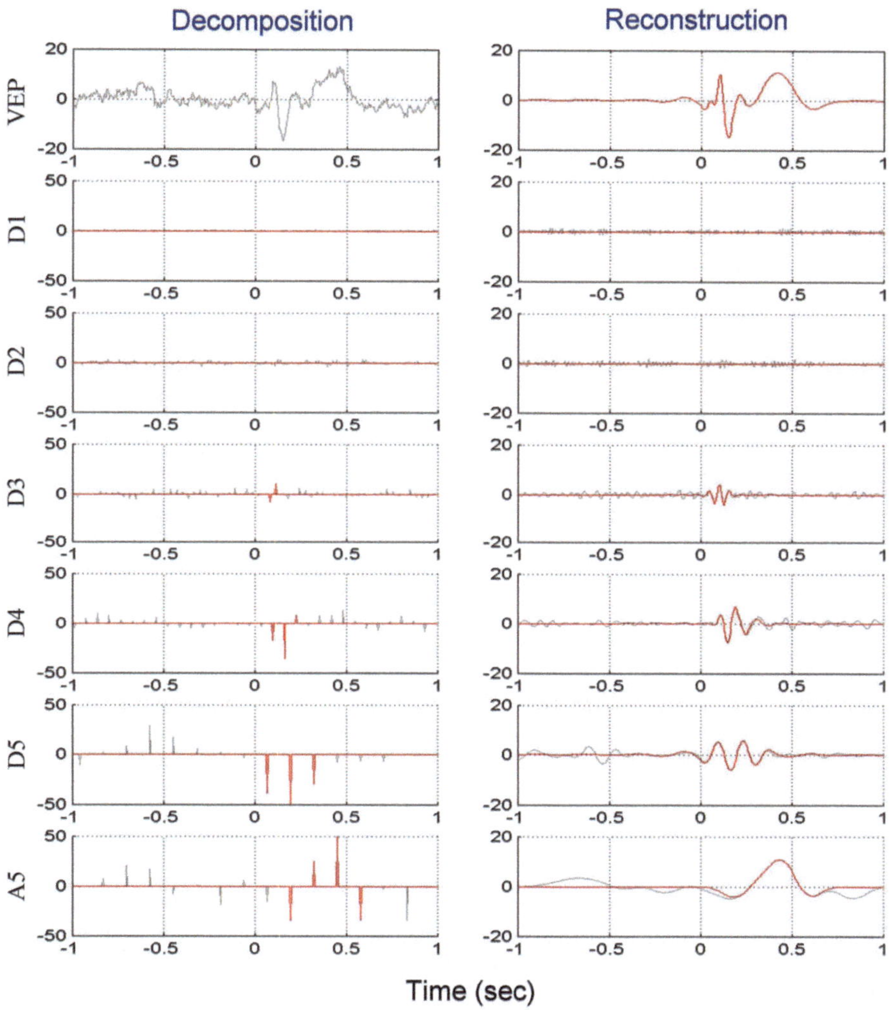

Fig. 5.2 In gray, multiresolution decomposition and reconstruction of the average evoked potential of Fig. 5.1. Note that the evoked responses are correlated with a few wavelet coefficients (in *red*) in the scales D4, D5, and A5. The *red traces* show the reconstruction of the average evoked responses from these coefficients

contour plots. In the denoised plot, we observe between 100 and 200 ms a yellow/red pattern followed by a blue pattern corresponding to the P100–N200 peaks. The more unstable and wider yellow/red pattern at about 400–600 ms corresponds to the P300. Noteworthy, all these responses are more difficult to be recognized in the original signal.

In line with the previous arguments, an analysis with simulated data – where the denoising performance could be quantified and compared to other methods – showed that denoising significantly improved the signal-to-noise ratio as well as the estimation of the amplitude and latency of the single-trial responses. Furthermore, results

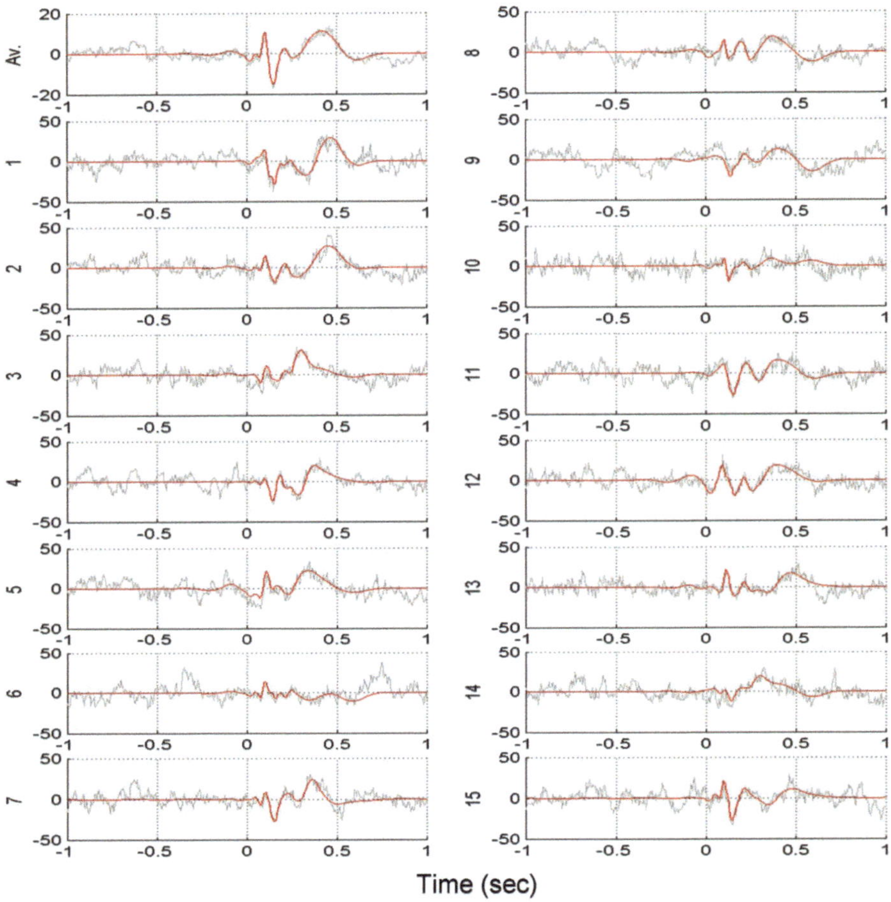

Fig. 5.3 Average evoked potential and first 15 single-trial responses with (*red*) and without (*gray*) denoising, corresponding to the data of the previous figure. Note that the single-trial responses are much easily identified after denoising

were significantly better than those obtained with Wiener filters, even though these are designed to provide optimal filtering in the mean square error sense (Quian Quiroga and Garcia 2003).

In summary, the wavelet denoising method for obtaining single-trial evoked potentials consists of the following steps[1]:

1. The activity of the average evoked potential is decomposed using the wavelet multiresolution decomposition.
2. The wavelet coefficients correlated with the evoked responses are identified and the remaining ones are set to zero. The chosen coefficients should cover a time range in which the single-trial evoked potentials are expected to occur.

[1] A simple implementation of the method in Matlab as well as tutorials, help files, and sample data sets can be obtained from www.le.ac.uk/neuroengineering.

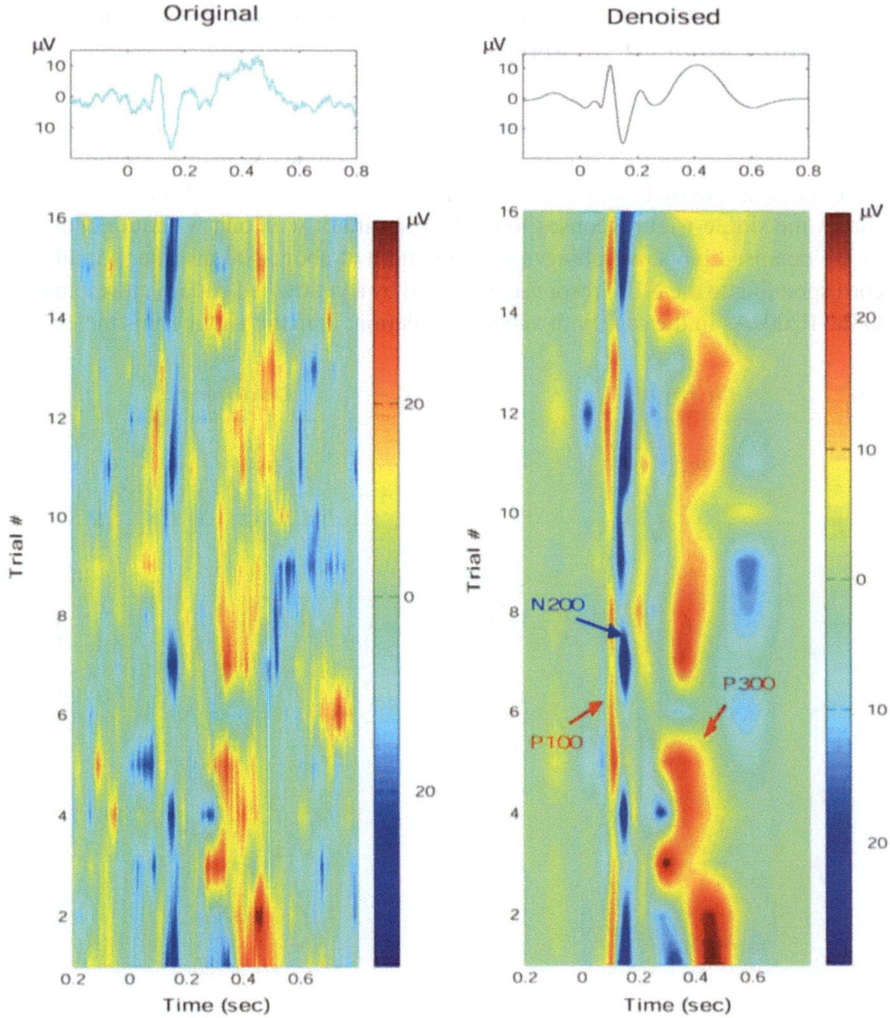

Fig. 5.4 Average evoked potentials with and without denoising (*top*) and contour plots of the single-trial responses (*bottom*) corresponding to the data of Fig. 5.3. Single-trial responses are clearer after denoising. Note the latency and amplitude variability of the P300

3. The inverse transform is applied, thus obtaining a denoised average evoked potential.
4. The denoising scheme defined by the previous steps is applied to the single trials.
5. The validity of the method can be checked by applying the same procedure to ongoing EEG test signals, taking, for example, pre-stimulus data.

5.4 Application to Auditory Evoked Potentials: Selective and Latency-Corrected Averages

Figure 5.5 shows the average auditory evoked potentials of a typical subject (492) obtained from electrode Cz using an oddball paradigm. Nontarget stimuli (75%) were tones of 1,000 Hz and target stimuli (25%) were tones of 500 Hz. The figure shows the single-trial responses to nontarget stimuli with and without denoising. In the denoised plots, we observe a blue pattern at about 100 ms after stimulation corresponding to the N100 response, followed by a yellow/red pattern corresponding to the P200. As in the case with visual stimulation, in many single trials the evoked

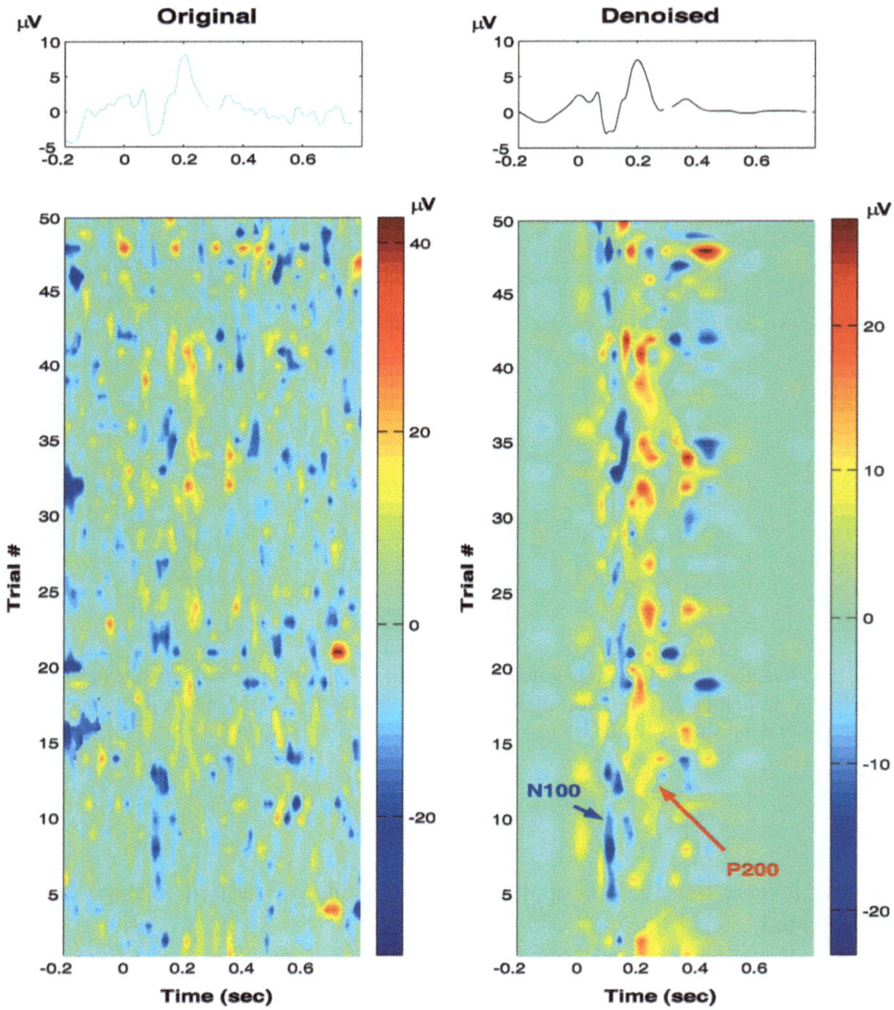

Fig. 5.5 Average (*top*) and single-trial (*bottom*) auditory evoked potentials with and without denoising. As for the case of visual evoked potentials in the previous figure, denoising improves the visualization of the single-trial responses

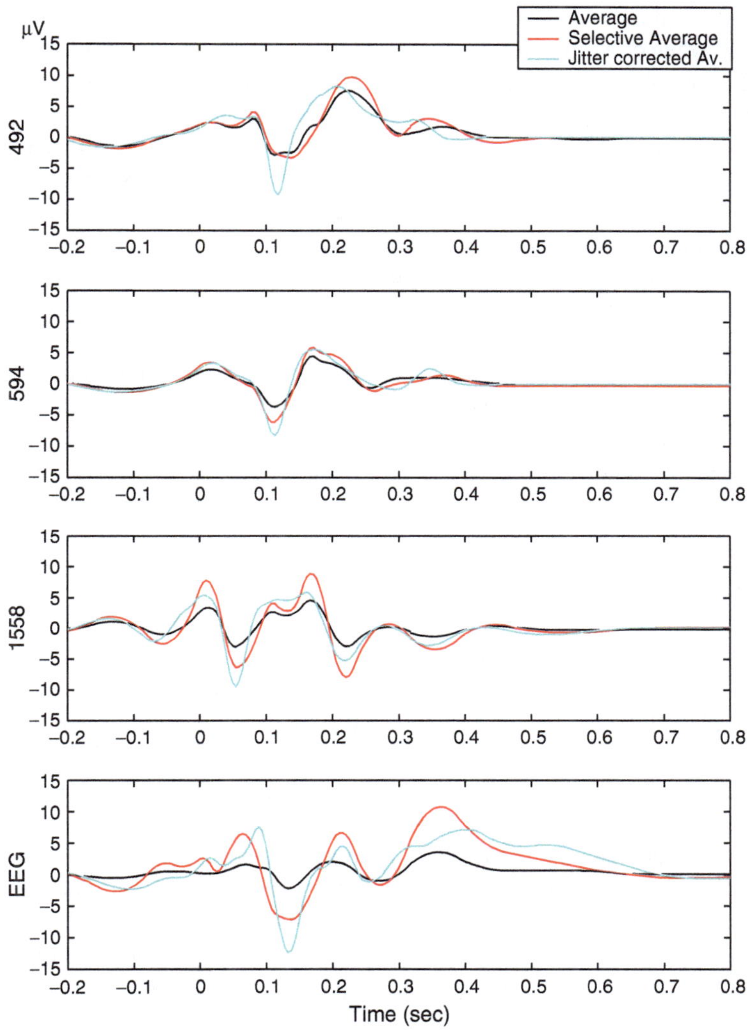

Fig. 5.6 Denoised, selective, and jitter-corrected averages for three subjects and control EEG data. The data for subject 492 corresponds to the one shown in Fig. 5.5

responses are not well defined. Moreover, the latency of the N100 has some variability. Therefore, two ways of improving the averages are by (1) selecting only those trials with good evoked potentials (selective averaging) and (2) correcting for latency jitters between trials. For the selective average, we calculated the cross-correlation between the denoised average and the denoised single trials to select the trials with a cross-correlation larger than a certain value (0.4). From these trials, we then calculated the jitter-corrected averages by aligning the maximum of the N100 peaks to the maximum of the average (see Quian Quiroga 2000; Quian Quiroga and Garcia 2003).

Figure 5.6 shows the average evoked potentials, the selective averages, and the jitter-corrected averages for three subjects and for a control signal obtained by averaging segments of ongoing EEG data. For subject 492, the average is much improved when correcting for latency jitters, due to the latency variability shown in Fig. 5.5. For subject 594, the average is mainly improved by the selective average and not by the jitter correction (i.e., meaning that the latencies of these single-trial responses were more stable compared to subject 492). Note also that the jitter-corrected averages are sharper than the original evoked potentials. This observation brings us back to the point of whether or not an average evoked potential is representative of the single-trial responses (see Sect. 5.1). The third subject (1558) should be treated with more care because no clear evoked potentials are recognizable in the average. This can be due to a lack of responses in the single trials or due to a high latency jitter. Although both the selective and the latency-corrected averages seem to better resolve some components, they also increase background oscillations, meaning that rather than improving the visualization of real evoked responses, we may be just aligning ongoing EEG oscillations. This suspicion is confirmed by the fact that a similar result is obtained when applying the same procedure to the test EEG signal.

5.5 Habituation and Sensitization

Figure 5.7a shows the grand average evoked potentials of 13 rats to auditory click stimuli of 1 ms duration, measured in the vertex (for details, see Quian Quiroga and van Luijtelaar 2002; de Bruin et al. 2001). We observe two positive components, at 13 and 20 ms, and 4 negative ones at 18, 24, 38, and 52 ms, respectively. Besides giving the amplitude and latency of these components, there is really not much more we can say with this "classic" analysis of average evoked potentials. However, we could expect systematic changes in these responses during the recording session due to habituation, that is, a response decrease given by the stimulus repetition, which can typically be fitted by a negative exponential function (Sokolov 1960). Moreover, we could have increases of the responses in the first trials due to sensitization processes, that is. the system getting prepared to better process a given stimulus (Thompson and Spencer 1966; Groves and Thompson 1970).

In order to study any possible habituation and sensitization processes, for each of the evoked potentials of Fig. 5.7a, we analyzed the single-trial amplitudes in the first 100 trials. For each rat, the single-trial amplitudes of each peak were automatically defined from the maximum (minimum) value within an appropriate time window after wavelet denoising (for details, see Quian Quiroga and van Luijtelaar 2002). Figure 5.7b shows the amplitude variations of the different evoked potentials as a function of trial number. There is a clear exponential amplitude decay for the first four peaks with trial number, which stabilizes after 30–40 trials. Moreover, for the P13, P20, N24, and most markedly for the N18, the responses to the first trial were smaller than for the following trials, thus showing sensitization. The two late components (N38 and N52) did not have a slow exponential decay as the earlier responses. However, there was a fast amplitude decay for the first three trials pointing toward a fast habituation process (Quian Quiroga and van Luijtelaar 2002).

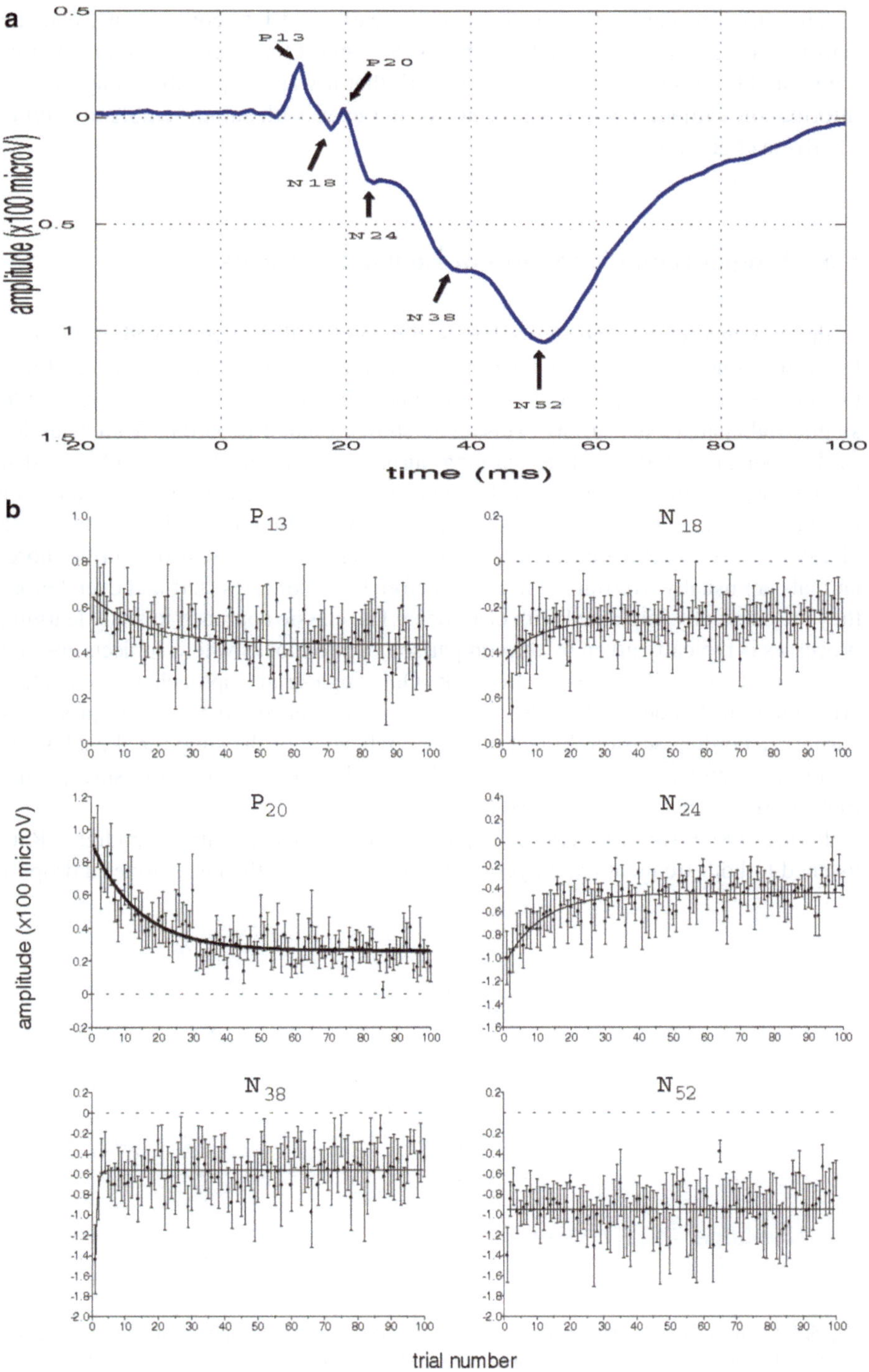

Fig. 5.7 Grand average auditory evoked potential for 13 rats (*top*). Amplitude variations with trial number for the six evoked potentials observed in the grand average (*bottom*). There are clear systematic changes with trial number due to habituation and sensitization processes

The different habituation patterns of the early and late components, and the sensitization observed especially for the N18, suggest that these evoked responses are related to different functions. We remark that it was not possible to assess this information without a single-trial analysis, not even with sub-ensemble averaging (de Bruin et al. 2001).

5.6 Single-Trial Correlates of Learning in Rats

In the previous section, we showed evoked potentials that were related to different brain functions, but we did not have any behavioral measure to specify which functions these were. Perhaps one of the most interesting cognitive processes related to single-trial changes is learning. This again shows a major departure from "classic" evoked potential studies, where learning during an experiment is typically avoided because it introduces changes in the evoked responses that are not desirable for obtaining good averages. To avoid this variability, in some cases the first trials are discarded, or paradigms are practiced before starting a recording session. Furthermore, animals are usually overtrained in a task before the recordings start and even before they are surgically implanted with electrodes. The classic approach to study learning processes is typically given by the comparison of two blocks where subjects are first naïve and then well trained in the task. Our single-trial approach is completely orthogonal to this design because we actually do want to have learning processes during a recording session. In fact, the key application of the single-trial analysis is to correlate any learning process, assessed with different behavioral measures, to the trial-by-trial changes of the evoked responses.

Figure 5.8 shows a typical example of a two-block learning paradigm. Rats listened to both frequent and target tones, and only upon the target tones a drop of

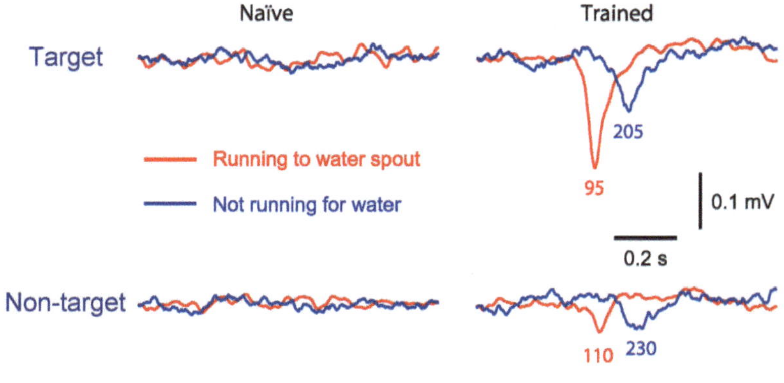

Fig. 5.8 Grand average evoked responses in the dentate gyrus for the naïve and trained rats upon target and nontarget stimuli. Note the appearance of a negative potential at about 100 ms for the trained rats for the correct target trials (i.e., with the rat running to the waterspout) (Adapted from Talnov et al. 2003)

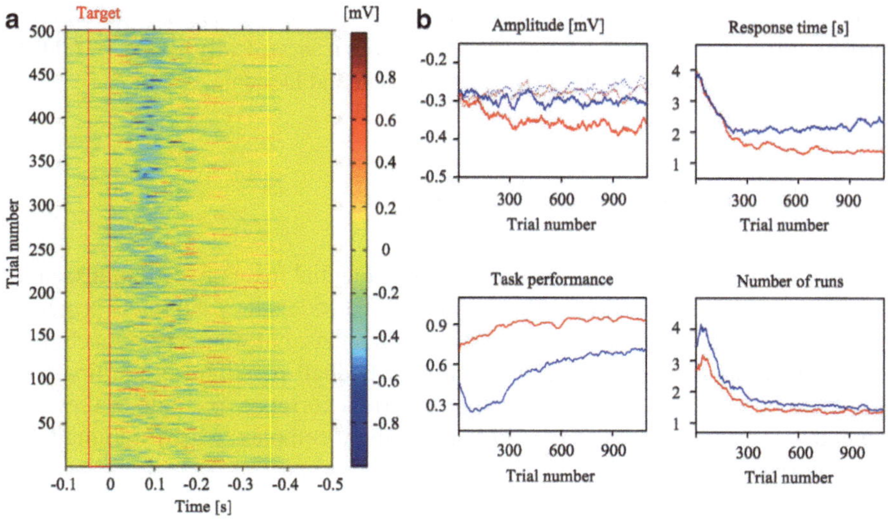

Fig. 5.9 (**a**) Single-trial evoked responses to target stimuli for 500 trials in a representative rat. Note the appearance of an N100 at about trial 200. (**b**) Grand average amplitudes of the single-trial evoked responses for the correct target (*solid red*) and nontarget (*solid blue*) stimuli. *Dotted lines* correspond to the same analysis performed with pre-stimulus data, as control. The other plots correspond to three different behavioral measures for the target (*red*) and nontarget (*blue*) stimuli obtained with a moving average of n = 50 trials. From the behavioral measures, we observe that the rats learned the task at about trial 200–300 in agreement with the appearance of the N100 response in the single-trial plots (Adapted from Talnov et al. 2003)

water was delivered in a waterspout. Learning took hundreds of trials and, following the "classic" approach, in Fig. 5.8 we see the comparison of the evoked responses in the naïve and trained rats. We observe that only for the trained rats and for the correct target trials (i.e., when the rats run to the waterspout) there was a negative component at about 100 ms (N100) in the dentate gyrus.

From the difference between the naïve and trained rats, it can be inferred that the N100 is correlated to the learning of the oddball paradigm. But can we follow how this component appeared on a trial-by-trial basis and correlate it to behavioral measures of the learning that took place? For this, using wavelet denoising we first identified the N100 component in the single trials, as shown in Fig. 5.9a. In this figure we observe a negative (*blue*) pattern at about 100 ms appearing after trial 200. The upper left plot of Fig. 5.9b shows the grand average (across seven rats) single-trial peak amplitudes, which were automatically obtained as the minimum value in a proper time window. In agreement with the pattern observed in Fig. 5.9a, there is a clear amplitude decrease (peak becoming more negative) only for the correct target trials (*solid red line*). The appearance of this N100 was correlated to several behavioral measures – response time, task performance, and number of runs – as shown in Fig. 5.9b (see Talnov et al. 2003 for details). After learning, the N100 responses could be switched on and off by changing the stimulus probability: the N100 appeared whenever the target probability was less than ~30%, and disappeared whenever the target probability was above 50%. This modulation with stimulus probability is very reminiscent of the P300, which is typically found in oddball paradigms with human subjects (Duncan-Johnson and Donchin 1977).

5.7 The Learning Oddball Paradigm

In the previous section we saw how an evoked potential in rats, very reminiscent of the P300 in humans, appeared with the learning of an oddball paradigm. In principle, it would be interesting to do the same study in humans, but we are way much more clever than rats and we immediately learn an oddball paradigm in the first trial. So, there is really no learning curve that we could correlate to single-trial evoked potentials. What we need is a paradigm much more complex, where learning takes several trials.

The P300 amplitude is modulated by target probability (Duncan-Johnson and Donchin 1977) and also by inter-stimulus and inter-target intervals (Croft et al. 2003; Fitzgerald and Picton 1981; Gonsalvez et al. 1995; Gonsalvez and Polich 2002). Moreover, sequence (Squires et al. 1976) and expectancy effects (Donchin 1981; Jentzsch and Sommer 2001; Jongsma et al. 2005; Polich and Kok 1995) have a strong influence in the evoked responses. In particular, we expect that the P300 will increase with the unexpectedness of a target stimulus; that is, the P300 is lower if we can predict that there is a target stimulus coming. In a typical oddball experiment, infrequent target stimuli (appearing about 25% of the times) are randomly interleaved in between standard (nontarget) stimuli. We could then change the stimulus expectancy – without changing the mean target probability, inter-stimulus and inter-target intervals – by switching from a random oddball sequence to a regular one, where the targets appear at regular intervals (e.g., after four nontargets). The predictability of the targets in the regular sequence should lead to a decrease of the P300 response, and we can then measure the learning of such repeated pattern – which is typically unconscious – by tracking the single-trial changes in the P300 amplitude. The advantage of this "learning oddball" paradigm is that the learning process is measured directly from the brain signals, without typical confounds of motor response issues, as when pattern learning is measured from reaction times (Seger 1994).

Figure 5.10a shows the grand average auditory evoked responses (24 subjects) to the learning oddball paradigm, where in each of six consecutive blocks a random oddball sequence with eight targets was switched to a fixed sequence with the same number of targets. We observe that the random sequence (*dotted lines*) generated a larger N200 and P300 components. Another evoked response that depends strongly on stimulus expectancy – but with exactly the opposite behavior given that it increases with stimulus predictability – is the contingent negative variation (CNV; see Sect. 1.5), shown in the leftmost plot of Fig. 5.10a. The CNV was higher for the regular sequence, where the target stimuli were (at least unconsciously) expected. Panels B and C show the amplitude of the CNV and the P3-N2 (the P300 minus the N200) for each of the eight random and the eight regular targets, averaged across

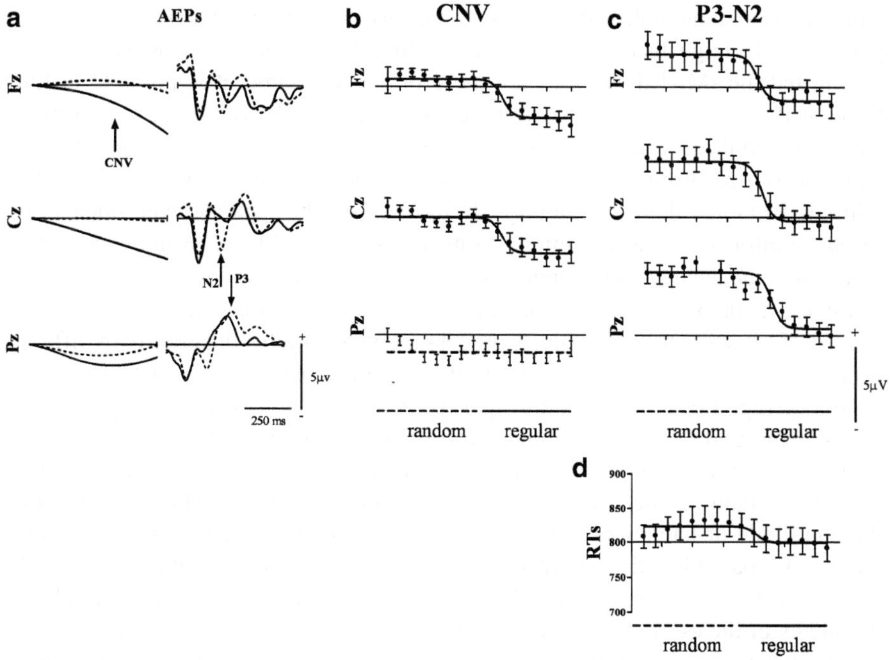

Fig. 5.10 (**a**) Contingent negative variation (CNV) and evoked responses for the random (*dotted lines*) and regular (*solid lines*) oddball sequences. (**b**) CNV single-trial amplitudes for the random and regular sequences. (**c**) Same for the single-trial P3-N2 responses. Both for the CNV and the P3-N2 there is a change in response amplitude when the sequence switches from random to regular. This learning effect is not clear from the reaction times (**d**) (Adapted from Jongsma et al. 2006)

blocks. Amplitudes were obtained from the single-trial responses after wavelet denoising. Note that the switch to the regular sequence led to a dramatic change in both components, which was correlated to the implicit learning taking place in the experiment. These changes could be fitted by sigmoid functions. As mentioned above, reaction time measurements mix learning and motor effects, and consequently the learning process is not clearly observed from the reaction time curves (Fig. 5.10d). Interestingly, when analyzing each block separately it was observed that the onset of the sigmoid learning curves decreased with each block, thus showing a higher order learning effect (Jongsma et al. 2006).

5.8 Simultaneous EEG and fMRI Recordings: Role of Single-Trial Analysis

In the last years, several works reported the use of simultaneous EEG and functional magnetic resonance imaging (fMRI) recordings. The basic idea is to combine the good temporal (but bad spatial) resolution of the EEG with the good spatial (but bad temporal) resolution of the fMRI. These simultaneous recordings can be challenging, in part due to the large distortions introduced to the EEG signals by the fMRI recording. Given that the fMRI distortions are more or less the same, it is possible to filter these artifacts using different commercially available algorithms. But even if we can deal with these artifacts, the key question is whether it is really necessary to do simultaneous recordings or whether the same experiment could be done separately with the EEG and the fMRI settings. In most cases, the separate experiments give the same information as the simultaneous recordings, and it seems not necessary to go through the technical challenge of doing the recordings simultaneously. There is, however, a major exception, and this is when single-trial information is used. Indeed, if for some reason subjects process a given stimulus differently in different trials (e.g., due to learning or varying degrees of attention), this same variability cannot be reproduced in a second experiment with the other recording system. The only way around this is to do both measurements simultaneously.

The idea to join the EEG and fMRI data is to correlate the evoked responses in each trial (obtained after wavelet denoising), with the voxel fMRI activity. If either the amplitude or latency of the evoked responses covary with certain voxels, then we get the spatial localization from these voxels and the time localization from the latency of the evoked responses. We only need enough variations in the amplitude or latency of the evoked potentials to be able to asses these covariations, but this is exactly what we get with the learning oddball paradigm described in the previous section.

Using the learning oddball paradigm with simultaneous EEG and fMRI recordings, it was indeed shown that the P300 elicited activations in the frontal, temporal, and parietal regions, mainly in the right hemisphere (Eichele et al. 2005). Other evoked components that were modulated by the learning oddball paradigm, the P2 and N2, had different spatial localizations. Note that these different localizations get mixed when considering solely the fMRI data because it has a temporal resolution of a few seconds.

Perhaps the P300 is not the most interesting evoked response to localize with the fMRI data given that it seems to involve many generators (Picton 1992; Polich and Kok 1995). But the important message here is that the single-trial analysis described in this section offers a new opportunity to combine the temporal information of evoked responses with the spatial information from the fMRI recordings.

5.9 A New Mechanism of Sleep-Induced Learning Revealed by Single-Trial Analysis

In the previous sections, we described the oddball paradigm, which basically involves distinguishing standard from target stimuli. This can take hundreds of trials for rats but such distinction is done from the first trial by human subjects. To make the task more difficult, in order to follow how learning develops trial by trial, we used complex sounds that were formed by a sequence of eight consecutive and

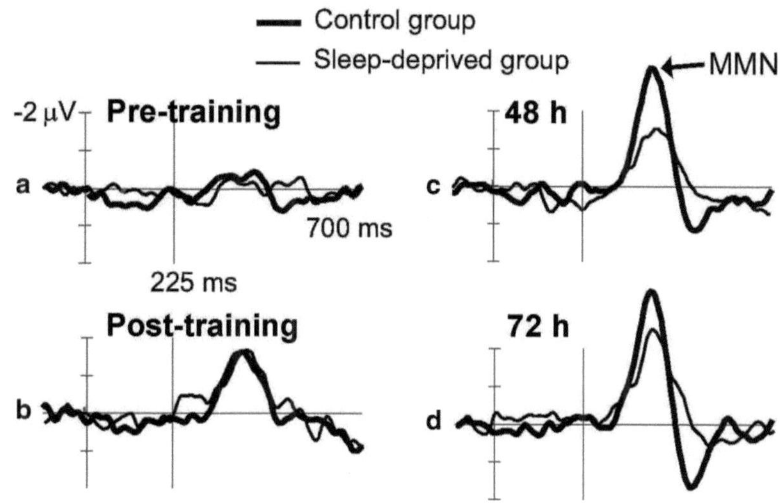

Fig. 5.11 Grand average (10 subjects) responses in a frontal-central electrode (Fz) for the sleep-deprived and the control group. For both groups, a mismatch negativity (MMN) response appeared immediately after training and only for the control group it increased further in amplitude in the recordings performed 48 and 72 h after training. To visualize the MMN responses, to each target stimulus we subtracted the standard stimulus following it, thus stressing the differential responses between targets and standards (Adapted from Atienza et al. 2005)

very fast tones, altogether lasting 365 ms, in which the frequency of the sixth tone (225 ms after the first one) was slightly changed for the target stimuli. Subjects were asked to respond to the targets as accurately and quickly as possible and required between 3 and 6 blocks of 240 sound presentations (with 36 targets in each block) to learn the paradigm (for details, see Atienza et al. 2004).

It is already well established that sleep following a training session contributes to the consolidation of learning. At this respect, it has been shown that the learning of an oddball paradigm with the abovementioned complex sounds was correlated with a mismatch negativity component (MMN; see Sect. 1.5) that appeared after training. This MMN was significantly lower for subjects that were sleep deprived after learning the task compared to control subjects (Atienza et al. 2004). Evoked potentials in the sleep-deprived subjects and in the control group were measured 48 and 72 h after learning to avoid differences in the arousal levels between both groups, as corroborated by awareness tests (Atienza et al. 2004).

Figure 5.11 shows the grand average MMN in a frontal electrode (Fz) immediately before and after training, as well as at 48 and 72 h post-training for the control and the sleep-deprived subjects. No clear responses are observed before the subjects learnt the task, but after training subjects were able to automatically detect the target stimuli, something that was correlated with the appearance of the MMN component 200 ms after the deviant tone in the complex sound (i.e., 425 ms after stimulus onset). This MMN has been suggested to reflect the triggering of an automatic shift of attention toward the deviant sound (Näätänen 1992). Interestingly, for the control

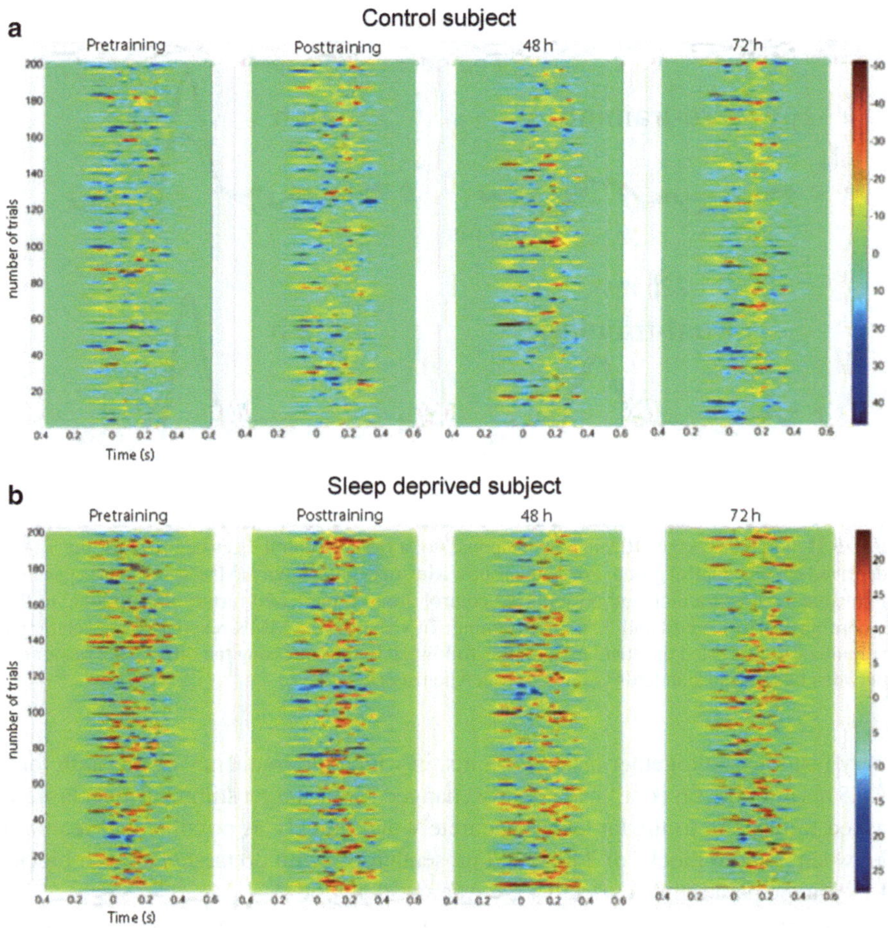

Fig. 5.12 Contour plot of the denoised single-trial target responses for a control and a sleep-deprived subject. Only for the control subject there is a clear time locking of the responses in the 48 and 72 h post-training sessions (Adapted from Atienza et al. 2005)

group the MMN showed an additional amplitude increase in the following post-training sessions. This change was not apparent for the sleep-deprived subjects, thus suggesting an influence of sleep for the consolidation of learning.

The neural mechanisms underlying the MMN increase for the control group cannot be elucidated from the previous analysis of average responses. However, it was postulated that the difference between the control and the sleep-deprived subjects could be due to a recruitment of more neurons or a strengthening of synchronization during sleep rehearsal of the task (Atienza et al. 2004). To get further insights into these possible mechanisms, a single-trial analysis with wavelet denoising was performed and, surprisingly, none of these explanations were correct!

Figure 5.12 shows the contour plots of 200 trials after denoising for one control and one sleep-deprived subject, across the different recording sessions. Note that

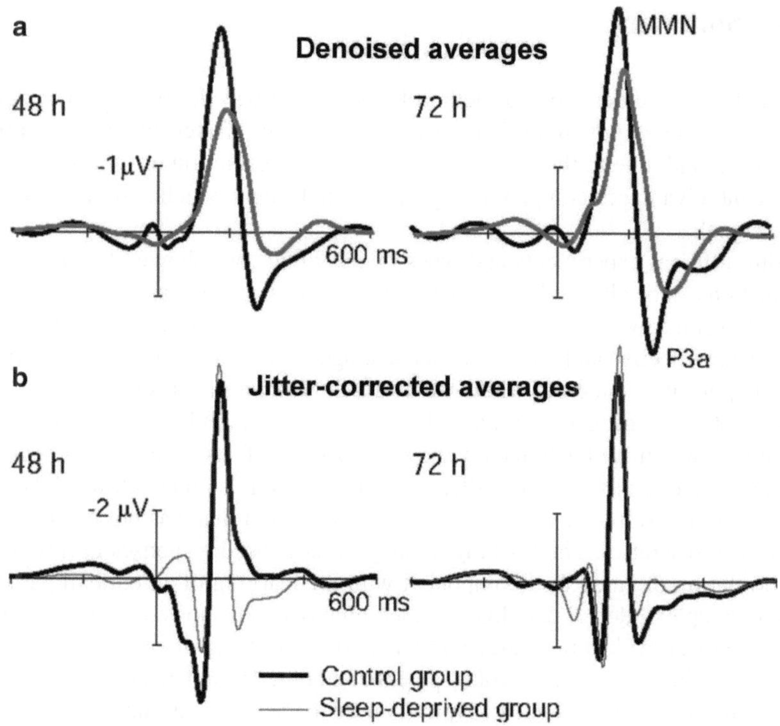

Fig. 5.13 (**a**) Denoised grand average responses for the control and sleep-deprived groups in the 48 and 72-h post-training session. (**b**) Same as (**a**) but after correcting for latency jitters. Note that after jitter correction, the MMN amplitude difference between the two groups disappeared

for the control subject the MMN (the *red* pattern at about 200 ms) becomes time locked (i.e., has a small latency variability) in the recordings performed 48 and 72 h after training, an effect that is not present for the sleep-deprived subject.

From the previous single-trial analysis, it seems that the difference between the control and the sleep-deprived group was not due to a recruitment of more neurons or a larger synchronization, but rather due to a reduction in the variability of the single-trial MMN latencies (Atienza et al. 2005). To show this, in Fig. 5.13 we display the denoised grand average responses during the 48- and 72-h post-training sessions and the jitter-corrected averages. The latency-jitter correction was done by aligning the latencies of the single-trial MMN responses to the one measured for the average MMN. The MMN amplitude differences between the two groups clearly disappeared after correcting for latency jitters, and we can then conclude that precise timing, namely, a decrease in the variability of the MMN latency, accounts for post-training sleep-dependent enhancements of the auditory MMN. In other words, sleep rehearsal induced a reliable and automatic processing of the task that led to time-locked responses.

5.10 Summary

In this chapter we described a method based on the wavelet transform to denoise single-trial evoked potentials. The method is very fast (faster than the fast Fourier transform), and, due to the optimal time-frequency resolution of wavelets, it gives significant advantages compared to typical digital filters which assume stationarity of the signal.

With different experiments and types of recordings, we showed that much more information can be obtained from a single-trial analysis in comparison to the standard study of average evoked potentials. In particular, with auditory evoked potentials we showed that selective and jitter-corrected averages gave improvements that may have clinical applications in cases where the available number of trials is limited (e.g., for evoked potentials triggered by painful somatosensory stimuli). With evoked potentials in rats, we studied habituation and sensitization effects. The varying degrees of habituation and sensitization of different evoked responses pointed toward the correlation with diverse functional processes, an information that was not available from the average responses. The tracking of single-trial evoked responses in rats allowed the correlation of an evoked component with the learning of an oddball paradigm. A more complex oddball paradigm, the learning oddball, showed systematic single-trial changes of evoked responses correlated to the learning of an oddball sequence. Such correlation between the evoked potential amplitudes and behavior was not clear from the reaction times, which are typically used to measure learning effects. By using single-trial analyses, we also showed a sensible approach to merge the time information of the EEG with the spatial information of fMRI recordings, a methodology that has large potential for further applications. Finally, we showed how a detailed single-trial analysis of MMN responses demonstrated that the learning of a task through sleep consolidation was related to a time locking of the responses, in contrast to what was postulated from the analysis of average responses.

More than favoring a particular method for denoising single-trial evoked potentials or a particular application, the main goal of this chapter was to introduce a radical change in the way experiments are designed and the data is analyzed. Typically, single-trial variations are avoided in order to obtain nicer averages. However, these variations can carry the most interesting information, as they may be correlated to different cognitive processes occurring during the experiment. The possibility of identifying evoked responses in the single trials, and the correlation of these single-trial responses with behavioral measures, then, opens a fresh window to new experimental designs and scientific questions.

References

Atienza M, Cantero JL, Stickgold R (2004) Posttraining sleep enhances automaticity in perceptual discrimination. J Cogn Neurosci 16:53–64

Atienza M, Cantero JL, Quian Quiroga R (2005) Precise timing accounts for posttraining sleep-dependent enhancements of the auditory mismatch negativity. Neuroimage 26:628–634

Bartnik EA, Blinowska KJ, Durka PJ (1992) Single evoked potential reconstruction by means of wavelet transform. Biol Cybern 67:175–181

Başar E (1980) EEG-Brain dynamics. Relation between EEG and brain evoked potentials. Elsevier, Amsterdam

Bertrand O, Bohorquez J, Pernier J (1994) Time-frequency digital filtering based on an invertible wavelet transform: an application to evoked potentials. IEEE Trans Biomed Eng 41:77–88

Brandt M, Jansen B, Carbonari J (1991) Pre-stimulus spectral EEG patterns and the visual evoked response. Electroen Clin Neurophysiol 80:16–20

Croft RJ, Gonsalvez CJ, Gabriel C, Barry RJ (2003) Target-to-target interval versus probability effects on P300 in one- and two- tone tasks. Psychophysiol 40:322–328

Lopes da Silva F (1993) EEG Analysis: Theory and Practice. In: Niedermeyer E, Lopes da Silva F (eds) Electroencephalography: Basic principles, clinical applications and related fields. Williams and Wilkins, Baltimore, pp 1097–1123

De Bruin NMWJ, Ellenbroek BA, Cools AR, Coenen AML, van Schaijk WJ, van Luijtelaar ELJM (2001) Sensory gating of auditory evoked potentials in rats: effects of interstimulus interval and repetitive stimulation. Biol Psychol 55:195–213

Donchin E (1981) Presidential address, 1980 Surprise!…Surprise? Psychophysiol 18:493–513

Doyle DJ (1975) Some comments on the use of Wiener filtering for the estimation of evoked potentials. Electr Clin Neurophysiol 38:533–534

Duncan-Johnson C, Donchin E (1977) On quantifying surprise: The variation in event-related potentials with subjective probability. Psychophysiology 14:456–467

Effern A, Lehnertz K, Grunwald T, Fernandez G, David P, Elger CE (2000a) Time adaptative denoising of single trial event-related potentials in the wavelet domain. Psychophysiol 37:859–865

Effern A, Lehnertz K, Fernandez G, Grunwald T, David P, Elger CE (2000b) Single trial analysis of event related potentials: non-linear de-noising with wavelets. Clin Neurophysiol 111:2255–2263

Eichele T, Specht K, Moosmann M, Jongsma M, Quian Quiroga R, Nordby H, Hugdahl K (2005) Assessing the spatio-temporal evolution of neuronal activation with single-trial ERP-fMRI. Proc Nat Acad Sci USA 102:17798–17803

Fitzgerald PG, Picton TW (1981) Temporal and sequential probability in Evoked Potential studies. Can J Psychol 35:188–200

Gonsalvez CJ, Polich J (2002) P300 amplitude is determined by target-to-target interval. Psychophysiol 39:388–396

Gonsalvez CJ, Gordon E, Anderson J, Pettigrew G, Barry RJ, Rennie C, Meares R (1995) Numbers of preceding nontargets differentially affect responses to targets in normal volunteers and patients with schizophrenia: a study of event-related potentials. Psychiat Res 58:69–75

Groves P, Thompson R (1970) Habituation: A dual-process theory. Psychol Rev 77:419–450

Jentzsch I, Sommer W (2001) Sequence-sensitive subcomponents of P300: Topographical analyses and dipole source localization. Psychophysiol 38:607–621

Jongsma M, van Rijn C, Quian Quiroga R, Schijk W, Dirksen R, Coenen A (2000) Influence of the power spectrum of the pre-stimulus EEG on the consecutive auditory evoked potential in rats. In Lehnertz K, Elger CE, Arnhold J, Grassberger P (eds) Chaos in Brain? World Scientific. Singapore

Jongsma MLA, Eichele T, Quian Quiroga R, Jenks KM, Desain P, Honing H, van Rijn CM (2005) The effect of expectancy on omission evoked potentials (OEPs) in musicians and non-musicians. Psychophysiol 42:191–201

Jongsma M, Eichele T, van Rijn C, Coenen A, Hugdahl K, Nordby H, Quian Quiroga R (2006) Tracking pattern learning with single-trial even-related potentials. Clin Neurophysiol 117:1957–1973

Näätänen R (1992) Attention and Brain Function. Hillslade. Erlbaum, NJ

Niedermeyer E, Lopes da Silva F (eds) (1993) Electroencephalography: basic principles, clinical applications and related fields. Williams and Wilkins, Baltimore, pp 1097–1123

Picton TW (1992) The P300 wave of the human event-related potential. J Clin Neurophysiol 9:456–479

Polich J, Kok A (1995) Cognitive and biological determinants of P300: an integrative review. Biol Psychol 41:103–146
Quian Quiroga R (2000) Obtaining single stimulus evoked potentials with wavelet denoising. Physica D 145:278–292
Quian Quiroga R (2006) Evoked potentials. In: Webster JG (ed) Encyclopedia of medical devices and implementation. Hoboken, John Wiley
Quian Quiroga R, Garcia H (2003) Single-trial event-related potentials with wavelet denoising. Clin Neurophysiol 114:376–390
Quian Quiroga R, van Luijtelaar ELJM (2002) Habituation and sensitization in rat auditory evoked potentials: a single-trial analysis with wavleet denoising. Int J Psychophysiol 43:141–153
Quian Quiroga R, Atienza M, Jongsma M (2007) What can we learn from single-trial event-related potentials? Chaos and Complexity Letters 2:345–365
Regan D (1989) Human brain electrophysiology. Evoked potentials and evoked magnetic fields in science and medicine. Elsevier, Amsterdam
Seger CA (1994) Implicit learning. Psychol Bull 115:163–196
Sokolov EN (1960) Neuronal models and the orienting response. In: Brazier MA (ed) The central nervous system and behavior III. Macy Foundation, New York
Squires KC, Wickens C, Squires NK, Donchin E (1976) The effect of stimulus sequence on the waveform of the cortical event-related potential. Science 193:1142–1146
Talnov A, Quian Quiroga R, Meier M, Matsumoto G, Brankack J (2003) Enthorinal imputs to dentate gyrus are activated mainly by conditioned events with long time intervals. Hippocampus 13:755–765
Thakor N, Xin-rong G, Yi-Chun S, Hanley D (1993) Multiresolution wavelet analysis of evoked potentials. IEEE Transl Biomed Eng 40:1085–1094
Thompson RF, Spencer WA (1966) Habituation: a model phenomenon for the study of neuronal substrates of behavior. Psychol Rev 73:16–43
Walter DO (1969) A posteriori "Wiener filtering" of average evoked responses. Electr Clin Neurophysiol (Suppl) 27:61–70

Chapter 6
Basic Concepts for Spatial Analysis

6.1 Introduction

In order to embark on the study of so complex organ as the brain, we select, observe, and measure one of the various forms of energy produced and used by the brain—electric, magnetic, chemical, thermal, and metabolic—and a hierarchical level of analysis—microscopic, mesoscopic, and macroscopic—each with its characteristic space-time scales. By choosing to read our book, we infer that readers have chosen to analyze the electroencephalogram from the scalp (EEG), the electrocorticogram from cortical surfaces (ECoG, Fig. 6.1), and the local field potentials from the depth of the brain (LFP) in any or all accessible forms and locations (Lopes da Silva 1993; Basar 1998). Then we characterize and classify the phenomena that we want to analyze and understand. We have begun with time series analysis of single channel recordings; now we undertake the spatial analysis of signals from arrays of channels. To that end, we require some basic concepts that we introduce in this chapter, with references to detailed treatments in other chapters.

Our approach is to regard the recording surfaces of scalp and cortex as screens, across which flicker the electric potentials we sample as seemingly random dots, like the snow of late night television. Yet we know the dots are not wholly random. They contain patterns that, when we learn to find and read their images, will give us a rich source of information with which to understand how brains create knowledge from sensory information. We conceive cortical input of sensory information and output of knowledge as patterns that have spatial dimensions. The existence and general locations of cortical patterns are already suggested by various techniques of brain imaging that include fMRI, BOLD, EEG, MEG, and intravital optical dyes. In the following chapters, our aim is to present high-resolution images that we obtained from multichannel EEG and ECoG signals by using electrodes in dense arrays. The images display the *spatial textures* of emergent patterns in the ECoG and EEG. We postulate that such textures may we carry the subjects' knowledge about the sensory stimuli, instead of representing the features of the stimuli.

W.J. Freeman and R. Quian Quiroga, *Imaging Brain Function With EEG: Advanced Temporal and Spatial Analysis of Electroencephalographic Signals*, DOI 10.1007/978-1-4614-4984-3_6, © Springer Science+Business Media New York 2013

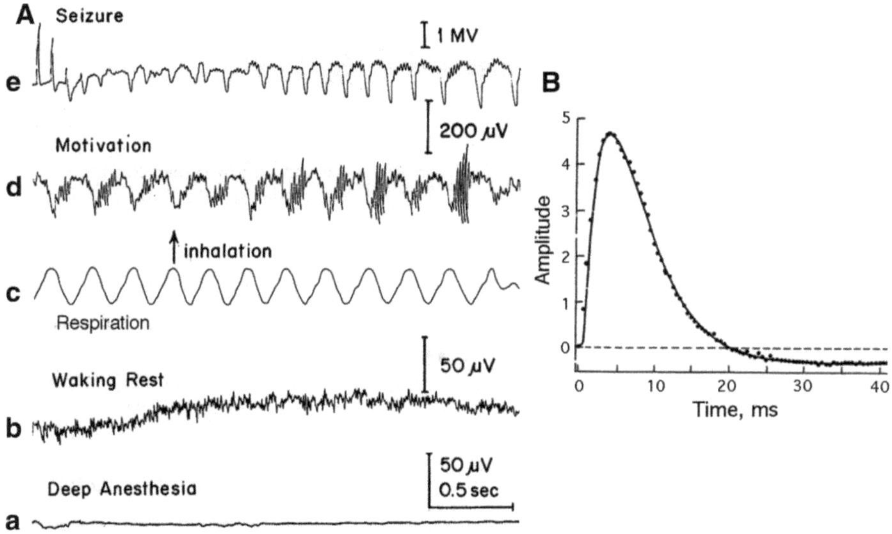

Fig. 6.1 (**a**) Examples are shown of some types of ECoG activity recorded from the rabbit olfactory system: (*a*) flat ECoG under deep anesthesia induced by intravenous pentobarbital; (*b*) awake and resting; (*c*) respiration recorded with an elastic tube around the chest (a pneumograph); (*d*) state of working; (*e*) complex partial seizure (petit mal epilepsy) induced by intense electric stimulation (From Freeman 1987). The spatial image of a seizure spike is shown in Fig. 7.8, Sect. 7.5. (**b**) The impulse response of the olfactory cortex (*dots*) under deep anesthesia that suppresses oscillations (*a*) is a multi-neuronal excitatory postsynaptic potential (*EPSP*). The subsequent downward overshoot is a dendritic after-potential, which reflects the fact that after every dendritic and axonal event, the ionic concentrations that have been diminished are restored by dissipation of metabolic energy (Chap. 11) (From Sect. 2.5.2 in Freeman 1975)

We interpret and explain the data by using brain theory. Einstein is reputed to have stated, "Make everything as simple as possible, but not simpler." Our premise is that cortical dynamics should not only focus on the level of microscopic neural networks. We need to define a more complex macroscopic level of masses of interactive neurons, into which the sensory networks inject their comparatively few microscopic pulses and from which issue macroscopic clouds of organized pulses in the billions. Our approach in this book is to record, analyze, and explain the dendritic potentials accompanying the synaptic currents that regulate the pulses. Each ECoG and EEG signal reflects the sum of contributions from an interactive mass of neurons. The interactions create neural populations with properties that differ from the firing of individual neurons in isolation or in sparse networks such as Hebbian assemblies (Sect. 8.4) (Freeman and Vitiello 2006). In the simplest description, we conceive sensory cortex as a self-regulating, self-stabilized system of neuron populations. It modulates and is modulated by other parts of the brain, but it does so on its own terms. The aim of EEG and ECoG analysis is to discover what those terms are.

Whereas cortical neural networks do microscopic dynamics using discrete pulses at precise time intervals, populations give rise to macroscopic dynamics implemented by *pulse and wave densities*, which we model as continuous state variables in space and time (Sect. 6.2). The activity densities form vector fields (Sect. 9.4) of propagating clouds of pulses, which are indirectly manifested in the scalar fields of the ECoG and EEG (Fig. 9.10, Sect. 9.6.1). It is not enough to know the amplitude and spectral distribution at each point in time and space. We also need to measure or infer the gradients and rates of change at each point. While ECoGs usually appear nearly random (Sect. 6.3), we find that the oscillations in the beta-gamma range (Sect. 3.10) occurring in brief epochs have spatial patterns of amplitude that are briefly steady state (Sect. 6.4). In order to locate the patterns more precisely, we introduce the Hilbert transform, which gives the high temporal resolution needed to display the transitions between successive images (Sects. 6.4.1 and 9.3). In those stationary images, we show that the two major operations of normal cortical dynamics—dendritic integration of waves and axonal transmission of pulses—are executed in near-linear domains (Sect. 6.5). Conformance to superposition justifies our use of the tools of linear analysis. This brings to the fore the necessity for understanding the roles in image formation of the state-dependent pulse-wave conversion at synapses and the nonlinear wave-pulse conversion at trigger zones (Sect. 6.6). With this platform we describe the interactive mechanisms that produce the broadband oscillations in the ECoG and EEG background activity at rest (Sect. 6.7), from which images emerge as narrow-band oscillations in cortices at work. The images have the form of finely textured spatial patterns of amplitude modulation of *carrier waves* in the beta-gamma range (Sect. 6.8). We introduce the concept of criticality (Sect. 6.9), which we use to explore the cognitive process of creating perceptions from sensations. We conclude with a summary (Sect. 6.10).

6.2 State Variables and Their Interrelations: Gains

6.2.1 State Variables: Axon Pulses Versus Dendritic Waves

The most basic concept is the neural state variable. Most neurons have two forms of activity, each with a specific site and function (Fig. 6.2a). Their axons *transmit* information by generating trains of action potentials (pulses, "spikes"). Their dendrites *integrate* information by generating synaptic currents, because synapses act like switches that briefly turn on a chemical battery. As shown in every basic text of neuron physiology, currents always flow in closed loops in and out of neurons (*dashed lines* in Fig. 6.2) because of charge conservation. When the dendritic currents flow across the fixed resistances of cortex, skull, and scalp, they generate voltage potential differences ($V=IR$) that we record as the EEG and ECoG (Sect. 7.1). Fluctuations are solely due to changes in current, I, and not resistance, R. Passing a fixed alternating current at 1 KHz across the cortex while measuring the ECoG gives an invariant 1 KHz sine wave, which proves that the fluctuations of the ECoG

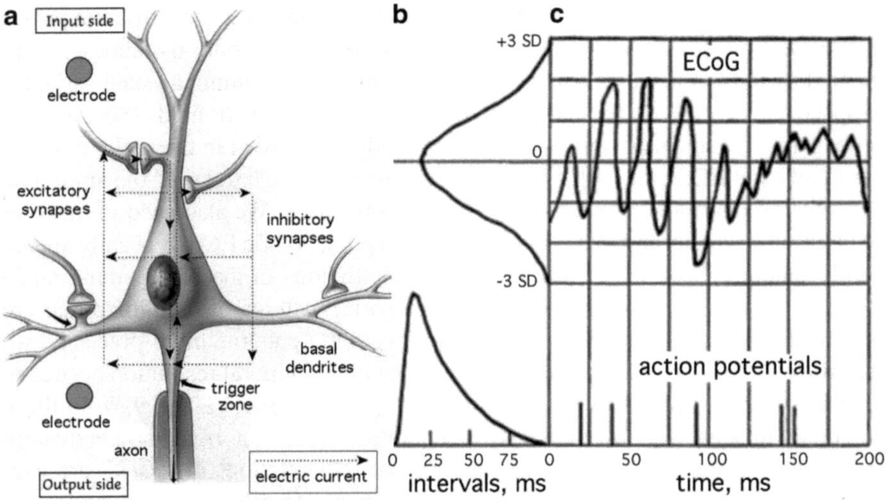

Fig. 6.2 (**a**) The parts of a neuron are schematized. The *dashed lines* show the currents that always flow in closed loops due to charge conservation. Each synapse drives current in at excitatory synapses and out at inhibitory synapses. The synapses transmit to the trigger zone by the loop currents and not by pulses along the dendritic shafts. Pulse-wave conversion at synapses is denoted $G_d(p_{in})$. Wave-pulse conversion to output p_{out} at trigger zones is denoted $G_a(v)$. (**b**) A pulse train of a representative neuron was recorded with a microelectrode near the cell body, from which an interval histogram was calculated. The ECoG was recorded with a surface macroelectrode with respect to a depth electrode (as shown) or to a distant macroelectrode outside the skull. The amplitudes in the Gaussian amplitude histogram were expressed in units of the ECoG standard deviation (SD). (**c**) The pulse counts in each amplitude bin of 0.1 SD were divided by the number occurrences of amplitude in that bin to calculate the pulse probability conditional on ECoG amplitude (Sect. 6.6) (From Fig. 5 in Freeman 2001)

are not caused by fluctuations in cortical specific resistance. They are caused by fluctuations in dendritic currents. Moreover, ECoGs are not sums or envelopes of action potentials.

The axon expresses the information it transmits in short time windows by the precise time intervals between the pulses and in long time windows by the average interval or pulse frequency. The axon state variable is discrete. Pulses are all-or-none. An axon cannot add pulses, but it can multiply them by the number of its terminal synapses. The pyramidal cortical neurons have only one axon, which forms synapses on 10^4 other neurons but none on itself (Braitenberg and Schüz 1998). Each synapse converts each pulse it receives to a wave of ionic current. The current of excitatory synapses flows in at the synapse and out everywhere else in a closed loop of current. The outflow at the trigger zone of the axon depolarizes the membrane and increases the tendency to fire pulses. The current of inhibitory synapses flows in a closed loop out at the synapse and in everywhere else. The inflow at the trigger zone hyperpolarizes the membrane and reduces the tendency to fire pulses. The dendritic state variable is the continuous sum of the potential differences (not of the ionic currents, which are carried by whatever anions and cations are available). The vast surface area of the dendrites supports synapses from on average 10^4 other neurons; the competition for space on the surface is intense. The dendritic tree gives every synapse direct access to the trigger zone by the flow of its loop current, which explains the tree-like structure of dendrites and their function, which is integration by summation of innumerable synaptic potentials at the trigger zone. Note that the synaptic potentials can sum to zero but not the ionic currents and their cost in metabolic energy (Fig. 6.1b). The widespread cancelation of excitatory and inhibitory synaptic potentials (not energy) is one reason why intelligence (i.e., brain function) is so expensive (Raichle and Mintun 2006).

State variables are used to describe how each neuron performs four sequential operations (Fig. 6.2). At the microscopic level, the dendrites transform incoming pulses to synaptic potentials. Then they integrate them. At the trigger zone, the axon converts the sum of dendritic potentials to a train of pulses. Then it transmits without attenuation but with delay. At the macroscopic level, the synaptic currents of neurons in local cortical columns generate extracellular potential differences observed as the ECoG, which approximates the amplitude of the wave activity. The instantaneous sum of pulses from active neurons in the column determines the dendritic wave amplitude, and the wave amplitude at each instant determines the output pulse intervals of the train that are sent on average to 10^4 other neurons, some but not all of which will fire. Two reciprocal conversions are required: pulse-wave conversion at synapses where afferent axons end and wave-pulse conversion at trigger zones where efferent axons begin.

6.2.2 *State Variables: Microscopic Bits Versus Macroscopic Densities*

The next basic concept is the distinction between the state variables of the neuron *versus* the neural population. The neuron is microscopic (Fig. 6.2). Its pulse variable is pulse frequency measured with a microelectrode placed near the cell body, and its wave variable is transmembrane potential measured with a microelectrode in the cell body or main dendritic shaft. The population consists of upward 10^4 neurons that interact by chemical synapses at long ranges and electrical synapses (e.g., gap junctions, electric synapses and ephapsis (Section 11.2; Anastassiou et al. 2009)) at short ranges, thereby constraining each other into macroscopic functional populations (Section 11.2). The dendritic state variable of the population is the wave amplitude, $v(t)$, with mean v_o. The axonal state variable is a pulse density, $p(t)$, with mean density p_o. For example, the density of neurons is $1 \times 10^5/\text{mm}^3$ of cortical volume (Braitenberg and Schüz 1998) and $3 \times 10^5/\text{mm}^2$ surface area from a mean neocortical depth of 3 mm. All neurons must generate pulses continually, or they atrophy and die. No one knows what the average rate is. If the mean pulse frequency were a modest 1/10 s, the pulses would total 3×10^4 p/s/mm^2 of cortical surface area, with each overlapping pulse lasting ~1 ms.

Whereas in a network the pulse and its synaptic potential can be represented by a binary digit in a logical device, in the population the element of integration must be represented by an increment in wave density, $\Delta v(t)$, and the element of transmission is pulse density, $\Delta p(t)$, in a time interval, Δt. In the construction of a differential equation to describe the macroscopic dynamics (Freeman 1975), the time interval is allowed to approach zero so that the wave and pulse variables, $v(t)$ and $p(t)$, are sequences of infinitesimals, $dv(t)$ and $dp(t)$. The conversion of incoming pulse density to wave density at synapses on dendrites is expressed by a nonlinear function, $G_d(p)$, and the conversion at trigger zones is given by a nonlinear function, $G_a(v)$. At each time step, t, the dimensionless forward gain k_j of the population (Freeman 1979) is defined by the product of the two conversions: $k_j = G_a(v) \cdot G_d(p)$ (Fig. 6.10). The forward gain is evaluated by measuring the input and output and calculating the ratio, $k_j = p_{out}/p_{in}$. The forward gain of an excitatory population is denoted k_e; that of an inhibitory population is k_i.

Both the axonal gain, $G_a(v)$, and the dendritic (synaptic) gain, $G_d(p)$, have small-signal, near-linear ranges. However, the near-linear range for $G_d(v)$ is so much wider than that of $G_a(p)$ that the conversion can be linearized by replacing $G_d(t)$ with a coefficient, k_e or k_i, which can be modified to represent changes with the processes of learning, arousal, and normalization. This change of functions greatly simplifies modeling neurodynamics because it leaves only the nonlinear function of the trigger zones, $G_a(v)$, following dendritic integration and preceding axonal transmission, and in piecewise linearization (Freeman 1975) the function can be replaced by the tangent at an operating point (Fig. 6.10a, b, Sect. 6.5).

The distinction must be clearly drawn between the microscopic level, at which each neuron interacts with a sparse selection of neurons in networks such as in a Hebbian assembly (Fig. 8.8, Sect. 8.4; Sect. 11.2; Amit 1995), and the macroscopic level, at which each neuron interacts with a subpopulation in its column or surround. When an excitatory neuron fires, it excites some among 10^4 other neurons. When those fire they excite some among 10^4 others by feed forward connections. The numbers affected subthreshold by percolation (Kozma et al. 2005) increase geometrically, 10^4, 10^8, 10^{12}.... Each neuron is embedded in the mass and receives feedback from its own action. The feedback path has been modeled by topologically dividing each population into a receiving subset from which a transmitting subset is continually renewed (K-sets, Chap. 5 in Freeman 1975; Freeman and Erwin 2008). The two subsets form a feedback loop for which the *loop gain* of the functional connectivity is defined by the dimensionless product of the two forward gains, $k_{ee} = k_e k_e$, and comparably for an inhibitory population which is also modeled as positive feedback, $k_{ii} = k_i k_i$. The interaction of excitatory and inhibitory populations constitutes negative feedback with loop gain, k_n, given by the product $k_{ei} k_{ie}$. The modal loop gain represents quantitatively the intensity of interactions among the neurons creating and sustaining the population activity.

Evaluation of loop gain is not as simple as it is for forward gain. It is evaluated in steps: first, constructing a differential equation modeling the feedback (Chap. 5 in Freeman 1975); second, solving the equation for impulse input; third, fitting the waveform of the impulse response to an evoked potential; and fourth, calculating the feedback gain from the characteristic frequencies and decay rates of the population (Fig. 6.12, Sect. 6.7; Fig. 8.5a, Sect. 8.3). In the steady state, the loop gain must be unity; $p_{out} = p_{in}$ at every instant. When the functional connectivity is weak, the loop gain is less than unity, and the responses to test input decay exponentially. When it exceeds unity, the test responses grow exponentially (Fig. 8.8, Sect. 8.4). The difference in the sign of the exponent (Fig. 9.13c, Sect. 9.7.1), which is determined by the loop gain, underlies the qualitative difference between the microscopic pulse frequency in a network and the macroscopic pulse density of a fully interactive cortical population. In summary, with gain <1 the impulse response decays; with gain $=1$ the steady state prevails; with gain >1 the impulse response grows exponentially; and the cortex may transit from sparse activity to high pulse density in a state of coherence, which as we will see can be manifested in phase locking in the carrier frequency pass band (Sect. 11.2).

In the weakly interactive, loosely coupled background state, we conceive that cortical neurons are receptive to extracortical input. In the strongly interactive, condensed state, we conceive that the neurons coordinate their firing by interactions with each other. Neurons may form both short and long connections in a power-law distribution of functional connection density among cortical neurons (Freeman and Breakspear 2007) (Sect. 6.6) so that every cortical neuron may interact within very few synapses, as predicted by random graph theory (Freeman and Kozma 2010) with every other cortical neuron. What we propose is that the immense number of neurons, the power-law connectivity, and the continuity of the neural sheet over vast areas of cerebral cortex (Sect. 8.1) can explain how cortex intermittently sustains independent coherent EEG oscillations simultaneously in multiple frequency bands over large areas in the ECoG and over the entire extent of the scalp in the EEG (Pockett et al. 2009 and Sect. 10.5).

6.2.3 State Variables: The Order Parameter

Textured images related to behavior are found only in the strongly coherent cortical states. Here differential equations are not merely a convenient approximation (Chap. 6 in Freeman 1975). The equations in pulse and wave density express the fundamental nature of cognitive processing in the cerebral cortex, by which microscopic information selects macroscopic patterns (Freeman and Vitiello 2010). The pulses form a macroscopic vector field, which is not a mean field because it is textured. It is a collective phenomenon well known in many-body physics (Vitiello 2001) by which interactions impose a degree of order. When we detect an image and measure its emergent pattern, we can define the feature vector as an *order parameter* (Haken 2002), that is, a field of force by which the neurons in a population impose order among themselves by synaptic interactions. The increased order is revealed not by the increase in firing rates or signal correlations but by the unique and intrinsic form of the sustained texturing that relates to a cognitive event. The $(1 \times n)$ feature vector that represents the spatial amplitude modulation (AM) pattern of these oscillations, where n is the number of channels, is a vectorial index of the order parameter, which is the closest we have come to numerical evaluation of the total strength of widespread synaptic connections that sustain the AM pattern. In further development below we define a scalar index of the magnitude of the vectorial order parameter ($H_e(t)$, Fig. 6.7a, Sect. 6.4.3), which we derive as an optimal measure of the order that is created during an act of perception (Fig. 11.1; Sect. 11.3).

Owing to the laminar geometry of the layers of neurons in cortex, the local amplitude of the ECoG is proportional to the local ionic current density and therefore to the local density of the pulse cloud vector field. We expect that the most fruitful outcome of learning to read the EEG and ECoG will be to infer the spatiotemporal *textures* in the pulse densities of the clouds, which were conceived by Sir Charles Sherrington (1940) in his metaphor of "an enchanted loom where millions of flashing shuttles weave a dissolving pattern, always a meaningful pattern though never an abiding one; a shifting harmony of subpatterns of points of light," points which have been handsomely simulated by Izhikevich and Edelman (2008). The challenge now is to detect, measure, and explain the patterns envisioned by Sherrington in the waking states of humans and animals.

6.3 Temporal and Spatial Textures: Gaussianity

The unfiltered background ECoG and EEG at rest (Sect. 7.3a) is monotonous and featureless in appearance (Fig. 6.3a). Spectral analysis of the ECoG (Sect. 3.4) shows broad distributions of the frequency components (b). It is useful to show temporal *power spectral density* (PSD_T) in log-log coordinates because it commonly shows a power-law, $1/f^{\alpha}$, form. Below a concave-downward inflection in the theta-alpha range, the PSD_T is flat, $\alpha=0$ (simulated in Fig. 6.12d, f, Sect. 6.7). Above, the $\log_{10}$

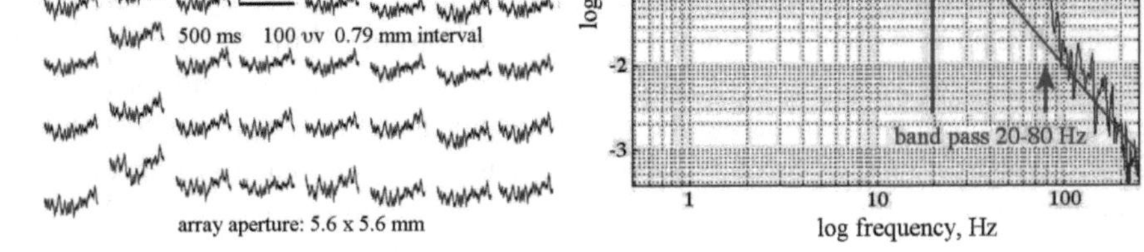

Fig. 6.3 (**a**) Multichannel recording from a high-density array fixed on the auditory cortex of a rabbit at rest. (**b**) The power spectral densities in the working ECoG were computed for all 64 signals and averaged. The power-law trend lines ($1/f^0$ and $1/f^{2.6}$) were drawn by hand to emphasize the multiple peaks of power in the theta- and beta-gamma ranges above the line, which were missing in the resting ECoG. We propose this graph as giving the canonical form of the PSD of resting EEG and ECoG (see simulated PSD_T in Fig. 6.12d, e) (From Figs. A1.1 and A1.2 in Freeman 2006)

power decreases approximately linearly with increasing $\log_{10}$ frequency in the beta-gamma range (12.5–80 Hz) with the exponent α between 2 and 4. In slow wave sleep the exponent averages near 3 (Freeman and Zhai 2009); in seizures it can go as high as 4 (Sect. 4.5). Above 75 Hz the slope either increases ($\alpha=4$, Kellis et al. 2009), or it flattens ($\alpha=0$) above a concave-upward, high-frequency inflection. The two inflections are used as criteria for sampling duration and interval (Sect. 7.2).

The multichannel ECoG at high spatial resolution discloses several important properties. The most prominent is that the waveforms are highly correlated but not identical; the first component of PCA (principal components analysis) applied to 64 signals incorporates 90–95% of the total variance. The signals have the same carrier frequency, but signals even from adjacent electrodes can differ strongly in amplitude and systematically in phase, differences that are not revealed by correlation or PCA. The differences show that the broad spatial correlation is not solely due to volume conduction or to activity at the reference lead electrode. The correlation is due instead to high-density coordination by synaptic interaction, which imposes a shared carrier frequency that is modulated in amplitude and phase by local variations of synaptic gains. The modulations are the source of the textures in the spatial images of the ECoG and EEG. The pair-wise correlations reveal the long distances across which synaptic interactions can sustain the coherence of carrier waves (Sect. 9.6, Table 9.1). The high correlation imposed by long-distance synaptic interaction often leads to overestimation of the effects of volume conduction and underestimation of the spatial resolution of the ECoG and EEG that can be achieved by array recording (Fig. 7.8, Sect. 7.5).

The tenfold fall in ECoG amplitude (square root of power) between the theta range (3–7 Hz) and the gamma range (30–80 Hz, Sect. 6.5) means that high-pass filtering is necessary in order to access high-frequency textural details. The presence of images emergent from the background activity is indicated by multiple peaks of power above the 1/f trend line in the beta-gamma range (arrows in Fig. 6.3b). Each peak reflects a brief epoch of narrow-band oscillation (Fig. 6.5a) in the longer segment of the ECoG. The center frequency in successive bursts varies randomly in the beta or gamma range (Chap. 9). The narrow-band bursts of oscillation can be appropriately measured with wavelets on single trials (Chap. 5).

Each digitized, filtered signal is expressed as a time series of numbers. The set of numbers from a recording epoch, for example, 10 s with sampling at 1,000/s giving 10^4 values, can be assembled into an amplitude histogram (Fig. 6.4) with the number of values in each bin on the ordinate and the range of values on the abscissa. In the example a set of 64 signals gives a matrix of 640,000 values. The distributions of ECoG and EEG amplitudes at rest conform closely to a Gaussian density distribution (A), but only within approximate limits of ±3 SD, because the tails do not go to infinity and, in practice, because of insufficient samples within stationary segments. As the cortex transitions from rest to work, the distributions deviate from Gaussian. Two examples are shown. Prolonged narrow-band oscillations usually in the gamma or alpha ranges give distributions with low kurtosis (B) that tend toward the U-shaped amplitude distribution of cosines. Low-voltage fast EEG and ECoG signals that are punctuated by episodic spikes give distributions with high kurtosis (C).

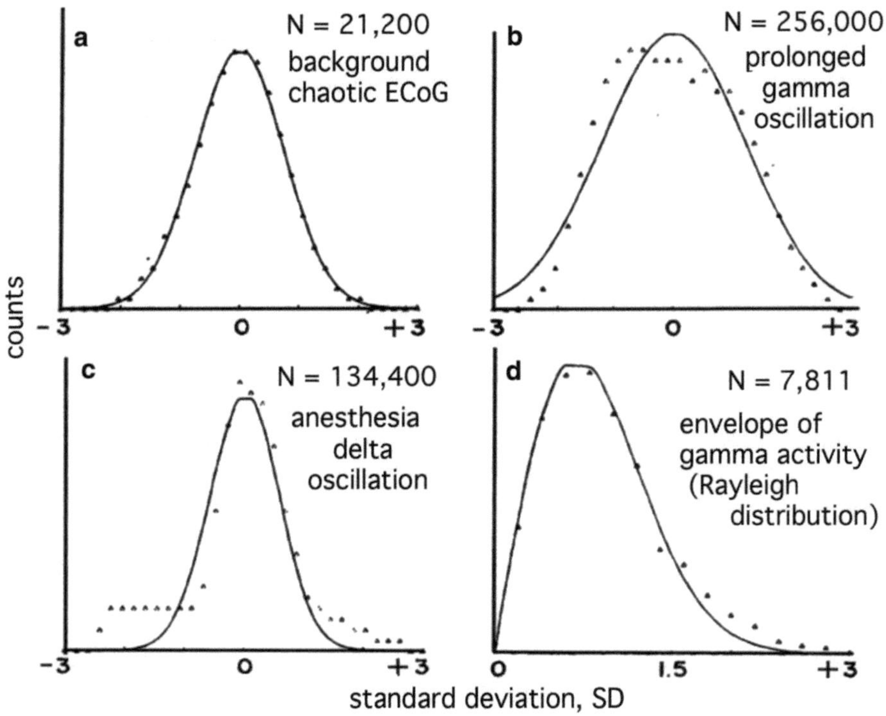

Fig. 6.4 Histograms of olfactory ECoG amplitudes in cat. (**a**) Normal density distribution at rest. (**b**) Platykurtosis with a burst of high-amplitude gamma activity. (**c**) Leptokurtosis during deep anesthesia with intermittent bursts elicited by barbiturate. (**d**) The envelope of band-pass-filtered ECoG noise (20–80 Hz) was fitted with the Rayleigh distribution, which is predicted for band-pass-filtered white noise (Sect. 9.1) (From Fig. 3.13, p. 148 in Freeman 1975)

The nearly Gaussian amplitude distribution from resting ECoG implied that the source of the spontaneous activity could be modeled with a noise generator. This was confirmed in a model of background ECoG based on positive feedback (mutual excitation) (Sect. 6.7) among pyramidal cells (Freeman and Zhai 2009). Deviations from Gaussianity and the power-law 1/f PSD (Fig. 6.5, Sect. 6.4.1; Sect. 9.3) are useful markers indicating the presence of nonrandom structures in the EEG and ECoG. Histograms of the envelope of band-pass-filtered EEG and ECoG give the Rayleigh distribution (d), which is characteristic of narrow band-pass-filtered white noise (Sect. 6.7), in which the filter imposes structure. This avenue is explored (Sect. 7.3) in terms of the spectral properties of types of $1/f^{\alpha}$ noise (white, $\alpha=0$; brown, $\alpha=2$; and black, $\alpha>2$) and in terms of the Rician (Rice 1950) and related distributions of extreme values of amplitude (Freyer et al. 2009), which may offer sensitive markers distinguishing resting from working ECoG and EEG (Sect. 7.4).

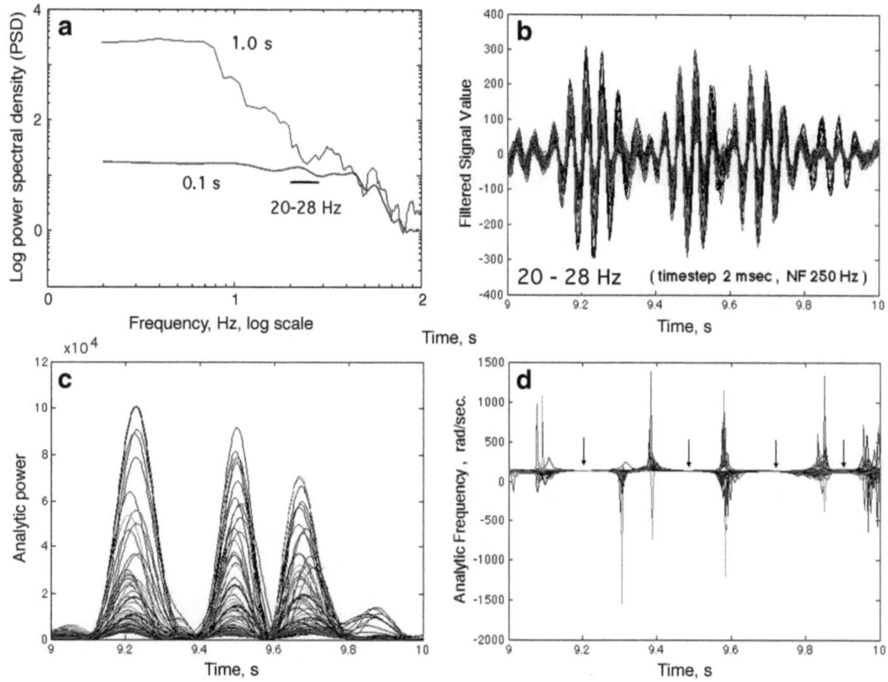

Fig. 6.5 (**a**) The PSD_T of the spatial ensemble average for a short segment contained a stationary burst of oscillation and for a longer segment contained several bursts. (**b**) The set of 64 ECoG were band-pass-filtered in the high beta range. We observed epochs with high phase-locked power, which were separated by epochs of low power resembling beats in Rayleigh noise (Fig. 6.4d). (**c**) The analytic power from the Hilbert transform (Sect. 9.5) varied widely in time and between the superimposed signals from the 64 channels. (**d**) The analytic phase, $f(t)$, and frequency, $w(t)$, remained steady during bursts, but varied widely between bursts and were briefly indeterminate between bursts. The arrows mark the locations where $SD_X(t)$ of the analytic frequency was minimal (Freeman 2012) (Adapted from Fig. 2 in Freeman 2009)

6.4 Stationarity of Spatial Patterns

A system is stationary when its statistical properties do not change. The meaning of the term stationarity depends on the context in which the properties are defined. It is customary to define a behavioral state and accept its statistics as the norm. In the context of training and testing animal and human subjects in discrimination learning, we conceive our subjects to be in a working state with two alternating substates. In this broad sense when sensory cortex is expecting, accepting, and preprocessing new microscopic information delivered to it by the senses, it is in a *receiving* state (Chap. 6 in Freeman 1975; Beggs 2008). When cortex is sending integrated macroscopic perceptual output, it is in a *transmitting* state. When cortex restricts itself to rapid switching between two of many possible states, it can be said to be *bistable* (Freyer et al. 2009). This conception of bistability differs from the concept of *metastability* (Kelso 1995; Bressler and Kelso 2001), in which cortex courses a trajectory among a collection of attractors that mold the trajectory but without capture into stationary states (Sect. 11.4). It differs also from the concept of *chaotic itinerancy* (Tsuda 2001), in which each attractor as the trajectory approaches collapses into ruins that influence the trajectory but do not capture it. In accord with the bistability of the working state, the cortex alternates between the chaotic state and the firm capture and binding by an attractor until the landscape collapses (Fig. 6.14, Sect. 6.9).

In the search for images in the EEG and ECoG, we first established an awake, working state, in which subjects learned to discriminate conditioned stimuli (CS) in reinforcement learning. Then we took samples by digitizing in search of invariant properties. A sample consisted of 40 trials with correct conditioned responses (CR) to 20 presentations of a reinforced CS+ giving the CR+ and 20 presentations of an unreinforced CS− giving the CR−. Each trial lasted 6 s with a 3 s control period and a 3 s tests period. Images were sought on the premise that perception occurs in frames, such as by sniff, saccade, whisk, and auscultation. By this hypothesis, the ECoG and EEG were expected to yield sequences of images, each image having relatively invariant carrier frequency, amplitude pattern, and phase distribution, alternating with periods of disorder and lack of structure.

6.4.1 Invariance of Analytic Frequency: The Hilbert Transform

The first step in searching for images was temporal band-pass filtering in order to remove the high-power, low-frequency components (Fig. 6.3b) and any high-frequency white noise. The optimal range for initial search for images was the beta-gamma range (20–80 Hz) (Sects. 8.5 and 9.6) because it was in this spectral range (Fig. 6.5a) that textured patterns in images have been found to have behavioral correlates (Freeman and Viana di Prisco 1986). The ECoG exhibited brief bursts of narrow-band oscillations at fixed center frequencies varying from each burst to the next in this range (Fig. 6.5b). Because the amplitude, phase, and frequency of the filtered ECoG and EEG could change rapidly and unpredictably, the Hilbert transform (Freeman 2007) was applied to the signals. The transform reexpressed the ECoG as instantaneous *analytic amplitude* (Fig. 6.5c) and instantaneous *analytic phase* at each time step. Without getting into technical details,[1] from each ECoG signal we got two time series: the analytic amplitude, $A(t)$, from which to derive AM patterns, and the analytic phase, $\varphi(t)$, from which to estimate the carrier frequency, $\omega(t)$, by dividing each phase step, $\Delta\varphi(t)$, in radians by the digitizing step, Δt (Fig. 6.5d), in seconds. We found that the Hilbert transform was well suited for EEG

[1] More details are given, Sect. 9.5 and in Freeman (2007). From a real ECoG signal, the Hilbert transform basically generates a complex signal. The original (real) signal together with its imaginary counterpart (obtained with the Hilbert transform) forms a complex *analytic signal*, in the sense that its Fourier transform is strictly positive. Using Euler's theorem, the analytic signal is plotted as a vector (Fig. 9.8c) rotating counterclockwise. The length of the vector gives the analytic amplitude; the angle from the real axis gives the analytic phase.

and ECoG analysis for several reasons. As a linear operator it was suitable for describing dendritic integration in its normal self-regulated range of cognitive function (Sect. 6.5). The real and imaginary components in quadrature (Fig. 9.8a, Sect. 9.5) adventitiously reflected the 90° phase lag between the excitatory and inhibitory populations generating the signals (Fig. 6.13, Sect. 6.8). The sum of squares of the real and imaginary components gave the analytic power for the combined energy dissipation of both neural populations and therefore provided the optimal electrophysiological correlate for measurements of blood flow using BOLD and fMRI (Logothetis 2008; Freeman et al. 2009). Most importantly, it gave high spatial resolution (Fig. 7.8d, Sect. 7.5) and high temporal resolution (Fig. 9.2c, d, Sect. 9.3) of changes in amplitude and frequency. The temporal resolution of the analytic amplitude, phase, and frequency of individual bursts was improved by searching the 1/f PSD_T of the signals in a narrow time window of 0.1 s stepped along the signal in search for a spectral peak (Fig. 6.5a, here 20–28 Hz). The flat segments of $\omega(t)$ in Fig. 6.5d of the 64 superimposed time series of the analytic frequency showed the degree of temporal invariance and the degree of spatial phase locking. The arrows indicate the minima of the spatial variance of the carrier frequency.

6.4.2 *Spatial Pattern Invariance of Analytic Amplitude*

Each EEG-ECoG signal from an electrode array specifies a coordinate axis. The collection of signals gives a set of axes that defines a state space. The set of 64 amplitudes specifies a point in 64-space. Considering the indefinitely large number of possible state variables and their broad ranges (as in epileptic seizures or spikes with extreme amplitudes (Fig. 7.8d, Sect. 7.5)), the brain state space is essentially unbounded. The ranges of variations in a set of normal multichannel EEG or ECoG signals (Fig. 6.2b) define a finite cortical state space. The state space for ECoG and EEG amplitudes is centered at zero by high-pass analog filtering in the amplifiers. The number of channels, here 64, sets the state space dimension, n. In the search for structure in the ECoG and EEG, we focus on a particular narrow frequency band that we select by finding a spectral peak (Fig. 6.5a) in the broad beta-gamma range (Fig. 6.3b).

The 64 analytic amplitudes, $A_j(t)$, (spatial sites $j = 1, \ldots, 64$) of the filtered signals at each digitizing time step, t, form a 64×1 vector, $\boldsymbol{A}(t)$, which is normalized by division of the 64 values by the spatial mean amplitude, $\underline{A}(t)$. The vector specifies a point in 64-space. Successive steps of the vector form a *trajectory* of points in the state space (Fig. 9.6a, Sect. 9.4.1). The Euclidean distance between successive points in this 64-space, $D_e(t)$, gives a measure of the rate of change in the spatial patterns along the trajectory between successive points in 64-space between $\boldsymbol{A}(t)$ and $\boldsymbol{A}(t-1)$ (Fig. 6.6a). The coincidence of low values of $D_e(t)$ with high values of mean analytic amplitude, $\underline{A}(t)$, shown in Fig. 6.6a, reveals a stationary state given by the persistence of a spatial image in the ECoG or EEG. $D_e(t)$ as a measure of pattern invariance should be distinguished from a measure of mean pattern amplitude, $\underline{A}(t)$; the spatial standard deviation of amplitude at each time step, $SD_X(t)$; or the temporal standard

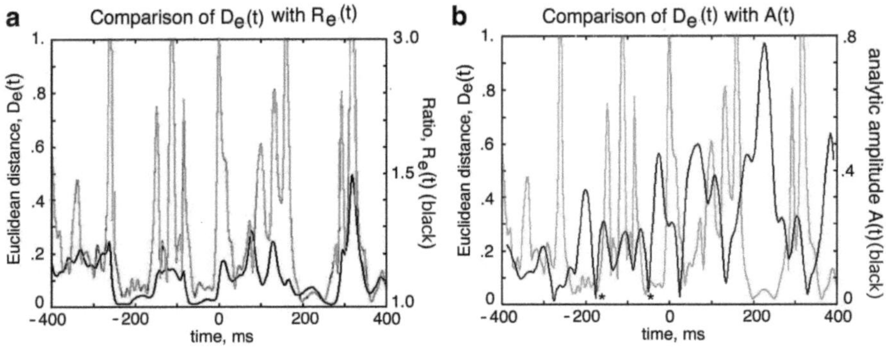

Fig. 6.6 The Euclidean distance, $D_e(t)$ (*gray curve*) between successive 64 digitized samples reveals epochs of low values that are nearly stationary. (**a**) During these epochs the 64 ECoG signals are synchronized, as shown by the reciprocal of the measure of ECoG synchrony, $1/R_e(t)$ (*black curve*). (**b**) During these epochs the measure of analytic amplitude, $A(t)$ (black curve), tends to maximal values. (From Fig. 1.3, Freeman 2004a)

deviation, $SD_T(t)$, of the spatial ensemble average of the 64 signals in a moving time window centered at time step t. $SD_T(t)$ is the average spatial standard deviation of the ensemble.

6.4.3 Estimating Broad-Spectrum Coherence

Narrow-band ECoG oscillations form by increases in synaptic interactions, which tend to lock the cortical populations into coherent oscillations. A measure of the degree of synchronization (Pikovsky et al. 2001) across ECoG signals from an 8×8 electrode array (Fig. 6.6b) is the ratio of the mean variance to the variance of the mean.[2] The ratio, $R_e(t) = SD_T(t)/\underline{SD_T}(t)$, ranges from unity with perfect correlation of all signals to a low value, $n^{-0.5}$, with perfect lack of correlation depending on the number of channels. This estimate of synchrony does not require measurement of the frequency or phase of the oscillations. They tend to change together. $1/R_e(t)$ evaluates the temporal covariance; $D_e(t)$ evaluates the rate of change in spatial AM pattern.

During a stationary epoch the mean power, $\underline{A}^2(t)$, of the 64 ECoG signals (the mean square of the amplitudes at each digitizing step) is high, and the rate of change, $D_e(t)$, is low. The ratio of the power (the rate of energy dissipation) to the rate of change in the spatial pattern, $H_e(t) = \underline{A}^2(t)/D_e(t)$, gives a scalar index of the vectorial

[2] We band-pass filter the 64 signals in the 20–80-Hz range and compute the spatial ensemble average signal $\underline{A}(t)$. We then construct a window twice the wavelength of the peak frequency in the mean PSD_T of each 6 s trial and step it along the 64 signals and the spatial ensemble average at intervals of the wavelength. At each step we calculate the mean SD of the 64 standard deviations, $\underline{SDT}(t)$, and the SD of the average waveform, $SD_T(t)$, in the window.

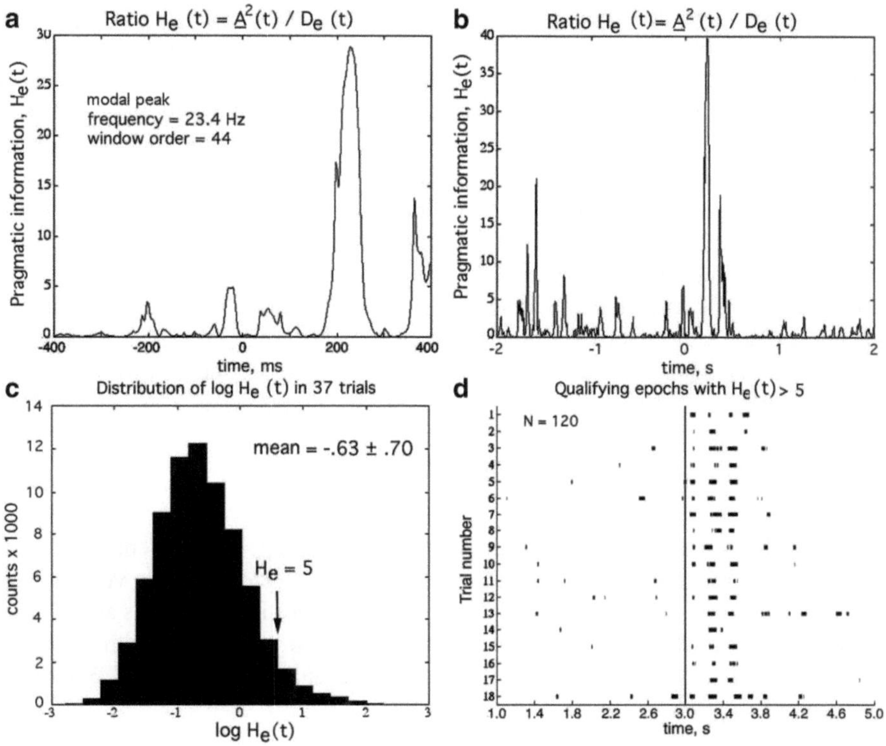

Fig. 6.7 (**a**) The pragmatic information, $H_e(t)=\underline{A}^2(t)/D_e(t)$, measures the degree of order in the 20–80 Hz pass band of the ECoG of a single trial. (**b**) The longer time scale shows brief peaks of high order following onset of a conditioned stimulus (CS) at 0. (**c**) The distribution of the scalar index has a long tail of infrequent high values. The threshold, here $H_e(t)=5$, is determined by repeating the classification of a set of AM patterns and constructing a tuning curve (Fig. 9.4a, Sect. 9.3.2; Fig. 10.3a, b, Sect. 10.3a, b). (**d**) Classifiable AM patterns are found in the segments following the CS where $H_e(t)>5$ (*black dashes* showing the durations of the segments and their time intervals on successive trials) (From Fig. 1.4 in Freeman 2004a)

order parameter, $\boldsymbol{A}^2(t)$, in a spectral band of the ECoG (Fig. 6.7a, b). The index is called the *pragmatic information*, $H_e(t)$, after Atmanspacher and Scheingraber (1990) (Freeman 2004a). An ECoG segment with values of $H_e(t)$ in excess of an empirical threshold (Fig. 6.7c) forms what we define as a *frame* as in a sequence of cinematic frames (Fig. 9.10, Sect. 9.6.1). An example (Fig. 6.7d) shows a sequence of stationary frames in a set of trials that occurs following delivery at time zero of a conditioned stimulus (CS). Examples of the spatial patterns that appear in stationary frames are shown, Sects. 8.5, 9.2, and 10.5. They resemble interference patterns seen in holograms (Pribram 1991) (Fig. 8.6 from allocortex; Fig. 9.5b from visual cortex; Fig. 10.5a from human cortex; Fig. 10.9 from scalp EEG).

We emphasize that the four manifestations of stationarity are constancy of the carrier frequency, increased $R_e(t)$ implying increased order, reduced $D_e(t)$ revealing increased stationarity of the AM pattern, and concomitant increase in the spatial mean and standard deviation of the fixed AM pattern (Fig. 6.5a, Sect. 6.3). That combination, explicitly formulated in $H_e(t)$, is the most sensitive index for evaluating ECoG and EEG images (Sects. 9.6 and 11.3). The scalar value of $H_e(t)$ indexes the intensity of the massive synaptic interactions by which the stationary AM patterns form and persist. The $(1 \times n)$ feature vector, $\boldsymbol{A}^2(t)$, reflects the spatial texturing of the synaptic gains. The repetitive increases in $H_e(t)$ following arrival of a CS (Fig. 6.7d) suggest the occurrence of qualitative changes in state from expectancy to processing and back again that resemble the condensation and evaporation of a raindrop from and to water vapor. If so, we propose to describe the onset of a stationary frame as a cortical *phase transition* from a noisy, disordered, gas-like receiving phase of cortex to an orderly, condensed, liquid-like transmitting phase, followed by return to the disorganized receiving phase (Sects. 8.4, 8.5 and 11.2).

6.4.4 Spatial AM Pattern Classification

Recognizable spatial AM patterns tend to recur upon repeated presentations of a CS. The recurrence is demonstrated by classifying the $(1 \times n)$ feature vectors that quantify each AM pattern. The local details of the images in themselves are of little interest. What gives the AM patterns meaning is their correlation with the CSs, which is shown by the clustering of points representing frames (Fig. 6.8) with

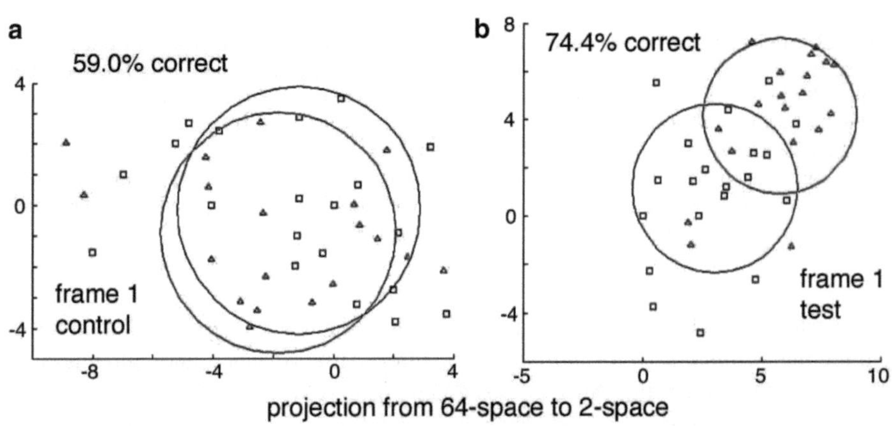

Fig. 6.8 The display of points is by nonlinear mapping (Sammon 1969), which projects the clusters from n-space into two space after rotating the clusters to give maximal cluster separation while preserving the relative distances between points. Classification is by finding the minimal Euclidean distance of each point to the nearest center of gravity. (**a**) Each point represents the geometric mean of a stationary frame in the 20–80 Hz pass band of ECoG signals from a rabbit visual cortex that were recorded in a control period. (**b**) The frames were extracted from trial sets during discrimination between two CSs, one reinforced (CS+, Δ), the other not (CS–, □). The *circles* show the standard deviation of the points in 2-space after projection from 64-space. The pre-stimulus control patterns were indistinguishable. The CS+ and CS– patterns recorded during the test period differed significantly (From Fig. 4 in Freeman 2005a)

respect to CSs that humans and animals have learned to discriminate. Frames with similar patterns on repeated presentations of a CS give a cluster of points in 64-space. The background control state also gives a distinctive cluster (Fig. 6.8a); presentations of the CS+ and CS− give overlapping but significantly different clusters (right). Each cluster is quantified by its geometric mean and its spatial SD_x in the plane of display to which the *n*-dimensional clusters are projected by nonlinear mapping (Sammon 1969). Examples of the categorization of spatial images with respect to cognitive behaviors are given in Chaps. 8, 9, and 10 (Figs. 8.7a, 9.9, 10.9).

6.5 Linearity: Additivity and Proportionality

Simply put, linear systems give output that is proportional to input, whereas nonlinear systems have little or no output in some domains of input yet disproportionately large outputs in other domains. Moreover, in linear systems the responses to multiple inputs add by superposition without changes in the responses from interactions. In order to test for linearity, we stimulate the cortex with a pulse of electric current. The pulse does not excite dendrites directly; it excites intracortical axons. Each axon gives an action potential but only for shock intensities above a threshold. Above threshold the response is all-or-none, not proportional. A suprathreshold shock given within 1–2 ms (the duration of the action potential) gives no response. That reveals the absolute refractory period. For 10–20 ms thereafter, the threshold returns exponentially to normal in the relative refractory period. The threshold and refractory periods show that the dynamics of single axons is nonlinear because the property of all-or-none violates the rule of proportionality.

However, a shock given to cortex can excite a number of axons in proportion to shock intensity. The current delivered by a pair of stimulating electrodes must penetrate the axons at one place (hyperpolarizing them near the anode) and exit the axons at another place (depolarizing them near the anode). Each increment in stimulus current can increase the number of axons that are depolarized beyond threshold. There is no interaction or integration among the axons. The most easily controlled input to cortex is by electrical stimulation of an afferent bundle, in which the axonal pulses elicit dendritic potentials monosynaptically. A stimulus pulse has intensity, *i*, and duration, *t*. The product of time and intensity, $i \bullet t$, specifies the amount of charge that is delivered during the brief flow. We record and measure a microscopic presynaptic action potential, and a macroscopic dendritic evoked potential in response to the square wave with a duration that is long enough to give both on and off responses. Then we decrease the pulse duration and increase the intensity while keeping the charge constant. Below some duration the response waveform no longer changes. The domain of input, below which the waveform of the evoked potential does not vary with duration, defines the range in which the input is an impulse (a Dirac delta function, $\delta(t)$). The utility of the impulse is that it puts all frequencies to the cortex viewed as a filter (Fig. 1.8, Sect. 1.6.2), and the cortical impulse response reveals all the characteristic frequencies of the cortex as it relaxes to its rest state (Fig. 6.9).

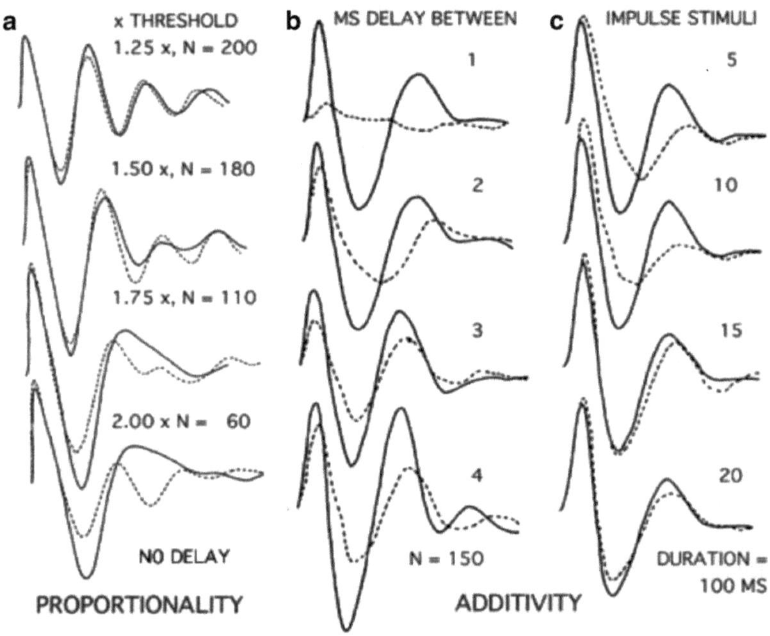

Fig. 6.9 Examples are shown of averaged evoked potentials (AEP) of cat olfactory cortex on electric stimulation (impulse perturbation) of the lateral olfactory tract (*LOT*) in testing for the domain of linearity. (**a**) *The dashed curves* show the single-shock control AEP at threshold intensity. *The solid curves* show the AEP when the stimulus intensity is increased in steps and the number of repetitions is decreased to give constant initial peak amplitude. (**b**) In paired-shock testing for additivity, the responses are stored to a conditioning shock at time zero and to a test shock without conditioning at the designated latency. We record the response to the pair of shocks, subtract the response to the conditioning shock, and superimpose the remainder (*dashed curve*) on the conditioning response (*solid curves*) (Adapted from Biedenbach and Freeman (1965))

In order to test whether cortex operates in a linear range, the test pulse must exceed the axonal threshold, and the interval between two test pulses must exceed the refractory periods of the stimulated axons (here ~15 ms). The examples of testing are evoked potentials from the olfactory cortex of a resting cat on stimulation of the lateral olfactory tract (LOT, Fig. 8.8a, Sect. 8.4). If the dynamics is linear, then when we vary the suprathreshold intensity, the amplitude of the evoked potential must vary proportionately but with no change in the waveform. This invariance holds (Fig. 6.9a) only when the peak-to-peak amplitudes of the evoked potentials do not exceed the self-regulated peak amplitude of the background ECoG because the limits of the background are imposed by nonlinearity (Fig. 6.10, Sect. 6.6). Because the evoked test "signal" is less than the background "noise," ensemble averaging is required (Sect. 1.4). When we give two or more impulses separated by an interval less than the duration of the response to the first impulse, the two responses must add without any change in waveform that would indicate interactions between the

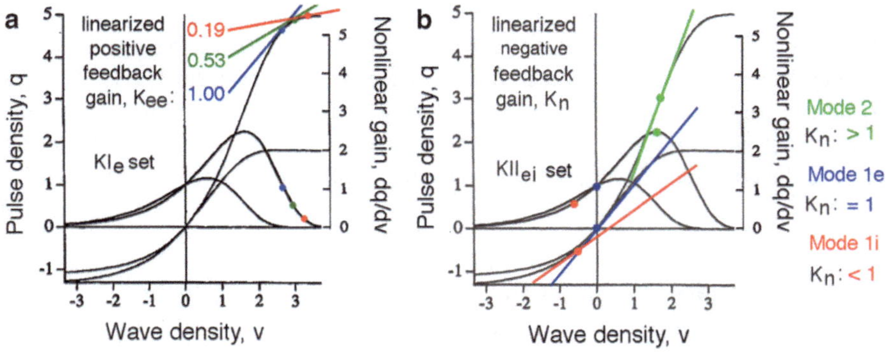

Fig. 6.10 The dependence of axonal pulse density output on dendritic wave density (Fig. 6.2, Sect. 6.2.2) conformed to an asymmetric sigmoid curve with bilateral saturation. (**a**) The right unity gain, $k_{ee} = 1$ (*blue tangent*), was found for a KI_e set. The slopes of two of the tangents (*green, red*) were evaluated by fitting sums of linear basis functions to impulse responses (Fig. 6.12a, b). (**b**) The left unity gain with $k_n = 1$ was found for a KII_{ei} set. The gains were evaluated by fitting sums of linear basis functions to impulse responses (AEP and PSTH as in Fig. 6.13) and plotted as tangents at three levels of excitatory bias from periglomerular cells to mitral cells: low (*red*), medium (*blue*), and high (*green*). (From Freeman 1979)

two responses. By these two criteria the cortical dynamics is linear within three constraints. The intensity of stimulation should be above threshold, the interpulse interval should exceed the duration of the refractory periods of the stimulated afferent axons, and the amplitude of the evoked activity should not exceed the range of the self-regulated ongoing background ECoG.

This test devised by Biedenbach and Freeman (1965) has been replicated for entorhinal cortex (Ahrens and Freeman 2001) and should be used for electric stimulation of areas of neocortex because it provides the experimental conditions for acquiring data suitable for modeling cortical dynamics with linear differential equations (Chap. 2 in Freeman 1975). The criterion for linearity is highly significant (Sect. 10.3) because it shows that normal dendritic integration is in a near-linear domain that is bounded by nonlinear axonal thresholds during excessive inhibition and axonal refractory periods upon excessive excitation or inhibition (Sect. 6.6). The proportionality and additivity of the first post-stimulus peak with respect to input intensity are robust proof that pulse-wave conversion at synapses occurs in a linear small-signal range, which justifies the modeling of pulse-wave conversion at synapses by a weighting coefficient that is adapted in learning and arousal but is otherwise time invariant. The major benefit of identifying the linear and near-linear domains is that multi-loop cortical feedback dynamics can be approximated with matrices of ordinary linear differential equations (Chap. 6 in Freeman 1975; Basar 1998). The solutions specify the linear basis functions to be used for measurement of evoked and background activity (sums of sines, cosines, exponentials), and the measured frequencies can be used to evaluate the feedback gains in the multiple types of loops (Fig. 9.13, Sect. 9.7.1). It is from these measured quantities that we can infer the existence of the point and limit cycle attractors (Fig. 9.13, Sect 9.7.1) that govern the cortical dynamics (Freeman and Kozma 2010). We postulate that after capture of cortical dynamics by an attractor, the sensory cortices rely on linear dynamics for the temporal band-pass filtering in feedback loops and the spatial transformation (Fig. 8.8, Sect. 8.4) in transmission and summation of AM patterns prior to the construction of gestalts (multisensory percepts, Sects. 10.3 and 11.2).

6.6 Ergodicity, Wave-to-Pulse Conversion, and Static Nonlinearity

The impulse responses (evoked potentials, Sect. 1.4) from electric shocks show that three of the four primary macroscopic operations occur within a small-signal, near-linear range. The first three are transmission of afferent pulse density $p_{\text{in}}(t)$ by axons, conversion of pulse density to wave density $v(t)=G_d[p(t)]$ by synapses, and integration of wave density $v(t)$ by dendrites. What keeps these operations within this simplifying range in feedback loops is the fourth operation, the conversion of wave density to efferent pulse density $p_{\text{out}}(t)=G_a[v(t)]$ at trigger zones (Fig. 6.2), which is nonlinear and asymmetric (Fig. 6.10b).

The experimental data were derived by calculating the pulse probability of single neurons conditional on ECoG amplitude (Sect. 3.3.3 in Freeman 1975). The amplitude-dependent sigmoid function was derived in normalized coordinates:

$$q(t)=\frac{[p(t)-p_o]}{p_o}, \tag{6.1}$$

where p_o was mean firing rate and, likewise, v replaced the ECoG (Fig. 6.2b). The function was a statistical mechanical generalization of the time-dependent Hodgkin-Huxley equations (Freeman 1979, reprinted as Chap. 10, Freeman 2000). The function had a single control parameter, Q_m, which was the asymptotic maximum of q:

$$q=\exp\left[\frac{v-(e^v-1)}{Q_m}\right]. \tag{6.2}$$

The forward axonal gain was given by the derivative of q with respect to v.

$$\frac{dq}{dv}=\exp\left[\frac{-(e^v-1)}{Q_m}\right]. \tag{6.3}$$

The value of the upper asymptote, here Q_m is 2 or 5, varied with the level of behavioral arousal from 2 under light anesthesia and rest (Fig. 6.1d) to 12 or more in a state of high motivation (Fig. 6.1d). The upper asymptote was set by the absolute and relative refractory periods (Fig. 6.11a). The lower bound was set by the

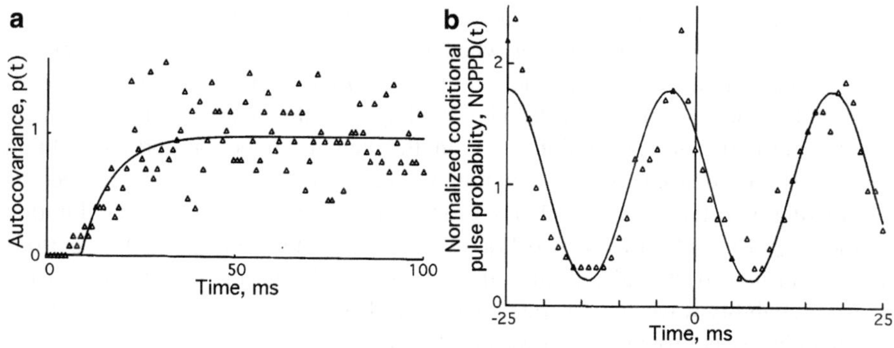

Fig. 6.11 (**a**) The autocovariance of the pulse train reveals the absolute and relative refractory periods (no firing followed by exponential recovery) but gives no evidence of gamma oscillation. (**b**) Calculation of the normalized pulse probability wave of the same pulse train conditional on time lag and ECoG amplitude reveals a modulation depth around the normalized mean firing rate exceeding 80% (From Sect. 3.3.2 in Freeman 1975)

threshold where $p=0$ (zero in Fig. 6.11b). The asymmetry of the function placed maximal gain, k_{max}, at normalized $v(t)=v_{max}=\ln(Q_m)$ to the excitatory side of the background steady state. This was due to the property that, when neurons are brought close to firing, their sensitivity to further input increases exponentially (e^v, Hagiwara and Tasaki 1958). The asymmetry gave two values of unity gain. The KI_e set is stabilized at unity gain (a), because an increase in wave density decreased the gain and limited the increase in pulse density output. This effect regulated the background activity. The KII_{ei} set stabilized at unity gain (b), which gave the possibility of limit cycle activity, because an increase in wave density increased both the pulse output and the feedback gain, k_n, in the range, $0<v<v_{max}$. The hypothesis holds that the asymmetry of the input-dependent gain is essential for the KII_{ei} set to undergo phase transition in response to ignition of a Hebbian assembly (Fig. 8.8a, Sect. 8.4; Fig. 9.13c, Sect. 9.7).

The derivation of this nonlinear function (Chap. 3 in Freeman 1975) is by calculating the probability of firing a pulse *conditional* on ECoG amplitude in the beta-gamma range (Fig. 6.2b). The question is asked at each digitizing time step and amplitude whether a pulse has occurred. The cumulative sum of pulses at each amplitude is divided by the number of times that amplitude occurred. The ratio is normalized by dividing it by the mean pulse rate, giving the normalized conditional pulse probability density on amplitude ($NCPPD_A$). The S-shaped *sigmoid* function has a small-signal, near-linear range but places limits by saturation on both ends of the range of the dendritic wave amplitude. The lower limit is imposed by the threshold of the neuron under strong inhibition. The upper limit is imposed by the absolute and relative refractory periods (Fig. 6.11a). The upper limit is far lower than the maximal frequency to which the neuron can transiently be driven by excitation, because it includes the elapsed time when the neuron is recovering from prior activity. Whereas the lower limit is a microscopic property, the upper asymptotic limit, Q_m, is a macroscopic property.

Pulse probability conditional on amplitude is extended to include time by further asking at each ECoG digitizing step whether a pulse occurred within ±25 ms before or after the current wave value. The conditional pulse probability density is normalized with respect to time ($NCPPD_T$) by dividing it by the mean, giving an estimate of the pulse density oscillation, *p(t)*, in the neighborhood or cortical column (Fig 6.11b). Evidence that the modulating wave, *p(t)*, is macroscopic is provided by comparing it with the autocorrelation of the pulse train averaged over the duration of the 10^4 pulses in the train (Fig 6.11a). The absolute and relative refractory periods prove that indeed the pulse train is from a single cell, but there is no sign of the pulse probability wave demonstrated in (b). Further evidence of the origin of *p(t)* in the population is provided by lowering the threshold for pulse detection so as to include pulses from several neurons in the neighborhood of the electrode in a multiunit record. The same sigmoid and pulse density wave emerges but with a much shorter period of observation.

The pulse probability wave has a frequency in the gamma range (20–80 Hz), whereas the individual neurons contributing have mean firing rates <10 Hz but with exponential interval histograms. Each neuron on average fires only once in several ECoG cycles, often only once in a burst of 3–5 cycles. These facts suggest that the neurons in each column of the population are *time multiplexing* in sustaining the background activity by randomly rotating their firing. Time multiplexing has several advantages. The foremost is decohering the background activity and dispersing the glial clean up by minimizing local buildup of potassium ions in the extracellular compartment. It extends the range of pulse density by increasing the number of neurons involved without requiring that single neurons be driven far outside their near-linear range and close to their upper limits. Time multiplexing of units in active states with gamma bursts can account for observations from multiunit microelectrode recordings showing that the neural firings are statistically phase locked to gamma frequencies (Sect. 6.8), even though the individual mean pulse firing rates are much lower. However, it is important to note that time multiplexing refers only to the macroscopic pulse densities in the background activity and in AM patterns. It coexists and works in tandem with precision coding in pulse trains at the microscopic level revealed by several types of category cells in perception and beyond: single-unit studies of *feature detector* neurons (Singer and Gray 1995) and neurons with high-level cognitive correlates, including face cells (Gross 2008), *mirror* neurons (Rizzolatti and Craighero 2004), hippocampal *place* cells showing precise precession with respect to theta waves in theta-gamma linkage (Buzsaki 2006), and *concept* cells (Quian Quiroga 2012) (Sect. 11.5).

In single neurons the microscopic wave-pulse conversion at the axon trigger zones from dendritic current amplitude to pulse frequency is well known to be linear (proportional and additive) over a range bounded at the lower end by thresholds and at the upper end by refractory periods (reviewed in Chap. 3, pp. 101–103, 154–159 in Freeman 1975). Within this range, wave-pulse conversion is represented by the ratio, $\Delta p/\Delta v$, which is the slope of a tangent line in the graph of p as a function of v. The near-linear, small-signal range is about $v=0$ in Fig. 6.10b. The ratio of output to input is a measure of the amplification, which is the *gain* of the operation (Sect. 6.2.1).

When the gain is the same over a range of amplitudes, the gain in that range is linear. In the steady state, the output equals the input, so the gain is unity. The upper steady state (Fig. 6.10a) is absolutely stable because increased wave input increases pulse output but decreases gain; the lower steady state (Fig. 6.10b) is conditionally stable because increased wave input increases both pulse output and gain.

In populations, the simplest way (though too simple) to conceive how wave density determines pulse density is that it is smoothed into a sigmoid curve by distributions of the thresholds and the refractory periods. An empirical way to derive the function is to graph the pulse density as a function of wave density. The ECoG gives a measure of the wave density because the local transcortical potential difference from superficial and deep electrodes is determined by the sum of extracellular potentials from many local neurons (Fig. 7.7b). However, as already mentioned there is no direct measure of the density of the pulse cloud. Therefore, we invoke the ergodic hypothesis. We assume that the time ensemble average of the wave-to-pulse relation for a single neuron over a period long enough to yield 10^4 pulses is equal to that for the spatial ensemble average of one pulse from each of 10^4 neurons. The wave-pulse conversion is evaluated experimentally by calculating the normalized probability of firing conditional on the amplitude of the ECoG (pp. 154–159 in Freeman 1975). The fitted curve is derived from two properties of neurons (Freeman 1979): the probability of firing increases exponentially with increasing depolarization (wave density), but when a pulse does occur, the probability briefly falls to zero followed by exponential return to steady-state gain, giving the double exponential term in the static nonlinear equation (6.2) for $G_a(v)$.

The normalized sigmoid function has a single parameter, the asymptotic maximum in pulse density, $Q_m = (p_m - p_o)/p_o$ (Fig. 6.10b). The analytic derivative, dp/dv, evaluates the axonal gain, $G_a(v)$, which is amplitude dependent and therefore nonlinear. The asymptotic maximum, Q_m, varies with behavioral arousal, which the cortex regulates for itself through neurohumoral nuclei in the brain stem (Panksepp 1998). Arousal increases cortical sensitivity and reactivity as manifested in the maximum of cortical pulse density. At all levels of arousal, any extreme deviation of wave density reduces the axonal gain asymptotically to zero by the thresholds during strong inhibition and by the refractory periods during strong excitation. Yet every level has a small-signal, near-linear range, which is the basis for modeling with piecewise linear differential equations.

It is important enough to reiterate that the axonal gain curve, $G_a(v)$, is asymmetric. At every level of arousal, the maximal axonal gain is displaced to the excitatory side of the steady state. The asymmetry is due to the exponential increase in sensitivity of neurons to excitation as they are brought to threshold. As a result, there are two values for wave density at which axonal gain is unity. The operating point of unity gain above the peak ($v_{max} = \ln \cdot Q_m$, in Fig. 6.10a) is the basis for the *spontaneous* background ECoG activity, which is robustly and unconditionally stable (Fig. 9.13b, Sect. 9.7.1), because any increase in wave density from input excitation decreases the gain and therefore the pulse density. Conversely any decrease in wave density increases the gain and the pulse density. Therefore, the upper operating point of unity gain is homeostatically maintained.

The operating point below v_{max} at $v=0$ is conditionally stable because any increase in wave density increases the interaction strength among the excitatory neurons. This property is necessary to explain cortical phase transitions (Fig. 9.13c, Sect. 9.7.1) in the formation of AM patterns (Fig. 8.7a) during perception. Such symmetric sigmoid functions as the logistic curve and hyperbolic tangent fail to explain the extremely selective sensitization to learned sensory inputs that cortices display in brains, seeking and attending specific information. We propose that the asymmetry of the nonlinear gain function shown in Fig. 6.10 can be a necessary property for focused arousal attention and (motivation, Fig. 8.1) because the destabilizing increase in feedback gain is only invoked by the amplification exerted by positive feedback in Hebbian assemblies that have been ignited by CSs (Fig. 8.8b, Sect. 8.4).

6.7 Positive Feedback, Stability, and Point Attractor

The background activity of cortex is essential to maintain cortical functions in a linear dynamic range. This principle was discovered 80 years ago by Nobel Laureate Haldan Keffer Hartline, when he demonstrated that the dynamics of spatial lateral inhibition in the eye of the horseshoe crab created spatial structure in the form of Mach bands and that the dynamics was linearized by background excitation, which was provided by illumination of the ommatidia (light receptors, Ratliff 1965). The neural dynamics was thereafter proven to be linear in the fly's eye (Reichardt 1962), in spinal motor systems (Houk and Rymer 1981), and in linear models of cerebral cortical dynamics (e.g., Basar 1998; Wright and Liley 1996; Liley et al. 1999; Wright et al. 2003; O'Connor and Robinson 2004), leaving unanswered the question of the origin of the background excitation in place of illumination.

Early on, mutual excitation among pyramidal cells was proposed as the basis for self-sustained activity, immediately after the discovery of neural networks by Rafael Lorente de Nó (1934) and the recognition of their significance for behavior by Donald Hebb (1949) in what we now call Hebbian assemblies (Fig. 8.8, Sect. 8.4), with follow-up by Daniel Amit (1995) in proposing reverberation as a mechanism for short-term memory, especially where mediated by NMDA receptors (Wang 2001). However, there has been widespread aversion to models using positive feedback as lacking in stability and reliability, except whereas negative feedback is associated with homeostasis and stability. In fact engineers are well aware that positive, regenerative feedback systems are not necessarily unstable, while negative, homeostatic feedback systems are not always stable. Both can be stable at low feedback gain and unstable at high feedback gain, where gain is defined as the ratio of output amplitude to input amplitude (Fig. 6.10b). The point of transition is where gain is unity. Below unity gain, responses to perturbation decay to the pre-stimulus state; above unity gain, the oscillatory responses can blow up (Fig. 6.1e); and the cortex can transit to a new state (Sect. 8.5).

Mutual excitation among excitatory cells persists as an especially attractive hypothesis because they comprise ~80% of neocortical neurons and they provide 90% of the synapses of cortical origin to each other. During embryological development cortical pulse density increases with connection density as neurons sprout axons and dendrites and form new synapses and gap junctions. At some threshold of density, each neuron gets more pulses than it transmits. Like a nuclear reactor the population goes *critical* and generates its own background activity, but, unlike a reactor, which requires external controls, cortex transits to self-stabilized activity at or near criticality (Kitzbichler et al. 2009) (Sect. 9.4). The transition has been extensively studied in simulations of cortex and neural tissue cultures with emergence of spontaneous bursting (Beggs 2008). The mechanism of stabilization of the background activity by refractory periods can be expected to hold for every population in every column of cortex because all neurons have refractory periods, which implies that there is no need for central regulation of the background activity other than the neuromodulatory control of the steady-state set point in relation to the degree of arousal (Freeman 2005b), which is quantitatively expressed in the parameter Q_m (Fig. 6.10b).

The modeling of wave-pulse conversion gives a definition and measure of gain, dq/dv, where $q(t) = p(t)/po - 1$, that enables us to calculate the gain of the excitatory feedback and show that in the resting state the gain is at unity. Demonstration of the steady state is necessary to postulate stability, but it is not sufficient. We must show that the system returns (is *attracted to*) its same steady state after perturbation. Stability, to reiterate, is predicted by the upward slope of the sigmoid curve and the downward slope of the gain curve at unity gain (*right dot*, Fig. 6.10b). Conditional stability (and instability) is predicted for unity gain where the sigmoid curve and gain curve both slope upwardly with increasing input, v (*left dot*).

An experimental demonstration of unconditional stability (Fig. 6.12a, b) shows the post-stimulus time histograms (PSTH) of the microscopic pulses from a single neuron in an excitatory population[3] to increased intensity of electric shocks. The averaging gives the impulse response of the macroscopic population (Sect. 6.6), which has a rapid rise and an exponential decay to the background firing rate. The decay rate, a, is proportional to response amplitude (Fig. 8.4f, Sect. 8.3). When the shock strength is decreased to threshold where response amplitude approaches zero, the decay rate also approaches zero (Fig. 6.12c), and the gain approaches unity. In dynamical terms, the asymptotic convergence of activity to the background after perturbation is evidence that a point attractor (Freeman and Breakspear 2007) governs the cortical background activity at unity gain (Sect. 5.2, pp. 285–305 in Freeman 1975). An attractor by definition is stable. Stability is demonstrated by the

[3] Experimental demonstration of the mechanism for stabilization of positive feedback in mutual excitation requires use of an excitatory population with no effective inhibitory neurons. The illustration is from periglomerular interneurons in the outer layer of the olfactory bulb (Section 5.2.3 in Freeman 1975). They are GABAergic and therefore mistakenly regarded as inhibitory. In fact they have a high intracellular concentration of chloride ions (Siklós et al. 1995). The action of GABA is to allow chloride ions to exit the neurons, causing depolarization and therefore excitation.

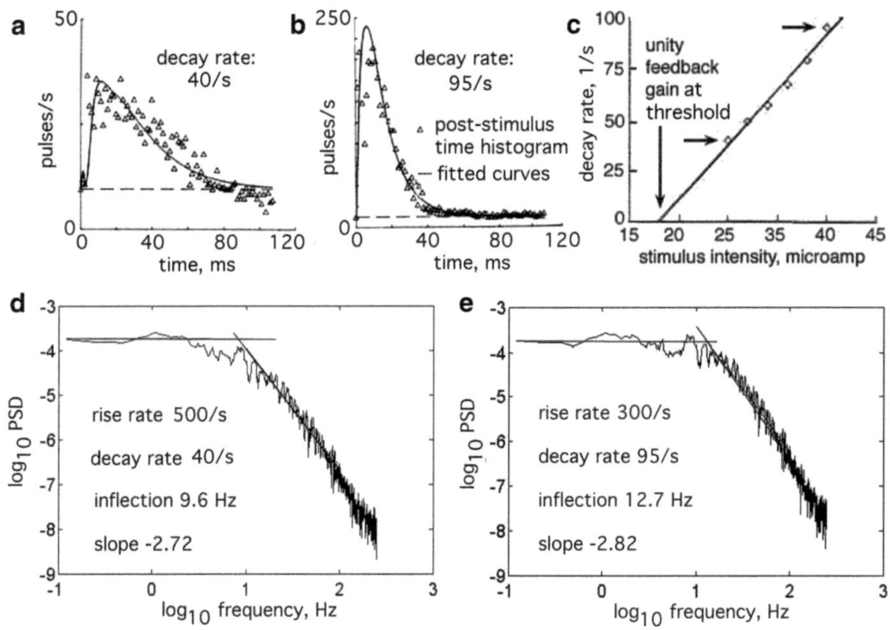

Fig. 6.12 (**a**) The symbols Δ show the post-stimulus time histogram (*PSTH*) of a representative neuron in response to a weak electric shock, which is fitted by nonlinear regression to fit the data with the solution to a fourth-order linear differential equation approximated by neglecting higher order terms with:

$$p(t) = p_o\left[1 + ab\left(e^{-bt} - e^{-at}\right)\right] \tag{6.4}$$

where p_o represented the mean pulse density, a = 500/s was the rise rate, and b = 40/s was the decay rate. Evaluation of the differential equation gave the feedback gain, k_{ee} = 0.59 (Fig. 6.10a). At steady state with no input, k_{ee} = 1.00 with zero decay rate. (b) Increased stimulus intensity increased the decay rate, b = 95/s, and decreased gain to 0.19. (c) The decay rate extrapolated to zero at threshold (zero amplitude of impulse response). We inferred that the prolonged discharge without inhibitory overshoot was reverberation due to mutual excitation (From Figs. 5.13 and 5.16 in Freeman 1975). (d, e) The decay rate, a, determined the inflection frequency of the power-law PSD_T. The rise rate, b, determined the exponent, α. The predicted range of the trend lines was $2 < \alpha < 4$ in $1/f^{\alpha}$, for a second-order filter, which we adopted as the range of black noise (Schroeder 1991). We propose that these two segments give the canonical form of the resting cortical PSD_T (From Figs. 5 and 7 in Freeman and Zhai 2009)

downward sloping gain function with the rising input-output function (Figs. 6.10b and 6.12).

Simulation of mutual excitation is by partitioning the population into a transmitting subset and a receiving subset (Sect. 6.2.2). Each subset is continually renewed from the other. The interaction is modeled with a positive feedback loop. The forward limb and feedback limb of each loop can each be minimally simulated with a second-order differential equation, giving the form of the open loop impulse response (Fig. 6.1b). The closed loop solution of the fourth-order equation for impulse input is shown by the black curve fitted to the data (Δ in Fig. 6.12a, b). The

PSD_T of the simulated background activity (Fig. 6.12d, e) gives the canonical form of the PSD_T of the resting ECoG, which has a flat low-frequency segment ($1/f^0$) and a downward high-frequency segment ($1/f^\alpha$) separated by a concave-downward inflection. Simulation has shown that the rise rate determines the exponent α of the PSD_T, and the decay rate determines the inflection frequency (Freeman and Zhai 2009). Both rate constants vary simultaneously with changes in the ratio of the intensity of the activity evoked by input and the pre-stimulus background firing rate (pp. 305–321 in Freeman 1975). The range of the exponent is predicted to be $2<\alpha<4$ from the fit of the solutions of the fourth-order equation to the PSTH. At one asymptote $\alpha=2$ is associated with coherent activity in an awake state, and at the other asymptote $\alpha=4$ is associated with complete disorder and disintegration of activity as in sleep and seizure. This range is consistent with experimental evaluation of PSD_T trend lines (Figs. 6.2b, 7.2 and 9.1).

6.8 Negative Feedback, Conditional Stability, and Limit Cycle Attractor

It is important to understand that the stability of ECoG and EEG background activity is not due to negative feedback by inhibitory interneurons. On the contrary, the negative feedback maintained by interactions among excitatory and inhibitory populations is manifested in beta-gamma oscillations, for which the steady-state gain is unity at zero normalized wave amplitude (Fig. 6.10b, $v=0$), but the slope of the gain curve is positive. This means that increases in wave density input not only increase pulse density output; they also increase the negative feedback gain. The increase in gain causes the impulse response to last longer with more cycles. That enhances the likelihood that the cortex can transit from the background state into an active state, in which the amplitude of oscillation *increases* exponentially, and the cortex enters sustained oscillation that is governed by a limit cycle attractor (Fig. 8.8b, Sect. 8.4; Fig. 9.13b, Sect. 9.7.1). The transition resembles that of a subcritical Hopf bifurcation (Fig. 6.30, p. 388 in Freeman 1975; Freeman 1987; Skarda and Freeman 1987).

The oscillations are intrinsically macroscopic because wide spatial divergence and convergence occur at all phases of the cycle (see Fig. 8.8a, Sect. 8.4). The participation of individual neurons can best be demonstrated by means of extended time averaging (ergodicity, Sect. 6.6), which supports the hypothesis of time multiplexing of neurons in cortical columns. Interaction strengths (gains) are only a part of the story. The chemistry of synapses contributes to the time constants of neurons, but that is outside the scope of our work. An insightful description of the cellular mechanisms for gamma oscillation at the microscopic level is based on the pharmacological properties of the several types of synapses involved (Traub et al. 1996; Whittington et al. 2000), which might tune the sensitivity of the populations to the expected frequencies of input pulse trains. The two levels of description of gamma activity are complementary. The microscopic description can be studied with pharmacological manipulations of the cortical neurons, while the macroscopic description can be studied with simulation of the large-scale neurodynamics of perception and higher cognition with solutions to matrices of differential equations. We propose that mass interactive dynamics at the macroscopic level governs the generation of beta-gamma oscillations and that the microscopic pharmacological properties determine the receptivity of neurons to beta-gamma input.

The populations of excitatory and inhibitory neurons in each part of the olfactory system oscillate at the same frequency. Regardless of frequency the phase of the inhibitory output lags the excitatory output on average by a quarter cycle (90°, $\pi/2$ radians), not zero lag synchrony (Traub et al. 1996; Whittington et al. 2000). The evidence is provided by simultaneous measurement of the impulse responses of the interactive populations and invoking the ergodic hypothesis (Sect. 6.4; Fig. 6.11).

Two olfactory areas stand in stark contrast. On the one hand, in the prepyriform cortex (and the neocortex, Chap. 9), the pyramidal cells generate the ECoG and the evoked potentials, which provide estimates of the wave density of the excitatory cells. Prolonged averages of the pulse trains of representative pyramidal cells provide estimates of the pulse density. The impulse response conforms to a damped cosine for both the evoked potential and the PSTH with on average zero phase lag between them (Fig. 6.13a), as predicted by the Hodgkin-Huxley equation, because both functions come from the same population, here superficial pyramidal cells.

The quadrature relation also holds for both the bulb and the cortex but with notable differences (Chap. 4 in Freeman 1975). In the bulb the deep inhibitory interneurons generate the ECoG giving $v_i(t)$ because they have the requisite axial geometry (Rall et al. 1966). They transmit by dendrodendritic synapses and as non-spiking neurons have no detectable pulses. The excitatory mitral cells contribute only to the deep LFP and not to the surface ECoG because they have radial geometry. Their pulse probability waves $p_e(t)$ *lead* the ECoG (Fig. 6.13a) by 90°. In the cortex the excitatory pyramidal cells generate the ECoG giving $v_e(t)$ because they have axial geometry. The interneurons generate only LFP in the depth because they have radial symmetry, but they do transmit by pulses $p_i(t)$ at axodendritic synapses. Their pulse probability waves *lag* the ECoG. The ECoGs of both the bulb and the cortex yield AM patterns that are classifiable with respect to CSs (Freeman and Viana Di Prisco 1986; Barrie et al. 1996). The excitatory and inhibitory populations are locked in spatial as well as temporal patterns; the utility of $v_e(t)$ and $v_i(t)$ for studies in perception is determined by their geometry, not by their sign of action.

In the search for macroscopic ECoG images, the contrasting wave-pulse relations found between the olfactory bulb and cortex show that description of the feedback mechanism of beta and gamma oscillations requires four state variables, $v_e(t)$, $p_e(t)$, $v_i(t)$, and $p_i(t)$, but only one of the four state variables gives classifiable images. In the olfactory bulb it is the wave density of the inhibitory population, $v_i(t)$; in the olfactory cortex and neocortex, it is the wave density of the excitatory population, $v_e(t)$. Despite the average 90° phase difference, ECoG images from both structures suffice for spatial AM patterns because of the tight coupling of the forward and feedback limbs. Verification is by time averaging of pulse trains from representative neurons and the dendritic potentials (Fig. 6.2), which requires stationarity, ergodicity, and Gaussianity.

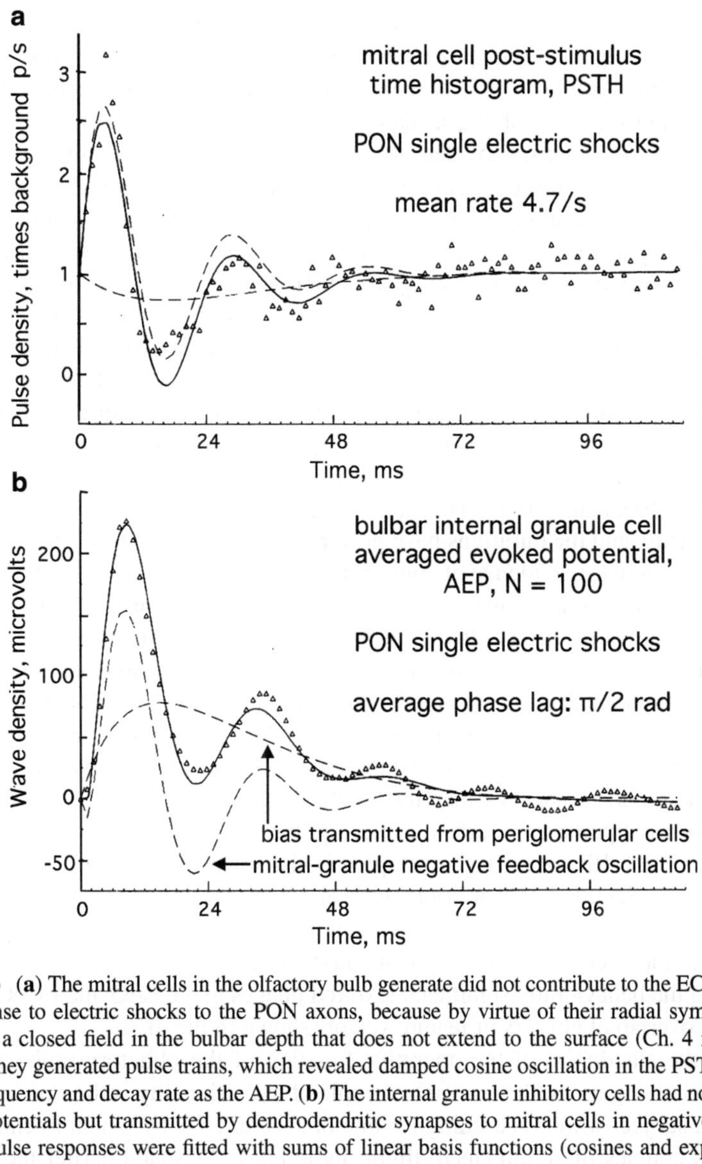

Fig. 6.13 (**a**) The mitral cells in the olfactory bulb generate did not contribute to the ECoG or AEP in response to electric shocks to the PON axons, because by virtue of their radial symmetry they generate a closed field in the bulbar depth that does not extend to the surface (Ch. 4 in Freeman 1975). They generated pulse trains, which revealed damped cosine oscillation in the PSTH with the same frequency and decay rate as the AEP. (**b**) The internal granule inhibitory cells had no detectable action potentials but transmitted by dendrodendritic synapses to mitral cells in negative feedback. The impulse responses were fitted with sums of linear basis functions (cosines and exponentials). The inhibitory population oscillated with on average a 90° *phase lag* behind the population ($\pi/2$ rad) in the evoked activity and the spontaneous waves of ECoG (Fig. 5.3, Sect. 5.4) and pulse probability waves (Fig. 6.11b). The two populations can be said to oscillate in quadrature, with deviations owing to varying intensities of k_{ee} and k_{ii} feedback gains (p. 315 and Chap. 6 in Freeman 1975)

6.9 Criticality: Choice by Means of an Attractor Landscape

Cerebral cortex manifests a special form of conditional stability, criticality (Sect. 9.4), which is defined in mathematics and physics as relating to or denoting a condition of readiness of a system to transit from one state to another. A widely cited model is self-organized criticality, in which a sand pile in the base of an hourglass is fed by a steady stream of sand at the apex (Bak 1996). The pile grows in height and steepness to a critical angle, which is thereafter maintained by avalanches. The size, frequency, and duration of the avalanches have power-law distributions, clearly suggesting that they are fractal. The cortices like all parts of the brain are open systems with continuous throughput of energy and information. Some of the metabolic energy is used to maintain the background activity, which is self-stabilized at set points relating to the level of arousal (Fig. 6.1). The ceaseless bombardment by sensory input drives the cortices away from their set points. The cortices continually relax toward the critical state in brief bursts of activity described as *neural avalanches* (Freeman 2004b; Plenz and Thiagaran 2007; Beggs 2008; Petermann et al. 2009), in reference to the concept of the sand pile (Sect. 9.4.2). The resulting background activity displays numerous overlapping bursts of activity, having no identifiable spatial AM patterns but having spatial patterns of phase (Figs. 8.6 and 9.2c) in the form of conic phase gradients (Freeman and Baird 1987; Freeman and Barrie 2000; Freeman 2004b). The locations and signs of the fitted cones (Fig. 9.5d) vary randomly, and the durations have power-law distributions (Figs. 9.3 and 9.4a). Despite the empirical evidence of conformance to self-stabilized, scale-free dynamics, the sand pile model fails to exhibit true criticality (Kozma et al. 2005; Kozma et al. 2012). Bonachela et al. (2010) describe it as "pseudo-critical" and suggest that we should "…look for more elaborate (adaptive/evolutionary) explanations, beyond simple self-organization, to account for this." In agreement, the small proportion of exceptionally large and long-lasting bursts (Fig. 9.7e) that do have classifiable AM patterns (Freeman 2005a, 2006) deviate strongly from randomness. The random small background avalanches manifest continual adjustments by cortex constantly striving for asymptotic approach to criticality, thereby achieving the steady state at unity gain (Fig. 6.10b). A stronger metaphor is visualizing cortical activity as a pan of boiling water (Sect. 9.4.2) that holds itself at a critical temperature (Freeman et al. 2006). It is this complex of multiple overlapping phase cones at differing carrier frequencies in the background activity that causes intractable problems in locating and measuring the behaviorally significant textured images in the EEG and ECoG, resulting in the overlap of clusters of feature vectors shown in Fig. 6.8b.

From the standpoint of the cognitive function of the olfactory bulb and other primary receiving areas, the relevant example of true criticality is the state of expectancy in a subject who is searching for one of several CSs, by which to choose one of several courses of action (feed, flight, fight, etc.). As stated earlier the state of expectancy is embedded in a hierarchy of states (Fig. 8.1a). In deep anesthesia the background ECoG is suppressed (a), the feedback gain is zero, and the impulse response (evoked potential) is not oscillatory; an exponential rise, a, and decay, b, are determined by the synapses and dendrites of the pyramidal cells (Fig. 8.1b). The cortical dynamics is governed by a point attractor (Fig. 6.14a). On recovery to waking rest the ECoG transitions to broad-spectrum chaotic activity (b), governed by a chaotic attractor, when a blue sky bifurcation replaces the point attractor by a point repeller (Abraham and Shaw 1983–1985).

On arousal from rest into the expectant state of searching, the chaotic background is amplified, and repeated bursts appear (Fig. 6.1d), which can be modeled

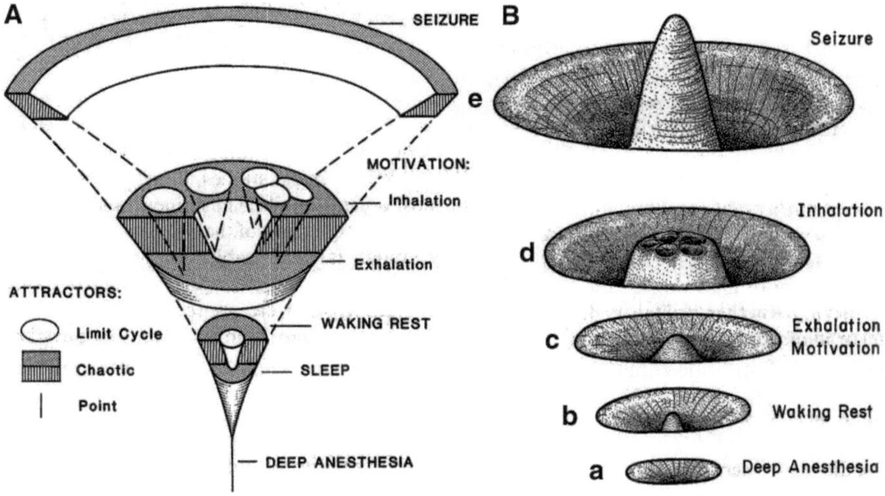

Fig. 6.14 (**a**) Bifurcation diagram of the hierarchical states of the olfactory bulb: (*e*) point attractor in deep anesthesia; (*d*) chaotic attractor (the non-convergent basin of attractor surrounding the central repeller); (*c*) aroused, expectant, motivated in the receiving state; (*b*) sniffing and processing in the transmitting state; (*a*) seizure state. (**b**) Hypothetical basins of attraction visualized in 3D (From Freeman 1987)

as transition by subcritical Hopf bifurcation. We view the olfactory bulb as sustaining *in potentia* a set of limit cycle attractors, one for each expected CS and each with its basin in an attractor landscape (Fig. 6.14a, b). The limit cycle attractors can be regarded as the implementation of a set of Bayesian priors (Doya et al. 2011), which are instructions based in memory from experience for constructing responses to expected CSs. The landscape is preselected from the cortical memory store by input from the limbic system in a process known as *preafference* (Kay and Freeman 1998). The size of each basin of attraction might represent either the breadth of the generalization gradient, or the likelihood assigned to each CS, or the degree of importance as in the sustained qui vive for a predator. The CS determines the choice of category for the CS by selecting a Hebbian assembly, which directs the cortical trajectory into a particular basin (Fig. 6.14). Convergence to the attractor directs the cortex to construct the selected AM pattern (Fig. 8.7). After completion of transmission of an AM pattern, the entire landscape collapses, releasing the trajectory from capture by the attractor, so that it can return to the high-dimensional state of expectancy. We conceive the expectant state as maintaining a trajectory through high-dimensional cortical state space, hovering over an attractor landscape, like an aircraft searching for a suitable field, with confinement to a low-dimensional space upon landing.

The landscape includes a basal attractor for the background input (which signals *no-go*) and an attractor for salient but unidentified novel stimuli (Fig. 8.3b, Sect. 8.2), so the sum of probabilities is unity. In the case of extreme input intensity, the cortex can be driven into a complex partial seizure (Fig. 6.1e) having high-intensity chaos (Freeman 1987), by which the limit cycle landscape is temporarily obliterated.

Criticality brings with it three properties that are crucial for ECoG and EEG image formation and interpretation. First, each sensory cortex maintains itself in a state of extreme sensitivity of readiness to condense into the transmission of a coherent pattern, but only upon receipt of sensory information that can direct the cortex into the basin of a limit cycle attractor that has been formed by reinforcement learning. This restriction indicates that exit from the critical state requires a transition energy, which we propose is provided by the ignition of a Hebbian assembly (Fig. 8.8, Sect. 8.4).

Second, the power-law form of the PSD_T and PSD_X (Sect. 7.2) indicates that within the beta-gamma range, the activity is scale-free (Sect. 8.1). As an immediate effect the appearance and statistics of the ECoG do not change when the pass band of the filter is changed, provided that the length of display and the spatial window of the time series are varied in proportion to the wavelength of the center frequency of the filter (Freeman et al. 2009). The scale-free property explains the significance of the random variation within the beta-gamma range of narrow-band carrier frequencies in successive bursts (Figs. 6.1d and 9.6a). The center frequency bears no relation to the cognitive content of the bursts, only to the size, duration, bandwidth, and intervals of bursts. It means that the signals from multiple sensory cortices can be integrated regardless of their differing carrier frequencies and phase relations (Chap. 10) because what is delivered to targets of transmission of AM patterns are the pulse densities, which we can best estimate from the vectorial analytic amplitude or power of the AM pattern.

Third, the correlation distance can be extremely large across which cortical populations can complete widespread phase transitions in very few milliseconds, regardless of their carrier frequencies (Sect. 10.5). This property may help to explain how images with beta-gamma carrier frequencies can condense and dissolve several times per second across distances that greatly exceed the modal lengths of the dendrites and most axons of the participating neurons (Table 1 in Freeman 2005a) and can include multiple cortical areas, even the entire scalp (Sect. 10.5). This property may also help to explain the multisensory integration that is required for Gestalt formation, which is a core cognitive property to be described and explained in Chaps. 8, 9, and 10. However, in pursuing the suggestion by Bonachela et al. (2010) quoted above, the search for the foundations of criticality in cortex should be extended into modern physics, specifically into random graph theory, nonequilibrium thermodynamics, and quantum field theory, as discussed in Chap. 11.

6.10 Summary

The potentials recorded in the ECoG are proportional to the sum of dendritic current density from active neurons in a cortical neighborhood (column), so they are macroscopic signals. Due to the high density of cortical neurons ($10^5/mm^2$) and synapses ($10^9/mm^2$), the state variable representing a local field potential is treated as a continuous wave density. The pulses by which cortex transmits its output are regulated by the wave density, $v(t)$, and collectively form a pulse cloud that is represented by a pulse density, $p(t)$. The neurons in the column contribute to the cloud by time multiplexing, that is, by random rotation of firing, so that in the absence of a very large sample of units, the ECoG is required to estimate instantaneous spatial patterns in the pulse cloud.

Evidence for the mutual dependence of pulse density, p, and wave density, v, is derived by time averaging in computing the probability of firing conditional on the ECoG amplitude. The demonstration requires the experimental condition of resting, in which cortical dynamics conforms to the assumptions of stationarity and ergodicity at the macroscopic level. The four operations of neural masses are confined to the self-imposed small-signal, near-linear range: pulse-wave conversion by synapses, $G_p(p_{in})$; integration of v by dendrites; wave-pulse conversion by trigger zones, $G_a(v)$; and pulse transmission by axons, p_{out}. Demonstration of linearity justifies measurement of ECoGs and evoked potentials (impulse responses) with sums of linear basis functions, including sines and cosines from the FFT.

In the resting state the background activity is Gaussian. Evidence for departure of cortex from resting to working is seen in deviations from Gaussianity by symmetry breaking. Peaks appear in the 1/f spectra of ECoG that signify bursts of narrow-band oscillations in the beta-gamma range. Bursts carry spatial patterns of amplitude and phase modulation (AM and PM) of a fixed carrier frequency. Some AM patterns are classifiable with respect to conditioned stimuli (CSs). Most PM patterns conform to a cone having no correlation with CSs but giving measures of the location, duration, and diameter of AM patterns. The n-values of instantaneous power at each digitizing step from n-electrodes define a ($n \times 1$) feature vector, which serves as an order parameter with which to estimate the strength and form of collective synaptic interactions that cortical populations exhibit in generating the AM patterns.

Linearity makes possible the description of the resting dynamics of multiple loops in cortical populations by defining feedback gain. The gain is unity in the steady state. The model is extended to the nonlinear working state by piecewise linear approximation (fitting tangents to $G_a(v)$ above and below unity axonal gain). The extended model serves to identify point, limit cycle, and chaotic attractors that govern the cortical dynamics in a hierarchy of states (Chap. 6 in Freeman 1975). The working state is characterized as critical by its alternation through phase transitions between a receiving phase governed by a chaotic attractor with a latent landscape of limit cycle attractors and a transmitting phase upon capture by a limit cycle attractor. The set of limit cycle attractors comprise Bayesian priors invested in memories of the knowledge of CSs embedded in each sensory cortex. The selection by a CS of one among them mediates the perception of a CS by a subject as its AM pattern emerges.

References

Abraham RJ, Shaw CD (1983–1985) Dynamics, the geometry of behavior. Ariel Press, Santa Cruz, 220 pp (part 1), 137 pp (part 2), 121 pp (part 3)

Freeman WJ, Ahlfors SM, Menon V (2009) Combining EEG, MEG and fMRI signals to characterize mesoscopic patterns of brain activity related to cognition. Special Issue (Lorig TS, ed) Intern J Psychophysiol 73(1):43–52

Ahrens KF, Freeman WJ (2001) Response dynamics of entorhinal cortex in awake, anesthetized and bulbotomized rats. Brain Research BRES 911/2, pp. 193–202
Amit DJ (1995) The Hebbian paradigm reintegrated: local reverberations as internal representations. Behav Brain Sci 18:617–657
Anastassiou CA, Perin R, Markram H and Koch C (2011) Ephaptic coupling of cortical neurons. Nature Neuroscience 14:217–223
Atmanspacher H, Scheingraber H (1990) Pragmatic information and dynamical instabilities in a multimode continuous-wave dye laser. Can J Phys 68:728–737
Bak P (1996) How nature works: the science of self-organized criticality. Copernicus, New York
Barrie JM, Freeman WJ, Lenhart M (1996) Modulation by discriminative training of spatial patterns of gamma EEG amplitude and phase in neocortex of rabbits. J Neurophysiol 76:520–539
Basar E (ed) (1998) Brain function and oscillations. Vol 1: Principles and approaches. Vol II: Integrative brain function. Neurophysiology and cognitive processes, Springer series in synergetics. Springer, Berlin
Beggs JM (2008) The criticality hypothesis: how local cortical networks might optimize information processing. Phil Trans R Soc A 366:329–343. doi:10.1098/rsta.2007.2092
Biedenbach MA, Freeman WJ (1965) Linear domain of potentials from the prepyriform cortex with respect to stimulus parameters. Exp Neurol 11:400–417
Bonachela JA, de Franciscis S, Torres JJ, Munoz MA (2010) Self-organization without conservation: are neuronal avalanches generically critical? J Stat Mech. doi:10.1088/1742-5468/2010/02/P02015
Braitenberg V, Schüz A (1998) Cortex: statistics and geometry of neuronal connectivity, 2nd edn. Springer, Berlin
Breakspear M, Stam CJ (2005) Dynamics of a neural system with a multiscale architecture, Philos Trans R Soc Lond B Biol Sci 1457:1051–1074
Bressler SL, Kelso JAS (2001) Cortical coordination dynamics and cognition. Trends Cogn Sci 5:2–36
Buzsaki G (2006) Rhythms of the brain. Oxford University Press, Oxford
Doya K, Ishii S, Pouget A, Rao RPN (2011) Bayesian brain. Probabilistic approaches to neural coding. MIT Press, Cambridge, MA
Freeman WJ (1975) Mass action in the nervous system. Examination of the neurophysiological basis of adaptive behavior through the EEG. Academic Press, New York, Posted in e-formats (2004) http://sulcus.berkeley.edu/MANSWWW/MANSWWW.html
Freeman WJ (1979) Nonlinear dynamics of paleocortex manifested in the olfactory EEG. Biol Cybern 35:21–37
Freeman WJ (1987) Simulation of chaotic EEG patterns with a dynamic model of the olfactory system. Biol Cybern 56:139–150
Freeman WJ (2000) Neurodynamics. An exploration of mesoscopic brain dynamics. Springer, London. http://soma.berkeley.edu/books/BD/MesoBrainDyn.html
Freeman WJ (2001) How brains make up their minds. Columbia UP, New York
Freeman WJ (2004a) Origin, structure, and role of background EEG activity. Part 1. Analytic amplitude. Clin Neurophysiol 115:2077–2088. http://repositories.cdlib.org/postprints/1006
Freeman WJ (2004b) Origin, structure, and role of background EEG activity. Part 2. Analytic phase. Clin Neurophysiol 115:2089–2107. http://repositories.cdlib.org/postprints/1486
Freeman WJ (2005a) Origin, structure, and role of background EEG activity. Part 3. Neural frame classification. Clin Neurophysiol 116(5):1118–1129. http://authors.elsevier.com/sd/article/S1388245705000064
Freeman WJ (2005b) NDN, volume transmission, and self-organization in brain dynamics. J Integr Neurosci 4(4):407–421
Freeman WJ (2006) Origin, structure, and role of background EEG activity. Part 4. Neural frame simulation. Clin Neurophysiol 117(3):572–589. http://repositories.cdlib.org/postprints/1480
Freeman WJ (2007) Hilbert transform for brain waves. Scholarpedia 2(1):1338. http://www.scholarpedia.org/article/Hilbert_transform_for brain_waves

Freeman WJ (2009) Deep analysis of perception through dynamic structures that emerge in cortical activity from self-regulated noise. Cogn Neurodyn 3(1):105–116
Freeman WJ, Baird B (1987) Relation of olfactory EEG to behavior: spatial analysis. Behav Neurosci 101:393–408
Freeman WJ, Barrie JM (2000) Analysis of spatial patterns of phase in neocortical gamma EEGs in rabbit. J Neurophysiol 84:1266–1278
Freeman WJ, Breakspear M (2007) Scale-free neocortical dynamics. Scholarpedia 2(2):1357. http://www.scholarpedia.org/article/Scale-free_neocortical_dynamics
Freeman WJ (2012) Movies of the filtered ECoG and the analytic amplitude and phase can be downloaded from http://soma.berkeley.edu/videos/?video=2
Freeman WJ, Holmes MD, West GA, Vanhatalo S (2006) Fine spatiotemporal structure of phase in human intracranial EEG. Clin Neurophysiol 117(6):1228–1243. http://soma.berkeley.edu/articles/EJ%20CLINPH02-28-06%20txt-fig.pdf
Freeman WJ, Erwin H (2008) Freeman K-set. Scholarpedia 3(2):3238. http://www. scholarpedia.org/article/Freeman_K-set
Freeman WJ, Kozma R (2010) Freeman's mass action. Scholarpedia 5(1):8040. http://www.scholarpedia.org/article/Freeman%27s_mass_action
Freeman WJ, Viana Di Prisco G (1986) Relation of olfactory EEG to behavior: time series analysis. Behav Neurosci 100:753–763
Freeman WJ, Vitiello G (2006) Nonlinear brain dynamics as macroscopic manifestation of underlying many-body field dynamics. Phys Life Rev 3:93–118. http://dx.doi.org/10.1016/j.plrev.2006.02.001, http://repositories.cdlib.org/postprints/1515
Freeman WJ, Vitiello G (2010) Vortices in brain waves. Int J Mod Phys B 24(17):3269–3295. http://dx.doi.org/10.1142/S0217979210056025
Freeman WJ, Zhai J (2009) Simulated power spectral density (PSD) of background electrocorticogram (ECoG). Cogn Neurodyn 3(1):97–103. http://repositories.cdlib.org/postprints/3374
Freyer F, Aquino K, Robinson PA, Ritter P, Breakspear M (2009) Bistability and non-Gaussian fluctuations in spontaneous cortical activity. J Neurosci 29(26):8512–8524
Gross CG (2008) Single neuron studies of inferior temporal cortex. Neuropsychologia 46(3):841–852
Hagiwara S, Tasaki S (1958) A study on the mechanism of impulse transmission across the giant synapse of the squid. J Physiol 143:114–137
Haken H (2002) Brain dynamics: an introduction to models and simulations, Springer series in synergetics. Springer, Berlin
Hebb DO (1949) The organization of behavior. Wiley, New York
Houk JC, Rymer WZ (1981) Neural control of muscle length and tension. In: Brookhart JM, Mountcastle VB, Brooks VB (eds) Handbook of physiology, Sect 1, the nervous system, Vol. II, Motor control, Part 1. American Physiological Society, Bethesda, pp 257–323
Izhikevich EM, Edelman GM (2008) Large-scale model of mammalian thalamocortical systems. Proc Natl Acad Sci USA 105(9):3593–3598
Kay LM, Freeman WJ (1998) Bidirectional processing in the olfactory-limbic axis during olfactory behavior. Behav Neurosci 112:541–553
Kellis SS, House PA, Thomson KE, Brown R, Greger B. (2009) Human neocortical electrical activity recorded on nonpenetrating microwire arrays: applicability for neuroprostheses. Neurosurg Focus 27:1–9
Kelso JAS (1995) Dynamic patterns: the self organization of brain and behavior. MIT Press, Cambridge
Kozma R, Puljic M, Bollobás B, Balister P, Freeman WJ (2005) Phase transitions in the neuropercolation model of neural populations with mixed local and non-local interactions. Biol Cybern 92(6):367–379
Kozma R, Puljic M, Freeman WJ (2012) Thermodynamic model of criticality in the cortex based on EEG/ECoG data. In: Plenz D (ed) Criticality in neural systems. Wiley, New York
Kitzbichler MG, Smith ML, Christensen SR, Bullmore E (2009) Broadband criticality of human brain network synchronization. PLoS Comput Biol 5(3):e1000314. doi:10.1371/journal.pcbi.1000314

Liley DTJ, Alexander DM, Wright JJ, Aldous MD (1999) Alpha rhythm emerges from large-scale networks of realistically coupled multicompartmental model cortical neurons. Netw Comput Neural Syst 10:79–92

Logothetis NK (2008) What we can do and what we cannot do with fMRI. Nature 453:869–878. doi:10.1038/nature06976

Lopes da Silva F (1993) EEG analysis: theory and practice. In: Niedermeyer E, Lopes da Silva F (eds) Electroencephalography: basic principles, clinical applications and related fields. Williams and Wilkins, Baltimore, pp 1097–1123

Lorente de Nó R (1934) Studies on the structure of the cerebral cortex. I. The area entorhinalis. J für Psychologie und Neurologie 45:381–438

O'Connor SC, Robinson PA (2004) Unifying and interpreting the spectral wavenumber content of EEGs, ECoGs, and ERPs. J Theoret Biol 231:397–412

Panksepp J (1998) Affective neuroscience: the foundations of human and animal emotions. Oxford University Press, Oxford

Petermann T, Thiagarajan TA, Lebedev M, Nicoleli M, Chialvo DR, Plenz D (2009) Spontaneous cortical activity in awake monkeys composed of neuronal avalanches. Proc Natl Acad Sci USA 106(37):15921–15926. doi:10.1073/pnas.0904089106

Pikovsky A, Rosenblum M, Kurths J (2001) Synchronization—a universal concept in non-linear sciences. Cambridge University Press, Cambridge

Plenz D, Thiagarajan TC (2007) The organizing principles of neural avalanches: cell assemblies in the cortex. Trends Neurosci 30:101–110

Pockett S, Bold GEJ, Freeman WJ (2009) EEG synchrony during a perceptual-cognitive task: widespread phase synchrony at all frequencies. Clin Neurophysiol 120:695–708. doi:10.1016/j.clinph.2008.12.044

Pribram KH (1991) Brain and perception: holonomy and structure in figural processing. Lawrence Erlbaum Associates, Hillsdale

Quian Quiroga R (2012) Concept cells: The building blocks of declarative memory functions. Nat Rev Neurosci 13:587–597. doi: 10.1038/nrn3251

Ratliff F (1965) Mach bands: quantitative studies on neural networks in the retina. Oxford University Press, Oxford

Raichle M, Mintun M (2006) Brain work and brain imaging. Annu Rev Neurosci 29:449–476

Rall W, Shepherd GM, Reese TS, Brightman MW (1966) Dendrodendritic synapses in the central nervous system. Exp Neurol 14:44–56

Reichardt W (1962) Nervous integration in the facet eye. Biophys J 2:121–143

Rice SO (1950) Mathematical Analysis of Random Noise - and Appendixes – Technical Publications Monograph B-1589. NY: Bell Telephone Labs Inc.

Rizzolatti G, Craighero L (2004) The mirror-neuron system. Annu Rev Neurosci 27:169–192

Sammon JW (1969) A nonlinear mapping for data structure analysis. IEEE Trans Comput C-18:401–409

Schroeder M (1991) Fractals, chaos, power laws. Minutes from an infinite paradise. WH Freeman, San Francisco

Sherrington CS (1940) Man on his nature. Oxford University Press, Oxford, pp 177–178

Siklós L, Rickmann M, Joó F, Freeman WJ, Wolff JR (1995) Chloride is preferentially accumulated in a subpopulation of dendrites and periglomerular cells of the main olfactory bulb in adult rats. Neuroscience 64:165–172

Singer W, Gray CM (1995) Visual feature integration and the temporal correlation hypothesis. Annu Rev Neurosci 18:555–586

Skarda CA, Freeman WJ (1987) How brains make chaos in order to make sense of the world. Behav Brain Sci 10:161–195

Traub RD, Whittington MA, Stanford IM, Jefferys JGR (1996) A mechanism for generation of long-range synchronous fast oscillations in the cortex. Nature 383:421–424

Tsuda I (2001) Toward an interpretation of dynamics neural activity in terms of chaotic dynamical systems. Behav Brain Sci 24:793–847

Vitiello G (2001) My double unveiled. John Benjamins, Amsterdam

Wang X-J (2001) Synaptic reverberation underlying mnemonic persistent activity. Trends Neurosci 24(8):455–463. doi:10.1016/S0166-2236(00)01868-3

Whittington MA, Faulkner HJ, Doheny HC, Traub RD (2000) Neuronal fast oscillations as a target site for psychoactive drugs. Pharmacol Ther 86:171–190

Wright JJ, Liley DTJ (1996) Dynamics of the brain at global and microscopic scales: neural networks and the EEG. Behav Brain Sci 19:285–295

Wright JJ, Rennie CJ, Lees GJ, Robinson PA, Bourke PD, Chapman CL, Gordon E, Rowe DL (2003) Simulated electrocortical activity at microscopic, mesoscopic, and global scales. Neuropsychopharmacology 28:S80–S93

Chapter 7
Image Sampling Based on Spectral Analysis

7.1 Introduction

The EEG and ECoG consist of collections of time series that are recorded from arrays of electrodes placed, respectively, on the scalp or surface of the cortex. Local field potentials, LFP, are recordings from penetrating electrodes deep in cortex and the brain. The surface potential differences are caused by flows of dendritic currents from cortical neurons that are oriented perpendicularly to the surface of recording. The dendritic currents are produced by chemical reactions in synapses that are activated briefly by pulses carried by incoming axons forming the synapses (Fig. 6.2, Sect. 6.2.1). The same currents regulate the probability of pulses on outgoing axons of the neurons integrating the pulse input. The macroscopic operations of cortex transform microscopic spatial patterns of incoming pulses to spatial patterns of outgoing pulse vector fields, which carry patterns that are revealed by images in the EEG and ECoG.

The spatial patterns of pulses reflect a *microscopic format*, in which the cortical information is expressed in the rates and intervals of pulses from neurons. The input activity is patterned in the cortical surface by the topographically organized input axons (Fig. 8.8a, Sect. 8.4). The spatial patterns of dendritic electric currents reflect a *macroscopic format*, in which the cortical information is estimated from ECoG and EEG potentials. Each sample is the sum of contributions from many thousand neurons in the neighborhood of each recording electrode. Accordingly, the dendritic potentials in each neighborhood provide an index of the macroscopic current density in mA/mm^2 (Sect. 6.2.1) that determines the local outgoing macroscopic pulse density ($pulses/s/mm^2$). When the cortex transmits a percept in a cloud of outgoing axonal pulses, we postulate that the information in the percept is carried as a spatial pattern of the pulse density, in which the contributions to local pulse density are shared among the many thousand local neurons by time multiplexing (Sect. 6.6), independently of any precisely structured microscopic activity. By our hypothesis, every neuron does double duty simultaneously at the microscopic and macroscopic levels. Sampling the microscopic pulse trains of a few neurons to represent a local

W.J. Freeman and R. Quian Quiroga, *Imaging Brain Function With EEG: Advanced Temporal and Spatial Analysis of Electroencephalographic Signals*,
DOI 10.1007/978-1-4614-4984-3_7,

pool of $\sim 10^4$ neurons is inadequate to reveal the macroscopic signal on single trials (Chap. 5), and time averaging is appropriate only in circumstances where stationarity and ergodicity may be assumed (Sect. 6.6). Among the several available methods for measuring macroscopic brain activity (time averaging of pulses, MEG, fMRI, BOLD), we focus on the EEG and ECoG.

The first step in testing the hypothesis of macroscopic formatting is to devise methods for effective spatial sampling of the textures of cortical and scalp potentials (Fig. 7.1). How wide should we make our arrays of electrodes that provide our windows of observation, and how closely should we space the electrodes? Texturing is familiar in the study of visual inputs, in which coarse patterns employ low spatial frequencies and fine textures require high spatial frequencies. It is formally identical to the temporal sampling problem that is encountered in the digitization of analog time series. The time digitizing interval must be small enough to avoid aliasing (Sect. 3.4) by undersampling of the high frequencies, yet the sampling windows must be long enough to capture the low frequencies. Therefore, spatial spectral analysis (Gonzalez and Wintz 1977; Freeman et al. 2000, 2003; O'Connor and Robinson 2004; Ramon et al. 2009) is needed to specify the optimal aperture size and interelectrode spacing of electrode arrays for ECoG and EEG recording in the search for images embodying a macroscopic format.

7.2 Search for the Spatial Grain of ECoG

The defining characteristic of the living brain is its spontaneous background activity (Sect. 6.7). Every part of the brain ceaselessly generates axonal action potentials and dendritic currents that are distributed through time and space and which by their extracellular potentials provide signs of the states of the brain and its ongoing cognitive activities. We infer that perceptual activity is minimal in states of rest, sleep, and anesthesia and that cognitively related spatial images are best sought in large-scale departures from the background activity. We begin our search and description of the macroscopic images with examination of the spectral properties of the background activity (Freeman 2004a, b, 2005, 2006).

At the outset, it is essential to keep in mind that the numbers of cortical components are very large. While the remarkably thin sheet averages ~3 mm in depth, the surface area in humans is $\sim 2 \times 10^5$ mm^2, a ratio of nearly 100,000. Each mm^2 overlies roughly 3×10^5 neurons with a total count of $1.37 \pm 1.6 \times 10^{10}$ neurons (Braendgaard et al. 2000) and roughly 10^{11} supporting glial cells. With each neuron giving and receiving 10^4 synapses, the estimated total number of cortical synapses ($\sim 10^{14}$) is still less than Avogadro's number (6×10^{23}). The high packing density of the neurons and synapses and the very low ratio of depth to surface area justify analyzing the macroscopic dynamics of the cortex as a continuous two-dimensional sheet. For this analysis, our model of the topology of cortex is that of a planar graph (Freeman et al. 2009b), in contrast with models of nuclei and reticular networks in 3-D.

Sampling begins by placement of a high-density array with spacing much closer than the finest texture expected. The rates of change with distance express the

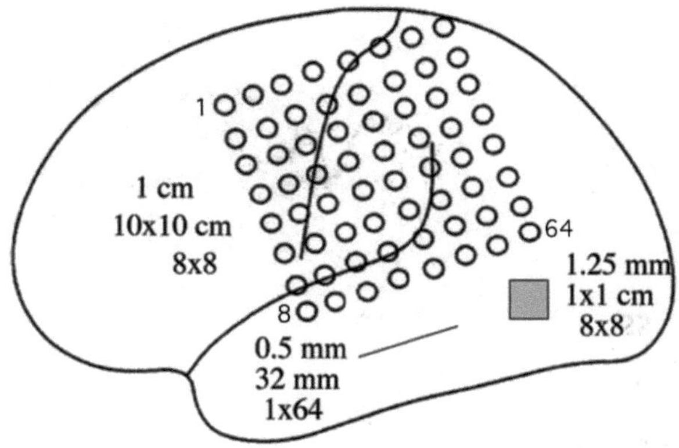

Fig. 7.1 Examples of human intracranial arrays, each with 64 electrodes, are compared with respect to interelectrode distances. The *large array* represents a standard neurosurgical array used for epileptic spike localization. The *short line* represents an array for spatial oversampling the ECoG. The *small rectangle* represents an optimized array for display of fine structure in ECoG (adapted from Menon et al. 1996)

coarseness or fineness of textures. The rates are expressed as spatial frequencies in c/m corresponding to the conventions of time-series analysis of temporal patterns in c/s (Hz) (Chaps. 2 and 3). A familiar example is an embroidered patchwork quilt, in which the low spatial frequency expresses the number of patches/meter and the high spatial frequency expresses the number of embroidery stitches/millimeter in each patch. In this analogy, the stitches would correspond to the spatial grain of columns and the patches to Brodmann areas. High resolution of dendritic potentials at the cortical surface is achieved by close spacing of electrodes. This rule contrasts with the need to use small tips to record action potentials, because the sources and sinks of action currents are very close together and wide electrode tips cannot separate them. The diameters of electrodes for resolving fine textures of dendritic potentials can be almost as large as the interelectrode distance in order to minimize electrode noise, which is proportional to electrode resistance and therefore to electrode size. High resolution of low spatial frequencies of the dendritic potentials requires a large array, which provides a wide aperture on the cortical or scalp surface.

The optimal spacing is determined by spectral analysis. In *temporal analysis,* the choices are the sample record length to accommodate the lowest frequency sought and the digitizing interval for analog-to-digital conversion to capture the highest frequency sought. In *spatial analysis,* the choices are the length or area of the array aperture, which is constrained by the size of the cortical surface of interest and its surgical accessibility and the interelectrode spacing. In practice, a major constraint is the number of available amplifiers (here 64), so a one-dimensional curvilinear array instead of a conventional two-dimensional surface array serves to span a wide spatial spectral range (Fig. 7.1). The available length of the contiguous surface of a gyrus (about 1–3 cm in human cortex) imposes a limit on array length so that the continuity of sampling is not broken by a sulcus. The interelectrode interval for sampling with

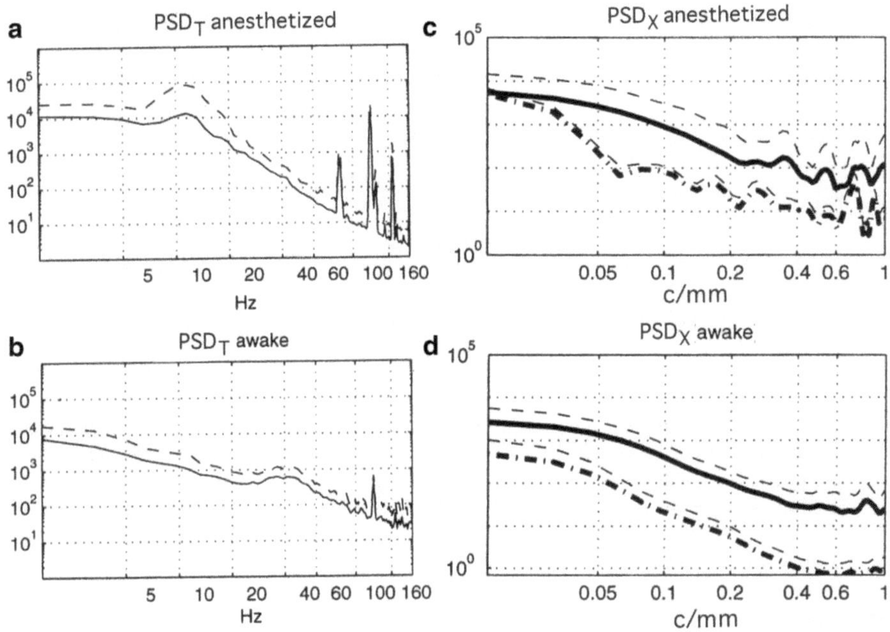

Fig. 7.2 ECoGs were recorded with a 1-D array placed on the exposed superior temporal gyrus in a subject undergoing surgical treatment for epilepsy. (**a**) Power spectral densities (PSD_T, *solid curves*) + SD_X (*dashed curves*) under anesthesia or (**b**) awake with eyes open. ECoGs were recorded with a 1-D array placed on the exposed superior temporal gyrus in subjects undergoing surgical treatment for epilepsy. (**c**) Power spectral densities (PSD_X, *solid curves*) + SD_X (*dashed curves*) under anesthesia or awake (**d**) with eyes open. Digitizing interval: 5 ms. Interelectrode interval: 0.5 mm. Temporal Nyquist frequency: 100 Hz. Spatial Nyquist frequency: 1.0 c/mm. The *lower curves* show the PSD_X and SD_X of the first component of PCA of the data, showing more clearly the flat high-frequency segment. The concave upward inflection of the curves in d at 0.4 c/mm give the spatial Nyquist frequency of 0.8 c/mm and the electrode interval of 1.25 mm. (From Freeman et al. 2000)

64 electrodes can then be calculated as 0.5 mm a linear array 32 mm long of 64 electrodes), which is 1/20 the interelectrode spacing of conventional neurosurgical intracranial arrays (Menon et al. 1996).

With a 1-D electrode array, the calculation of the spatial power spectral density (PSD_X) and the temporal power spectral density (PSD_T) were directly comparable (Fig. 7.2). The data were acquired from five neurosurgical patients (Freeman et al. 2000), four of them anesthetized and one awake under local anesthesia. A convenient spatiotemporal sample for temporal analysis was $N_T = 1{,}000$ points at 200 Hz for 5-s duration, which gave a 64×1000 matrix of ECoG amplitudes. Each signal was extended to the next highest power of 2 for the FFT, and a Hamming window was applied (Sect. 3.7). The temporal PSD_T was calculated (Fig. 7.2a, b) for the 1,000 time steps of each of the 64 signals and for the spatial ensemble average, $\underline{PSD}_T \pm SD_T$. The spatial PSD_X was calculated (Fig. 7.2c, d) at each time step, $N_X = 1{,}000$, and the average $\underline{PSD}_X \pm SD_X$ was calculated.

The most revealing displays of the PSD_T and PSD_X were in logarithmic coordinates: $\log_{10}$ power vs. $\log_{10}$ frequency (Fig. 7.2). The canonical form of the PSD_T and PSD_X from the ECoG *at rest* (Sect. 6.7) could be described in the beta-gamma range temporal frequency range as power-law between upper and lower flat segments: a linear decrease in log power with increase in log frequency with exponent a ($1/f^a$, giving the slope -a), which was modeled as filtered noise (Freeman and Zhai 2009). The search for signals was cast as the search for peaks rising above the trend (Fig. 6.3). Two such peaks were revealed in the PSD_T of the human ECoG (Fig. 7.2). One peak was seen during the anesthetized state in the alpha range in two of four subjects (A). The other peak was seen in the awake subject in the gamma range (B).

In temporal spectral analysis (Chap. 2), the lowest frequency different from zero was set by the duration of the signal and the highest by the sampling rate. Thus, the frequency of the PSD_T at the concave-downward inflection (typically 8–12 Hz) gave an estimate of the optimal duration (83–125 ms) of the window (temporal aperture) for FFT decomposition of bursts. The frequency at the concave-upward inflection in the higher frequency range gave an estimate of the optimal digitizing interval (spatial sampling rate). The reasoning was that cortical signals gave sloping PSD ($1/f^a$, $a>0$), whereas thermal and muscle noises gave flat PSD ($1/f^a$, $a=0$) comparable to featureless white noise. Signals and noises were additive, with signal greater than noise for low frequencies below the inflection but the reverse for high frequencies above the concave-upward inflection. Signal enhancement and noise abatement shifted the inflection to a higher frequency. The assumption was that useful information was available in spectral ranges with $a>0$, that sampling resources would be wasted in ranges with $a=0$, and that valuable information would be missed if sampling rates were not raised sufficiently to reveal the upper inflection. By this account, the middle segment gave the optimal range in which to seek temporal structure (beta = 13–30 Hz; gamma = 30–80 Hz; epsilon = 80–250 Hz). The sampling rate of 200 Hz was too low for reliable exploration of the high gamma and epsilon ranges. Future digitizing should routinely be at rates of 2 KHz or higher.

We explored the relation between temporal and spatial frequencies by applying filter banks with 7-Hz bandwidth to the ECoG before calculating the PSD_X. The same canonical form of the PSD_X was seen in all temporal 7-Hz pass bands across the temporal PSD_T centered from 5 to 50 Hz (Fig. 7.3) in awake and anesthetized states (Freeman et al. 2000). Two significant peaks deviated from the canonical form. The alpha band under anesthesia revealed power a full log unit above power in the awake state. The alpha peak was blocked in the awake state with eyes open, whereas the gamma peak contained greater power in the awake state. Both differences were consistent with the steeper slope of the PSD_T in the sleeping state than in the waking state (Sect. 7.3; Freeman et al. 2006b), which reflected the greater intensity of background activity in arousal compared with rest (Fig. 6.1) and the remarkable enhancement of power in the gamma range despite the dispersion relation, that is, the tendency for attenuation of signals by phase dispersion to increase in proportion to increasing carrier frequency (Sect. 2.3.3 in Chap. 3; Freeman 1975; Nunez 1981; Majumdar et al. 2006; Rudrauf et al. 2006).

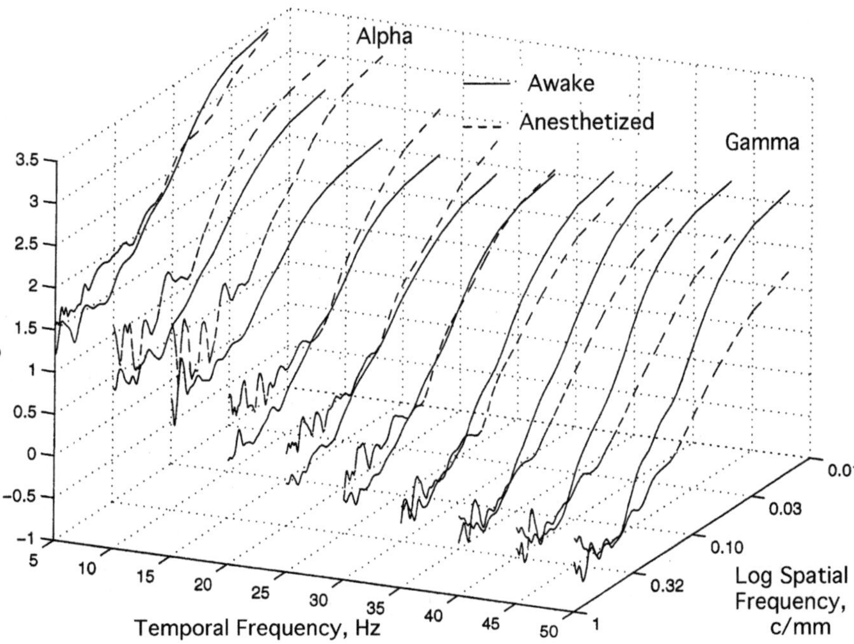

Fig. 7.3 The ECoG was filtered through 7-Hz pass bands before calculating the PSD_X. *Solid curves* from the awake subject showed more power in the gamma range, whereas the *dashed curves* from an anesthetized subject showed greater power in the alpha range. Similar features were found on spectral decomposition of the EEG (Fig. 7.12). (From Freeman et al. 2000)

In summary, we interpreted the spatial PSD_X at rest as approximating a canonical form with three segments. The segments at low and high spatial frequencies were relatively flat and could be modeled as white noise ($1/f^0$). The middle segment was approximated by a trend line with slope –a in the range of –2 to –4 (Fig. 6.12, Sect. 6.7). It defined the main search range for peaks indicating the presence of spatial images (0.04–0.4 c/mm). The concave-downward, low-frequency inflection in PSD_X corresponded to a minimal desired aperture wavelength (1 cm). The concave-upward, high-frequency inflection gave the optimal interelectrode sampling interval (1.25 mm).

7.3 Use of Macroscopic PSD_T and PSD_X to Define the Rest State

Spatial imaging of the background activity showed that when the background PSD_T of the ECoG closely conformed to $1/f^a$ with no significant peaks in the spectra (Fig. 7.4a), there were no recognizable patterns in spatial and temporal displays of the ECoG (Freeman et al. 2006a, b). The ECoG had the same appearance in many

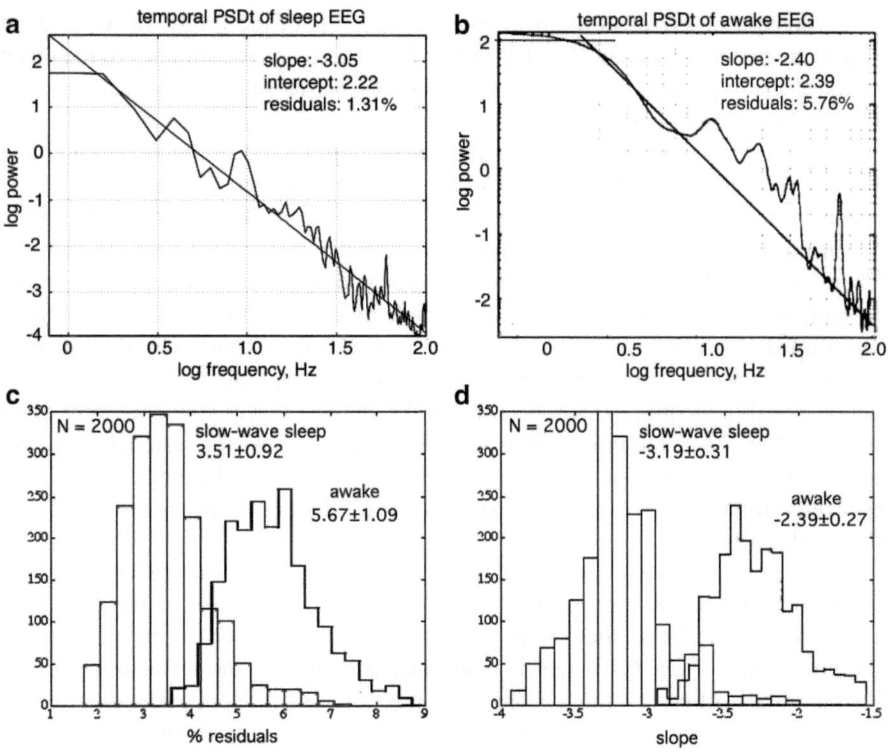

Fig. 7.4 (**a**) The form of the PSD_T ($\log_{10}$ power vs. $\log_{10}$ frequency) in deep slow-wave sleep was power-law with exponent a>3 (black noise, slope <–2, Schroeder 1991). The 1/f form revealed the absence of peaks above the 1/f trend line. (**B**) Upon arousal (eyes closed), the slope of the trend line flattened (a decreased), and multiple peaks appeared in the gamma range. It was accompanied by a peak in the theta range (3–7 Hz) manifesting theta-gamma linkage (Buzsáki 2006). (**c, d**) The slopes, –a, were estimated by linear regression to fit a line to the middle segment of the PSD_T with least squares deviation (6 s, N=2,000) (from Fig. 1 in Freeman (2006) and Fig. 4 in Freeman et al. 2006a)

locations, successive times, and different spectral pass bands, provided that the time base for display of the band-pass-filtered ECoG was contracted or expanded inversely in proportion to the center frequency of the pass band (Freeman et al. 2008). In dynamical terms, the ECoG was invariant on translation in time, space, and spectrum; the activity was approximately *scale-free* (Sect. 9.4) and *symmetric* in time, space, and frequency (invariant under translation and rotation and devoid of patterns) over the beta and gamma ranges (Freeman and Vitiello 2006). The emergence of structure in the form of recognizable spatial ECoG patterns was accompanied by peaks at one or more frequencies above the 1/f trend line in the PSD_T of the ECoG (Figs. 6.2 and 7.4b). We conceived the transition from formless background to structure as *symmetry breaking*, so the search for structure and signals was focused on deviations from uniformity and Gaussianity (Sect. 6.3) as simulated by random noise (Sect. 6.7) (Freyer et al. 2009).

Estimation of slopes in the beta-gamma segments of the PSD_T was done most simply by linear regression. The $1/f^a$ power-law relation gave estimates of the mean value of the exponent, a, (slope = −a) in sleep ECoG close to 3 (Fig. 7.4c), which was fairly precise when the residuals were low. The residuals were found by subtracting the fitted line from the data and calculating the SD of the deviations. The hypothesis was that the EEG and ECoG at rest conformed to random noise, so the spectrum was integrated near-white noise (Freeman and Zhai 2009). Peaks in the spectrum that deviated above the trend line revealed power that was not random. Therefore, the magnitude of the residuals indicated deviations from rest, and more meaningful estimates of the slopes were made by drawing trend lines that emphasized upward peaks (Fig. 7.4b), giving estimates of the exponent in the range of $2 < a < 4$ (*black noise*, Schroeder 1991) that conformed to prediction from a model of neural positive feedback (Sect. 6.7). In awake states, ECoG exponents were often below 2 (Fig. 7.4d), but those values were imprecise because the residuals by linear regression were large. The non-Gaussian peaks flattened the slope of the PSD_T and reduced estimates of a below 2 (brown noise, Fig. 7.5a), as the deviations increased the residuals.

Black noise is usually considered to result from nonstationarity (Sect. 6.4), such as the occurrences of rare events: epileptic seizures, epidemics, forest fires, etc. (Schroeder 1991), but in the case of the resting EEG and ECoG, the cause instead is the nonlinearity (Fig. 6.10) that is constantly imposed by the refractory periods. Refractoriness limits the microscopic firing rates not by an upper boundary but by the increased slope of the PSD_T across all frequencies. We propose that whatever the slope, the PSD_T can be used as one of the criteria by which to infer that an area of cortex is physiologically at rest (Freeman et al. 2009a) and not engaged in occult recovery from past activity, current maintenance, or prediction of future needs, while the subjects' overt behavior gives the appearance of resting (Raichle and Mintun 2006).

7.4 Null Spikes Revealed by Probability Distribution Function (PDF)

The strict 1/f form of the PSD_T may be a necessary condition for the resting state, but it is not sufficient, because the 1/f form is often seen in ECoGs from animals engaged in task performance when the duration of the sample is long (1–6 s). The short-time FFT (Chap. 3) from each step of a moving window (about 0.1 s) may yield one or more spectral peaks above a 1/f trend line (Fig. 6.3), which results from shifting of the carrier frequency within the beta-gamma range. We proposed in Sect. 6.3 that markers for nonrandom structures in the ECoG would be provided by subtle deviations from Gaussianity. In long segments of neocortical ECoGs, the distribution of amplitudes tended to appear Gaussian (Fig. 6.4a). The distribution of the envelope of the unfiltered ECoG (peak absolute values) conformed to the probability distribution function (PDF) for Rayleigh noise (Fig. 6.4d). We have made the

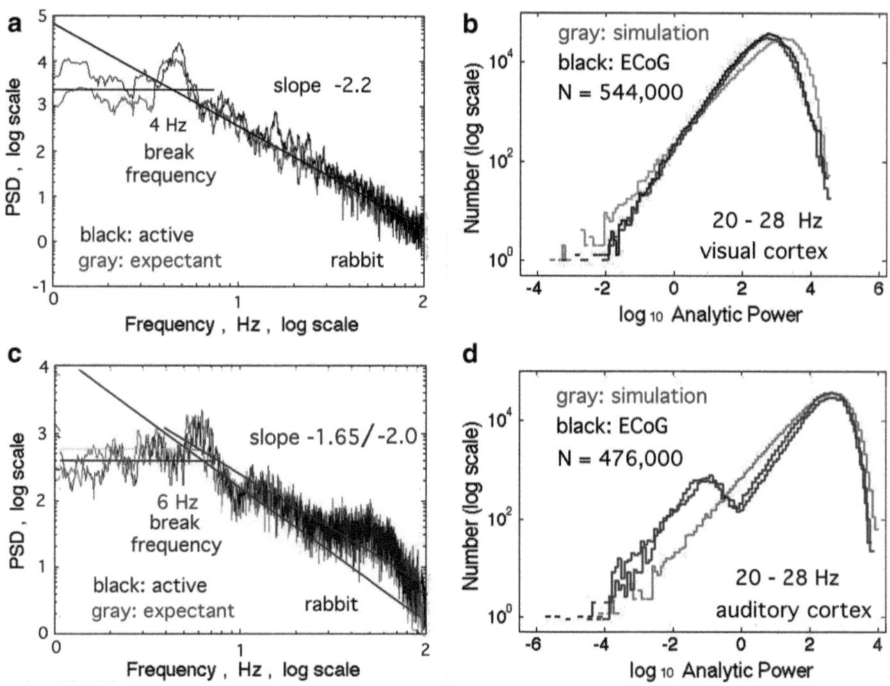

Fig. 7.5 Evidence is shown that neocortical ECoG usually conformed to black noise. (**a**) The PSD_T from the expectant and test states were indistinguishable from each other and from the canonical PSD_T for simulated black noise (Fig. 6.12, Sect. 6.7). (**b**) The frequency-specific probability distribution function (*PDF*) of the analytic power (Fig. 6.5c, Sect. 6.3) of the filtered ECoG in the high beta range is compared with the PDF from simulated black noise. (**c**) Deviations from noise gave peaks above the 1/f trend line (Fig. 6.3, Sect. 6.3). (**d**) In those cases, the PDF also deviated from that of noise. Freyer et al. (2009) fitted this type of PDF with a double exponential PDF, from which they inferred bistability of cortex. (From Freeman 2009)

case that Gaussian background activities at rest and at work were generated by mutual excitation in cortical populations (Fig. 6.12, Sect. 6.7) and that cognitive operations could give rise to deviations from normality. The background state was revealed form of the exponential PDF of analytic power, $A^2_{ij}(t)$,

$$P\left(A^2_{ij}(t)\right) = l\exp\left(-lA^2_{ij}(t)\right) \tag{7.1}$$

where l was the slope of the PDF in log-log coordinates. When the PSD_T conformed to 1/f (Fig. 7.5a), the PDF from the ECoG conformed to that of black noise (b). When peaks appeared in the beta-gamma range (c) of the PSD_T, deviations appeared in the PDF as well (d) (Freeman 2009). The same or similar deviations were found in the PDF of scalp EEG in the alpha range by Freyer et al. (2009) with an equivalent technique of decomposition using wavelets instead of the short-time FFT. The log-log PDF called our attention to extremely low values of log analytic power, $A^2_{ij}(t)$

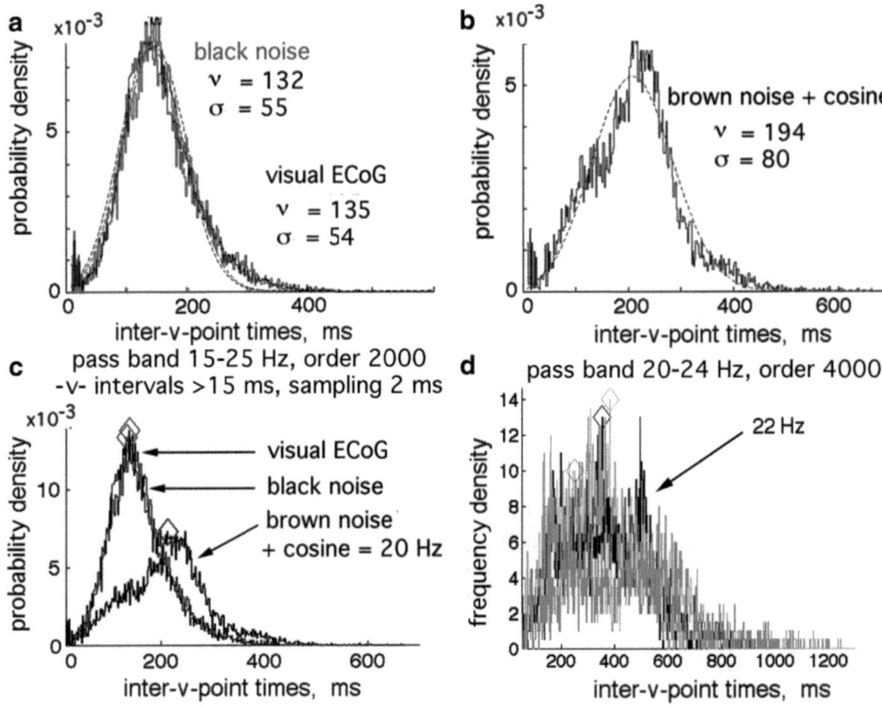

Fig. 7.6 (**a**) Superimposed PDFs of the time intervals between minima in analytic power of the filtered visual cortical ECoG and simulated black noise for the pass band 15–25 Hz were indistinguishable. *Dashed curves* were from (7.2). (**b**) Adding a 20-Hz cosine with amplitude 0.5, the SD of brown or black noise prolonged the intervals, "-v-", between the downward peaks of the filtered ECoG or noise. (**c**) The three density functions were superimposed. The small diamonds symbolized the modal values. (**d**) The same prolongation of intervals was found in ECoG when the filter pass band was centered on the carrier frequency of a burst. Black values were ECoG superimposed on black noise (*gray curve*) (Freeman 2009)

(Fig. 7.5d). Although they occurred infrequently, we came to invest them with special significance (Fig. 7.8d, Sect. 7.5; Fig. 9.12, Sect. 9.6.2) as *null spikes*.

More revealing was the PDF or the frequency density of the time intervals between minima (labeled "-v-", Fig. 7.6a) in the analytic power, $A^2_{ij}(t)$, of filtered ECoG, which were indistinguishable from the PDF of random noise, whether white, brown, or black, when the PSD_T conformed to 1/f. They were fitted with the function derived by Rice (1950):

$$P(x) = x / s^2 \exp\left\{\left(-\left[x^2 + n^2\right] / 2s^2\right)\right\} I_0\left(xn / s^2\right), \tag{7.2}$$

where $x = A^2_{ij}(t)$, s was the SD of the filtered ECoG, and n was a parameter that evaluated the deviation from randomness that was imposed on white, brown, or black noise by band-pass filtering, either endogenous from burst formation (Fig. 8.8b, Sect. 8.4) or imposed in data processing (Fig. 7.6b). When $n = 0$, the Rice distribution (concave upward to the modal value) reverted to the Rayleigh distribution.

Rice proved that the modal beat frequency in Hz depended only on the width of the pass band in Hz by a factor of 0.641 (Rice 1950), which we demonstrated for the ECoG (Freeman 2009) and EEG (Ruiz et al. 2010). Rice showed that the addition of a cosine signal to white noise increased the intervals between beats but only in the pass band of the added cosine. Simulations showed that adding a cosine to brown and black noise shifted the beat intervals as Rice predicted, but only in the center of the pass band (Fig. 7.6d, Sect. 7.4). Placing the pass band at the edges of the band shortened the beat intervals (Freeman 2009). The upward shift in modal intervals was found experimentally (d) with very narrow pass bands and high order of the filter, which decreased the sample size and increased the noise.

These findings are very preliminary and cannot support any conclusions, but they are important because they open an avenue to model the emergence of a narrow band burst in response to a CS (Fig. 8.8b, Sect. 8.4) as the addition of a cosine to the background ECoG noise. The Rice effect may explain the prolongation of bursts related to cognition (Fig. 9.7e, f, Sect. 9.4.2), and it may also explain their termination of a burst by intrinsic decoherence of the population in a beat (Fig. 9.12c, Sects. 9.7.1). Rice proposed using his PDF for detecting narrow band oscillations embedded in white noise. The simulations of brown and black noise (Freeman and Zhai 2009) show that band-pass filtering by itself causes beats to appear in the filtered ECoG, which increase the complexity of search for AM patterns. We predict that the use of extreme statistics as pioneered also by Freyer et al. (2009) will become a major source of new understanding of cortical dynamics.

7.5 Evaluating Spatial Resolution with the Point Spread Function (PSF)

The ECoG is generated by the flow of synaptic currents from layers of neurons packed in high density with their apical dendrites oriented perpendicular to the surface of the cortex (Fig. 6.2, Sect. 6.2.1). An example is presented in Fig. 7.7 of the properties of the ECoG generator in the olfactory bulb. Controlled activation is by single shock excitation of the lateral olfactory tract, which synaptically excites the granule cells in the bulb. During excitation (n1 in Fig. 7.7a), the synapses formed by the excitatory axons act like transmembrane batteries that drive ionic currents into the dendrites, creating an extracellular current sink (dashed curve in B). The intracellular currents converge to the axon at the cell body and flow out across the membrane, thereby causing them to fire by depolarization and incidentally creating an extracellular source. The extracellular dendritic current flows radially outward from the deep source toward the surface and radially converges to the superficial sink. The current converging to the sink establishes a bell-shaped distribution of negative potential on the bulbar surface. The source-sink pair from each neuron creates a scalar *dipole field* of potential that is surface-negative during the activation of excitatory synapses on the dendrites (n1 in Fig. 7.7a). The excited granule cells inhibit the mitral cells, which withdraw their background excitation of granule cells. Disexcitation reveals background inhibition. During inhibition, the direction of

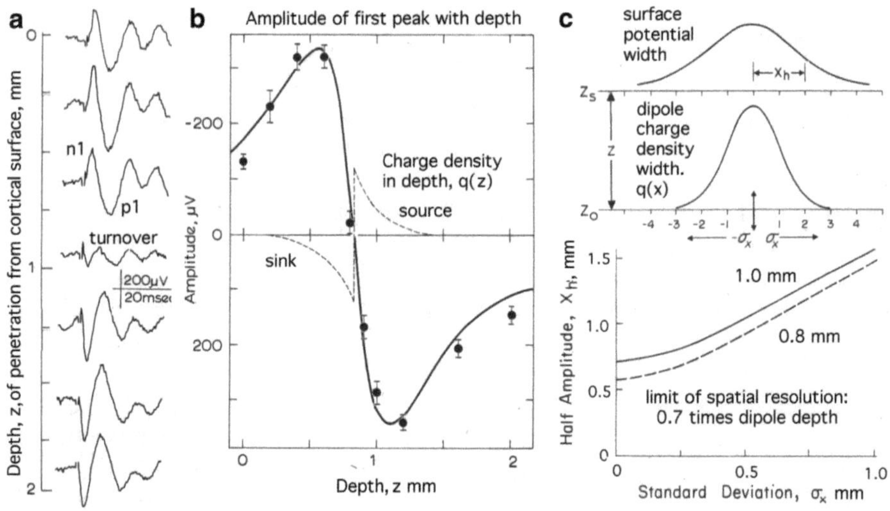

Fig. 7.7 (**a**) An example shows the change in evoked potential as the recording electrode is inserted into the bulb. Note the reversal in sign (−n1 to +p1) of the first peak at the turnover. (**b**) The *dashed curve* shows the source-sink charge density for the dendrites of the generating neurons alighted along the z-axis. The *symbols* show the mean ±SD of 100 summed evoked potentials at each step. The curve was calculated for the source-sink charge density in the volume conductor. This depth profile holds also for the depth LFP underlying the ECoG at the surface. (**c**) The example shows the bell-shaped distribution of surface potential (*upper frame*) for a Gaussian distribution of charge density (*middle frame*) at the depth of the turnover. The *lower frame* shows that the half-amplitude width of potential varies linearly with the SD of the dipole charge density for distances >0.5 mm. Extrapolation to zero SD gives the limit on spatial resolution of the ECoG imposed by volume conduction (from Figs. 4.28, 4.29, and 4.30 in Chap. 4 in Freeman 1975)

current is reversed, giving a surface-positive dipole field of potential (p1) with current radiating outwardly from the point of intersection of the dipole axis with the cortical surface (upper frame in Fig. 7.7c). The depth profile of the dipole field along its axis is calculated at each instant by expressing the current source-sink density as fixed charge density, calculating the centers of gravity for source and sink and the distance that separates them, and using Coulomb's law to evaluate the sum of potential created by the equivalent pair of +/− charge at each point along the dipole axis (solid curve in Fig. 7.7b).

Most cortical events involve co-activation of many dendritic dipoles, which give broadly distributed peaks of positive and negative potentials. The relation between instantaneous dipole width and peak width can be estimated by describing the dipole density as Gaussian and calculating the relation between the half-amplitude width (x_h in upper frame, Fig. 7.7c) and the SD of the dipole distribution (s_x in the middle frame). The relation is linear for distances on the surface >0.5 mm (lower frame). Extrapolation of the simulation to the dipole widths of cortical columns (Mountcastle 1998) and single dendritic trees (<0.5 mm) shows the limit of spatial resolution that is imposed by volume conduction. The width of the surgace peak of dendritic potential depends on the depth of the dipole layer. The theoretical limit is displayed in

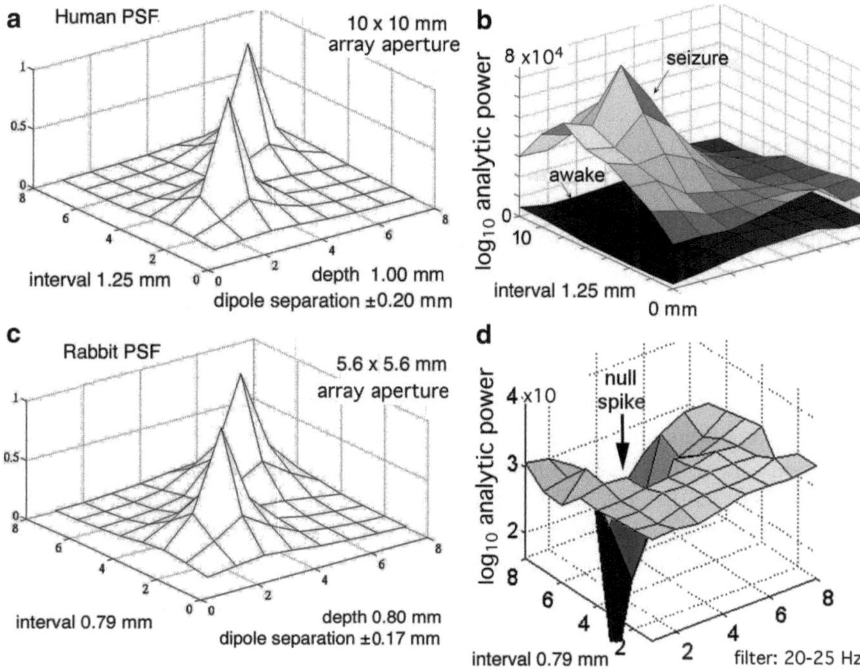

Fig. 7.8 (**a**, **c**) Simulated point spread functions (*PSF*) of two dipoles corresponding to two cortical columns or single neurons. The depth of the cortical dipole layer was estimated from measurements of the depth from the cortical surface of dipole turnover in human and rabbit. The "separation" refers to the distance between centers of gravity of source and sink (Fig. 7.7b). (**b**) An example of the spatial pattern of an epileptic spike like that in Fig. 6.1, A(e) but from human ECoG (Sect. 10.4), is compared with the calculated PSF (A). (**d**) An example of the spatial pattern of the $\log_{10}$ analytic power of rabbit ECoG compares the null spike (Sect. 9.6.2) with the PSF calculated for rabbit neocortex, which has approximately the same depth of generating layer as the olfactory cortex and the same waveform (Fig. 6.1e in Sect. 6.1). Such unusual events gave experimental evidence for the capability of the arrays illustrated to resolve fine details despite smoothing by volume conduction (from Fig. 2 in Freeman 2006)

graphs (Fig. 7.8a, c) of the surface distribution of potential of two cortical columns, which are 4.3 mm apart (three interelectrode steps in an 8×8 array). In analogy with the distribution of light from a point source, the spread of the peak or trough of potential at the surface is designated the *point spread function* (PSF). In the rabbit olfactory system, the average depth is 0.8 mm, which is approximately the depth of the mitral cell layer in the bulb and layer II of pyramidal cells in the olfactory cortex (Freeman 1975). The corresponding depth of layer II in human neocortex is approximately 1.0 mm (Mitzdorf 1987).

The simulation showed that the surface half-amplitude width of the PSF attributable to volume conduction is approximately 0.7 times the dipole depth mm, which for human is 0.7 mm and for rabbit is 0.5 mm. By this criterion, the interelectrode separations, respectively, of 1.25 and 0.79 mm suffice to resolve local peaks and troughs of ECoG activity in single columns separated by 1 mm or more. An example of a localized upward peak (Fig. 7.8b) is from human temporal lobe ECoG recorded during a complex partial seizure, which was comparable to a seizure induced in a rabbit olfactory cortex (Fig. 6.1e, Sect. 6.1). An example of a null spike (Fig. 7.8d) is from the visual cortical ECoG (Fig. 9.12b, Sect. 9.6.1). Such extreme spikes tended to recur singly or in clusters at theta rates after band-pass filtering in the beta or gamma range (Freeman 2009).

The most important result of the simulation (left frames) is the demonstration that the closely spaced arrays resolve spatial differences in amplitude and power of ECoGs from closely spaced cortical columns despite high temporal correlation between them. The reason for the discrepancy is that the entire array has phase-locked gamma oscillations in the same instantaneous frequency range (Fig. 6.3) but with amplitude modulation (AM). Correlation techniques including PCA (principal components analysis) extract the temporal coherence, whereas spatial mapping reveals the local textures of amplitude and power. We conclude that the shared frequencies and high correlations are mainly due to interactions among cortical neurons and they are not solely due to volume conduction or to activity at the reference lead in referential recording. Therefore, sampling intervals are chosen in accord with the spatial Nyquist frequency that we derive from the PSD_X of ECoG (Sect. 7.2).

7.6 Search for the Spatial Grain of EEG

Ideally, the EEG would be sampled simultaneously at high density over the entire head and displayed in the detail afforded by modeling with many thousand point sources (Izhikevich and Edelman 2008). Practical considerations constrain the sampling intervals for seeking textures using the PSD_X (Freeman et al. 2003; O'Connor and Robinson 2004; Ramon et al. 2009). At the time the observations described in what follows were done, the number of channels simultaneously recorded could not exceed 64; the array could not exceed the circumference of the head; the intervals had to be small enough to oversample the EEG in search of contributions, if any, from the gyri and sulci, owing to variations that they introduced in the distance from scalp to cortex and the orientation of the cortical surface with respect to the scalp surface (Fig. 7.9) (Freeman et al. 2009a). A practical compromise to adjudicate these constraints was a curvilinear 1-D array of 64 gold-plated pins (Freeman et al. 2003) that were glued in a strip of embroidery cloth with spacing of 3 mm between holes (rows of dots in Fig. 7.9). This interval met the engineering *practical* Nyquist criterion (three instead of two samples per gyrus of width 10 mm). The 189 mm length of 64 electrodes conveniently covered a substantial fraction of the circumference of the scalp in any orientation. The strip was bound to the head without electrode gel in view of the close proximity of the pins so as to avoid shunting of surface current by the gel. Broad temporal pass bands (e.g., 20–80 Hz) and the Hilbert transform (Sects. 6.4.1 and 9.5) were used to track temporal frequency modulation; PCA served to extract widespread waveforms on the assumptions of stationarity (Sect. 6.4) and linearity (Sect. 6.5).

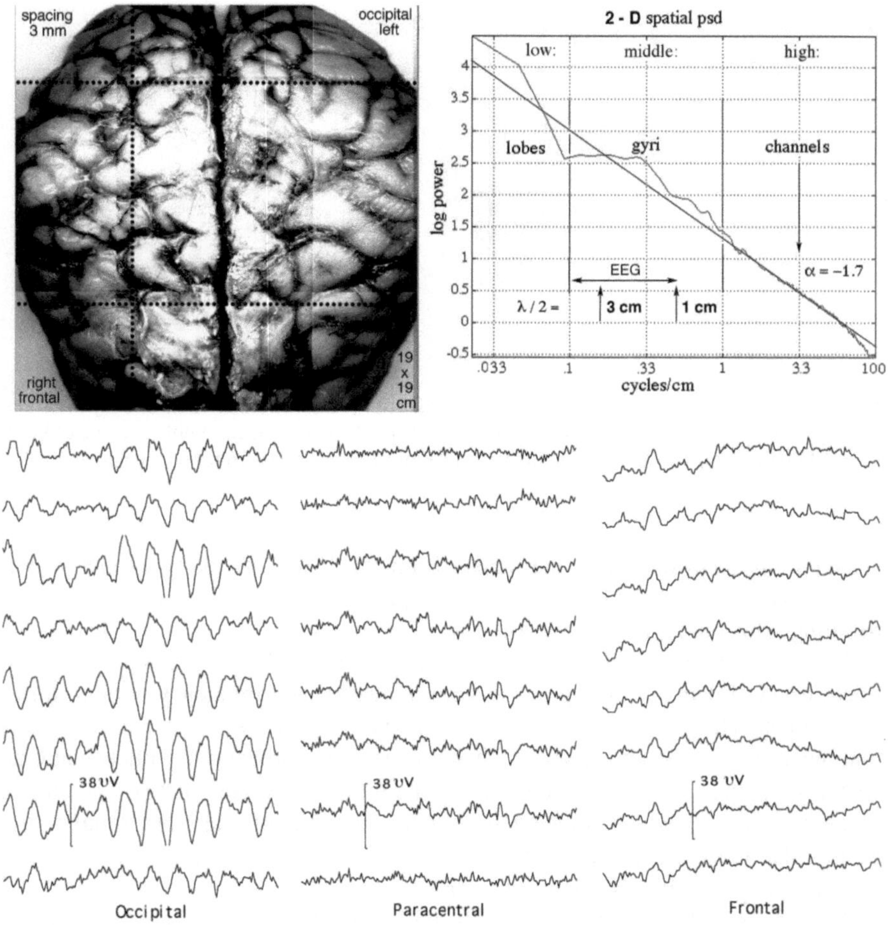

Fig. 7.9 *Above left*: photomontage to flatten the display of the brain surface as seen from above. The 64 × 1 scalp array is symbolized by the rows of dots superimposed in three orientations: frontal, occipital, and right parasagittal. The 64 dots are 3 mm apart. *Above right*: the spatial spectrum of the photomontage, showing a peak corresponding to the estimated width and length of the gyri. *Below*: examples of signals from 8/64 contiguous electrodes showing local and regional differences as well as similarities (from Freeman et al. 2003)

Examples of the temporal PSD_T from ten subjects (Fig. 7.10a) show nearly the same canonical form for the EEG as for the ECoG in log-log coordinates: a plateau in the delta-theta range and a concave-downward inflection to a trend of linear decrease in log power with increasing log frequency over the beta-gamma range. With eyes closed, there was a prominent peak in the alpha range near the inflection. There was no plateau at high frequencies, indicating that the sampling rate was too low (Sect. 7.2). The PSD_T from single channels differed in mean power but not in form. This was consistent with the finding by Pockett et al. (2009) of widespread intermittent synchrony in all bands (Chap. 10). The PSD_T slopes from the EEG calculated by linear regression (Sect. 7.3) were in the range of 1–2 (pink noise),

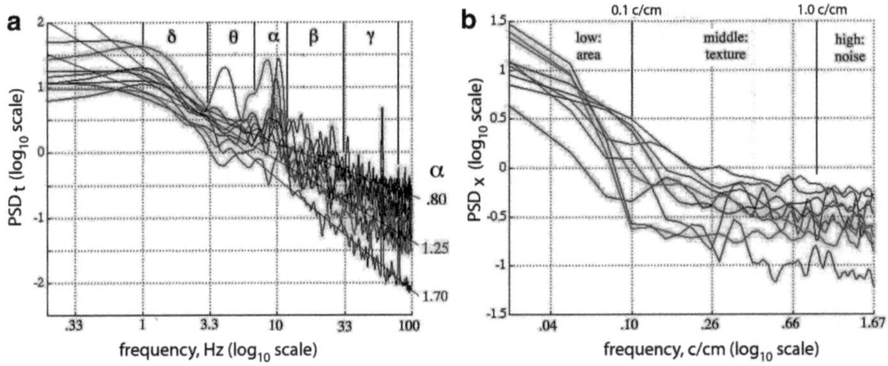

Fig. 7.10 (**a**) Spatial averages are shown of 64 PSD_T of normalized EEG from ten subjects at rest with eyes closed (2 s duration, 2-ms sample interval, 1,000 steps). (**b**) Ensemble average of PSD_X from 1,000 time steps, 3-mm sample interval, 189-mm 1-D aperture (from Freeman et al. 2003)

substantially shallower than the slopes from the ECoG. A likely explanation is contamination of the EEG by electromyographic (EMG) potentials (Sect. 7.8) contributing power at high frequencies.

Examples of the spatial PSD_X from the same subjects (Fig. 7.10b) showed a nonlinear trend of decrease in log power with increasing log frequency with no linear 1/f segment. There was no plateau in low spatial frequencies <0.1 c/cm (indicating that the array was too short, Sect. 7.2), but there was a broad, noisy plateau at high spatial frequencies >1.0 c/cm (confirming the intended oversampling). The middle range 0.1–1.0 c/cm was characterized by an irregular mix of peaks and troughs, suggesting that these are the spectral locations of significant deviations from randomness, possibly cognitively related structure.

Comparison of the EEG PSD_X with the ECoG PSD_X (Fig. 7.11) revealed the effects of four experimental problems that pose major difficulties in getting useful EEG images: (1) the long distance between the scalp surface and the generating cortical surfaces, compounded by the impedance barriers of the cerebrospinal fluid, skull, and scalp; (2) the lack of any site on the head for reference-free recording; (3) gyrification; and (4) muscle noise (electromyographic activity, EMG).

(1) The effect of the impedance barriers is seen in the leftward shift of the PSD_X (reduced power at all frequencies) of the EEG from the ECoG due to the combined effects of the distances between scalp and cortex and the shunting of dendritic currents by the low resistance of the scalp and cerebrospinal fluid on opposite sides of the high resistance of the skull (Ramon et al. 2009).

(2) The only bias-free reference for scalp recording is the sum of signals recorded densely over the entire body (Sect. 3.10.2) (Tucker 1993). Reference sites equidistant from a curvilinear array reduce the bias from incomplete coverage (Junghöfer et al. 1999) by common-mode rejection. The signals from parasagittal placements of the array can be referenced to the contralateral mastoid; exclusively frontal or occipital recordings can be referenced, respectively, to the inion and nasion.

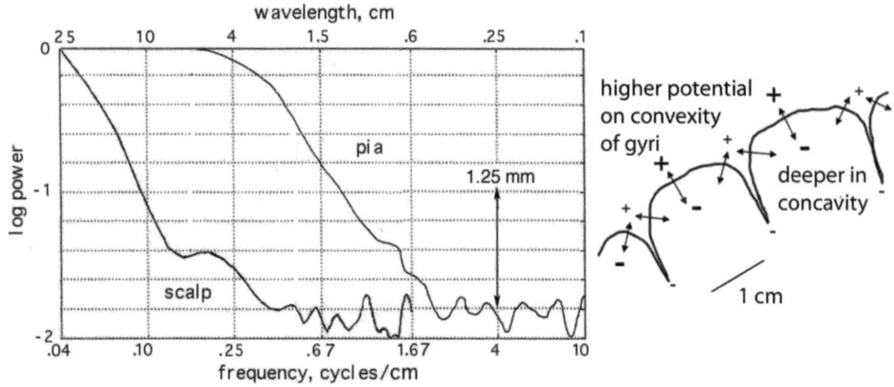

Fig. 7.11 Ensemble averages of PSD_X were from noninvasive scalp EEG in subjects at rest and the ECoG from the exposed surface (pial membrane) of a neurosurgical patient under local anesthesia. *Right inset*: the effects of gyri on the scalp EEG might explain peaks in the PSD_X in the middle range, on the condition that widely synchronized oscillations occupy a substantial area of cortex under the array, as shown by Pockett et al. (2009) and Ruiz et al. (2010). The lack of a flat plateau by which to demarcate a low-frequency concave-down inflection showed that the arrays were not long enough to encompass the low spatial frequencies (from Freeman et al. 2003)

Re-referencing is also done by subtracting the spatial ensemble average of the 64 signals in each frame from each signal, which preserves the AM pattern feature vectors (Sect. 6.4.3) but reverses the sign of the phase on every signal where the amplitude is less than the average (Fig. 10.9b). Once a reference is chosen, the fixed bias thus imposed has no significant effect on the textures of the spatial patterns of amplitude and their changes with behavior (Ruiz et al. 2009).

(3) A contiguous array of electrodes on the surface for the ECoG of a gyrus provides equal distances to the underlying dendritic generators. For the EEG, the distances from the electrodes to the cortical surface vary widely, especially over the midline sagittal and lateral Sylvian fissures. The variations in distances and orientations of the cortical surfaces with respect to the scalp surface (Fig. 7.11) impose fixed spatial differences in amplitude at all frequencies in the PSD_X, which are characteristic of each subject and can be related to the individual brain by use of structural MR images (Ramon et al. 2009). The characteristic 'signature' AM pattern (Sect. 8.2) is modulated by transient deviations that reflect behavioral state changes with fixed arrays. Decomposition of the PSD_X in 7-Hz temporal frequency bands shows that a small peak in the PSD_X occurs only in the beta and gamma ranges (Fig. 7.12a) and not in the high power ranges of theta and alpha (Fig. 7.12b). The evidence indicates that in some subjects, locations, and states, especially in the presence of strong alpha or theta, the spatial resolving power of the EEG extends to the spatial frequency range of human gyri and sulci. This inference has been supported by Ramon et al. (2009), using a 4-shell finite element model with 2,650 dipolar sources oriented normal to the local cortical surface. By calculating 51 parasagittal MR images, they demonstrated PSD_X peaks within the range of 0.32 to 0.9 cycles/cm (wavelength 3.125–1.11 cm/cycle). They concluded that "the practical Nyquist

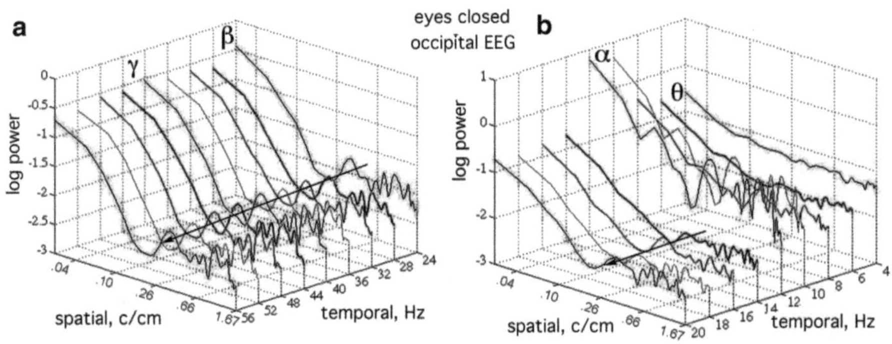

Fig. 7.12 The EEG was filtered in 7-Hz pass bands prior to calculating the PSD_X. The set of curves at right display the low temporal frequencies. The PSD_X for high temporal frequencies was rescaled to compensate for the low power. The notable feature was the spatial peak in the beta-gamma range, which was most prominent in conditions of high alpha or theta activity. As in the olfactory system, the bursts of beta and gamma oscillation recurred at theta and alpha frequencies (Fig. 8.3b, Sect. 8.2) (theta-gamma linkage). Similar features were found in the human ECoG (displayed from a different perspective, Fig. 7.3) (from Freeman et al. 2003)

frequency for sampling scalp EEGs should be 3.0 c/cm, and an optimal interelectrode spacing of about 3 mm is needed for extraction of cortical patterns from scalp EEGs in humans" (p. 191).

(4) Occult EMG potentials were presumed to be present in all scalp recordings (Whitham et al. 2007). For purposes of description and measurement of EMG, subjects were asked to tense their scalp muscles in order to add detectable noise to the signals (Fig. 7.13). EMG from movements of the eyes, jaw, and neck were treated as overt, not occult, and could be prevented by instruction or deleted by editing. The effect on the PSD_T and PSD_X was to increase the power not just above 20 Hz (Whitham et al. 2007) rather across the entirety of both spectra (Fig. 7.13). We suggest that the basis for the 20-Hz threshold is that the EEG PSD_T tends to $i/f^{2.5}$ exponent, whereas the EMG PSD_T tends to $1/f^0$ flat white noise, so the EMG dominates the high frequencies in the summed spectra.

7.7 PSD_X of EEG Plus EMG: Spatial Autocorrelation Function (SAF)

The broad distribution of power across the spatial spectrum suggests that occult EMG power is concentrated in spatial spikes, which are generated by muscle cells located under the skin and close to the array electrodes. This inference is supported by spatial autocorrelation (Fig. 7.14). Each EEG signal from the center 32 electrodes is lagged across the 1-D array for ±32 steps of 3 mm and cross-correlated with itself. The resulting spatial autocorrelation function (SAF) during sustained EMG has a sharp spike at the origin (Fig. 7.14a) and a decrease in amplitude with increasing

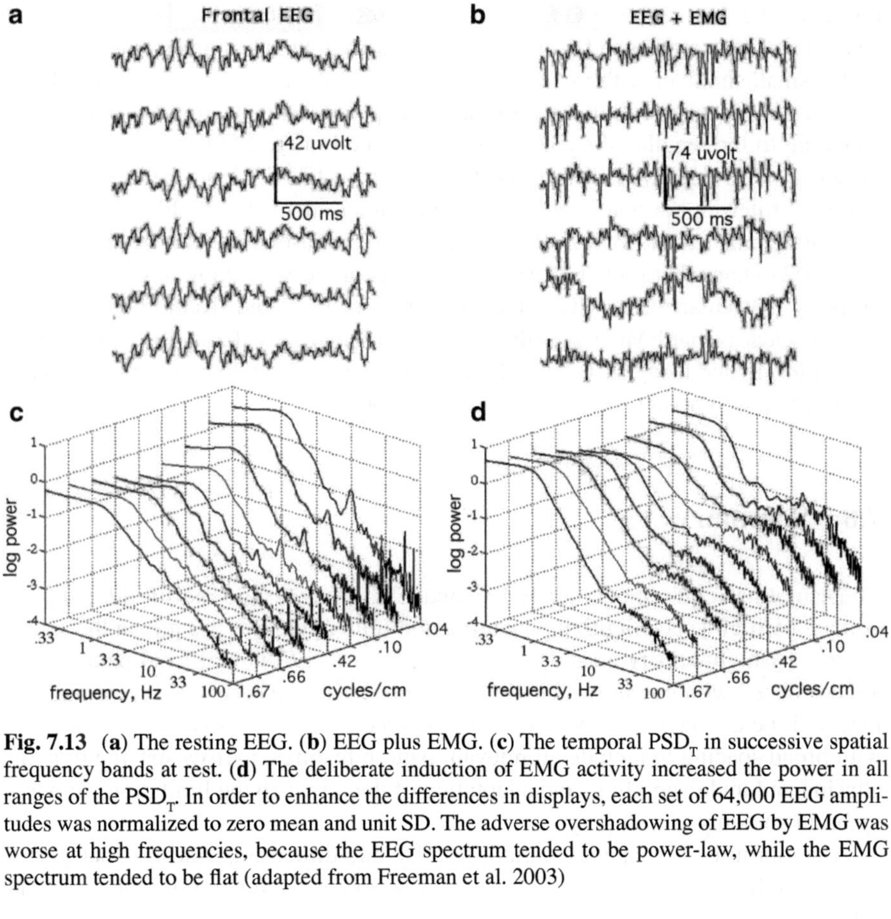

Fig. 7.13 (**a**) The resting EEG. (**b**) EEG plus EMG. (**c**) The temporal PSD_T in successive spatial frequency bands at rest. (**d**) The deliberate induction of EMG activity increased the power in all ranges of the PSD_T. In order to enhance the differences in displays, each set of 64,000 EEG amplitudes was normalized to zero mean and unit SD. The adverse overshadowing of EEG by EMG was worse at high frequencies, because the EEG spectrum tended to be power-law, while the EMG spectrum tended to be flat (adapted from Freeman et al. 2003)

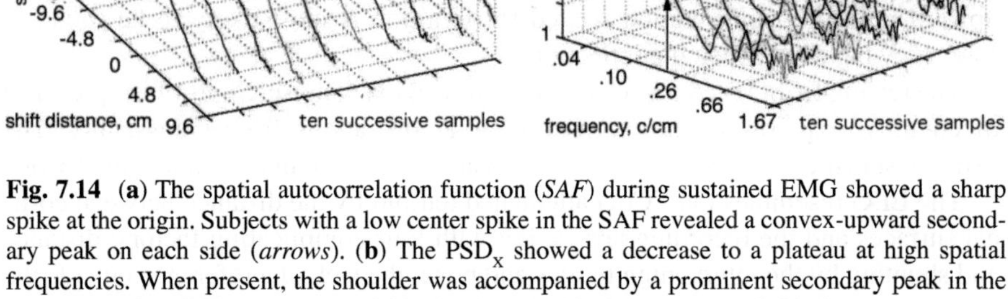

Fig. 7.14 (**a**) The spatial autocorrelation function (*SAF*) during sustained EMG showed a sharp spike at the origin. Subjects with a low center spike in the SAF revealed a convex-upward secondary peak on each side (*arrows*). (**b**) The PSD_X showed a decrease to a plateau at high spatial frequencies. When present, the shoulder was accompanied by a prominent secondary peak in the mid-range of the PSD_X, here about 0.22 c/cm (*arrow*) (from Freeman et al. 2010)

distance. The SAF of the EEG from subjects at rest with no visible EMG has a much smaller spike at zero lag and decreasing amplitude with increasing distance but often with a small shoulder with increasing correlation distance (arrows in A). The PSD_X reveals a secondary peak in the mid-range of spatial frequency (arrow in B) corresponding to the lengths of gyri, which is not found in the PSD_X of the SAF in the presence of strong EMG. It follows that the occult EMG cannot be attenuated by temporal filtering but that it might be significantly attenuated by high-pass spatial filtering (Gonzalez and Wintz 1977) of EEG from ultra-high-density scalp arrays of 256–1,024 channels (Tucker 1993; Junghöfer et al. 2004), taking advantage of the property that cutaneous EMG spikes are spatially localized (Ramon et al. 2009). What is clear is that EMG poses the greatest difficulty in realizing the full potential for exploring and using EEG spatial images for scientific and engineering purposes such as brain-computer interfacing (Freeman 2006).

7.8 Summary

The sampling requirements for ECoG and EEG imaging are met in the temporal domain by high-speed digitizing and time multiplexing. They are met in the spatial domain by high-density arrays to get the textures of ECoG and EEG patterns that are correlated with cognition. The limits of the size of the accessible recording surface and the available number of channels force a trade-off between aperture size with broad spacing to include all relevant pattern features, narrow spacing to avoid aliasing and undersampling, and adaptation of the arrays to the limitations imposed by brain anatomy and surgery.

A compromise for spatial sampling is by one-dimensional arrays that permit close sampling for high frequencies over long distances for low frequencies. The 1-D FFT of the ECoG in subjects at rest gives canonical forms of the PSD_T and PSD_X. In log-log coordinates, they conform to three line segments that are separated by two inflections. For long segments >1 s, the PSD_T and PSD_X in subjects at rest tend to conform to a power-law density distribution, $1/f^a$, $0 < a < 4$. The low-frequency and high-frequency segments tend to be flat as in white noise, $a = 0$.

The sampling rule is proposed that the optimal temporal or spatial aperture is set by the wavelength of the frequency at the lower, concave-downward inflection, and that the Nyquist sampling frequency is set by twice the high-frequency, concave-upward inflection. Information related to cognition is most likely to be found in frequency bands between the inflections.

The resting ECoG is characterized as Gaussian black noise. Signs of the presence of cognitive patterns in the ECoG appear as deviations from Gaussianity. The extreme statistics of Rice can help extract and explain singular events in the ECoG.

The EEG resembles the ECoG attenuated tenfold by the distance between cortical surfaces and the recording arrays, but with four complicating factors: distance and shunting by the intervening tissues and bone, the lack of a reference outside the EEG fields of potential, the folding into gyri and sulci, and electromyographic potentials (EMG) added to the EEG. All factors must be incorporated into simulations (Freeman 2006) of the PSD_X. The log-log PSD_X, point spread function (PSF), and spatial autocorrelation (SAF) of the EEG provide the tools needed to show that spatial resolution extends to the level of gyri and sulci and that occult EMG contributes noise to the entire ranges of the PSD_T and PSD_X, not only to the high beta and gamma ranges. Success in textural analysis of EEG in the gamma range will depend on solving the problem of removing the EMG potentials, using the new tools provided by high-density EEG recording: PSD_X, PSF, and SAF.

References

Braendgaard H, Evans SM, Howard CV, Gunderson HJG (2000) The total number of neurons in the human neocortex unbiasedly estimated using optical dissectors. J Microsc 157(3):285–304

Buzsáki G (2006) Rhythms of the brain. Oxford University Press, Oxford

Freeman WJ (1975) Mass Action in the Nervous System. Examination of the neurophysiological basis of adaptive behavior through the EEG. Academic Press, New York. Posted in e-formats (2004), http://sulcus.berkeley.edu/MANSWWW/MANSWWW.html

Freeman WJ (2004a) Origin, structure, and role of background EEG activity. Part 1. Analytic amplitude. Clin Neurophysiol 115:2077–2088. http://repositories.cdlib.org/postprints/1006

Freeman WJ (2004b) Origin, structure, and role of background EEG activity. Part 2. Analytic phase. Clin Neurophysiol 115:2089–2107. http://repositories.cdlib.org/postprints/1486

Freeman WJ (2005) Origin, structure, and role of background EEG activity. Part 3. Neural frame classification. Clin Neurophysiol 116(5):1118–1129. http://authors.elsevier.com/sd/article/S1388245705000064

Freeman WJ (2006) Origin, structure, and role of background EEG activity. Part 4. Neural frame simulation. Clin Neurophysiol 117/3:572–589. http://repositories.cdlib.org/postprints/1480/, http://dx.doi.org/10.1016/j.clinph.2005.10.025

Freeman WJ (2009) Deep analysis of perception through dynamic structures that emerge in cortical activity from self-regulated noise. Cogn Neurodyn 3(1):105–116

Freeman WJ, Vitiello G (2006) Nonlinear brain dynamics as macroscopic manifestation of underlying many-body field dynamics. Phys Life Rev 3:93–118. http://dx.doi.org/10.1016/j.plrev.2006.02.001, http://repositories.cdlib.org/postprints/1515

Freeman WJ, Zhai J (2009) Simulated power spectral density (PSD) of background electrocorticogram (ECoG). Cogn Neurodyn 3(1):97–103. http://repositories.cdlib.org/postprints/3374

Freeman WJ, Rogers LJ, Holmes MD, Silbergeld DL (2000) Spatial spectral analysis of human electrocorticograms including the alpha and gamma bands. J Neurosci Methods 95:111–121

Freeman WJ, Burke BC, Holmes MD, Vanhatalo S (2003) Spatial spectra of scalp EEG and EMG from awake humans. Clin Neurophysiol 114:1055–1060. http://repositories.cdlib.org/postprints/989

Freeman WJ, Holmes MD, West GA, Vanhatalo S (2006a) Fine spatiotemporal structure of phase in human intracranial EEG. Clin Neurophysiol 117:1228–1243

Freeman WJ, Holmes MD, West GA, Vanhatalo S (2006b) Dynamics of human neocortex that optimizes its stability and flexibility. Int J Intell Syst 21:1–21. http://repositories.cdlib.org/postprints/2385

Freeman WJ, O'Nuillain S, Rodriguez J (2008) Simulating cortical background activity at rest with filtered noise. J Integr Neurosci 7(3):337–344. http://repositories.cdlib.org/postprints/3373

Freeman WJ, Ahlfors SM, Menon V (2009a) Combining EEG, MEG and fMRI signals to characterize mesoscopic patterns of brain activity related to cognition. Int J Psychophysiol 73(1):43–52. http://repositories.cdlib.org/postprints/3386

Freeman WJ, Kozma R, Bollobás B, Riordan O (2009b) Chapter 7: Scale-free cortical planar network. In: Bollobás B, Kozma R, Miklös D (eds) Handbook of large-scale random networks.

Bolyai mathematical studies, vol 18. Springer, New York, pp 277–324. http://www.springer.com/math/numbers/book/978-3-540-69394-9
Freeman WJ, Ramon C, Holmes MD (2010) 1-D spatial autocorrelation function of EEG: a sensitive assay for occult EMG; a tool to reduce it. In: Abstract #738, Proceedings of 16th conference on human brain mapping, pp 6–10, June, Barcelona, Spain
Freyer F, Aquino K, Robinson PA, Ritter P, Breakspear M (2009) Bistability and non-Gaussian fluctuations in spontaneous cortical activity. J Neurosci 29(26):8512–8524
Gonzalez RC, Wintz P (1977) Digital image processing. Addison-Wesley, Reading
Izhikevich E, Edelman G (2008) Large-scale model of mammalian thalamocortical systems. Proc Natl Acad Sci USA 105(9):3593–3598
Junghöfer M, Elbert T, Tucker DM, Braun C (1999) The polar average reference effect: a bias in estimating the head surface integral in EEG recording. Clin Neurophysiol 110(6):1149–1155
Majumdar NS, Pribram KH, Barrett TW (2006) Time frequency characterization of evoked brain activity in multiple electrode recordings. IEEE Trans Biomed Eng 53(12):1–9
Menon V, Freeman WJ, Cutillo BA, Desmond JE, Ward MF, Bressler SL, Laxer KD, Barbaro NM, Gevins AS (1996) Spatio-temporal correlations in human gamma band electrocorticograms. Electroencephalogr Clin Neurophysiol 98:89–102
Mitzdorf U (1987) Properties of the evoked potential generators: current source-density analysis of evoked potentials in cat cortex. Int J Neurosci 33:33–59
Mountcastle VB (1998) Perceptual neuroscience: the cerebral cortex. Harvard University Press, Cambridge
Nunez PL (1981) Electric fields of the brain: the neurophysics of EEG. Oxford University Press, New York
O'Connor SC, Robinson PA (2004) Unifying and interpreting the spectral wavenumber content of EEGs, ECoGs, and ERPs. J Theor Biol 231(3):397–412
Pockett S, Bold GEJ, Freeman WJ (2009) EEG synchrony during a perceptual-cognitive task: widespread phase synchrony at all frequencies. Clin Neurophysiol 120:695–708. doi:10.1016/j.clinph.2008.12.044
Raichle M, Mintun M (2006) Brain work and brain imaging. Annu Rev Neurosci 29:449–476
Ramon C, Freeman WJ, Holmes MD, Ishimaru A, Haueisen J, Schimpf PH, Resvanian E (2009) Similarities between simulated spatial spectra of scalp EEG, MEG and structural MRI. Brain Topogr 22:191–196
Rice SO (1950) Mathematical analysis of random noise- and appendixes. Technical Publications Monograph B-1589. Bell Telephone Labs Inc., New York
Rudrauf D, Douiri A, Kovach C, Lachaux J-P, Cosmelli D, Chavez M, Adam C, Renault B, Martinerie J, Le Van Quyen M (2006) Frequency flows and the time-frequency dynamics of multivariate phase synchronization in brain signals. Neuroimage 31:209–227
Ruiz Y, Li G, Freeman WJ, Gonzalez E (2009) Detecting stable phase structures on EEG signals to classify brain activity amplitude patterns. J Zhejiang Univ 10(10):1483–1491
Ruiz Y, Pockett S, Freeman WJ, Gonzales E, Guang L (2010) A method to study global spatial patterns related to sensory perception in scalp EEG. J Neurosci Methods 191:110–118
Schroeder M (1991) Fractals, chaos, power laws. Minutes from an infinite paradise. WH Freeman, San Francisco
Tucker DM (1993) Spatial sampling of head electrical fields: the geodesic sensor net. Electroencephalogr Clin Neurophysiol 87(3):154–163
Whitham E, Pope K, Fitzgibbon S, Lewis T, Clark C, Loveless S, Broberg M, Wallace A, DeLosAngeles D, Lillie P (2007) Scalp electrical recording during paralysis: quantitative evidence that EEG frequencies above 20 Hz are contaminated by EMG. Clin Neurophysiol 118(8):1877–1888

Chapter 8
Allocortical ECoG Images Formed by Learning

8.1 Introduction

The olfactory system is by far the best place to start searching for the neural operations by which vertebrate cortices form spatial images. This is so for several reasons (Freeman 2001). Its function is clearly to transform sensations of odorant chemicals to perceptions of odors. In phylogenetic evolution of brains, olfaction was the first sensory modality to achieve full-scale cortical processing. It pioneered development of the vertebrate mechanisms of perception that were adopted and adapted by the predictive components of the systems for vision, touch, and hearing, all of which receive information from distant sources, thereby availing time for prediction (Zelano et al. 2011). A single synaptic layer separates the nasal receptors from input to the most intimate organs of the limbic system. The three-layered allocortex of the olfactory system is far simpler than six-layered neocortex (Freeman 2001). Beginning cortical studies with neocortex may end up being like trying to reverse engineer electronics by studies of VLSI instead of breadboard transistor circuits. Finally, all sensory and motor systems may have the same basic code in order to integrate information from multiple sources. In other words, any algorithm that is proposed to explain vertebrate brain function is likely to be compatible with the olfactory code for perception. Logically, the place to begin the search for the shared code is to explore the dynamics of the primordial archetype (Kozma et al. 2003).

Comparative neurologists (Herrick 1948; Roth 1987) have deduced that the surviving animal with the brain most closely conforming to that of the ancestor of vertebrates is the salamander (Fig. 8.1). Herrick (1948) remarked, "… the tiger salamander … is appropriately named, for within the obscurity of its contracted world it is a predaceous and voracious terror to all humbler inhabitants" (p.3). The olfactory bulb occupies the anterior third of each cerebral hemisphere. The medial third is devoted to spatial and temporal orientation provided by the hippocampal formation (Buzsaki 2006). The lateral third organizes motor functions (pyriform in Fig. 8.1).

W.J. Freeman and R. Quian Quiroga, *Imaging Brain Function With EEG: Advanced Temporal and Spatial Analysis of Electroencephalographic Signals*,
DOI 10.1007/978-1-4614-4984-3_8,

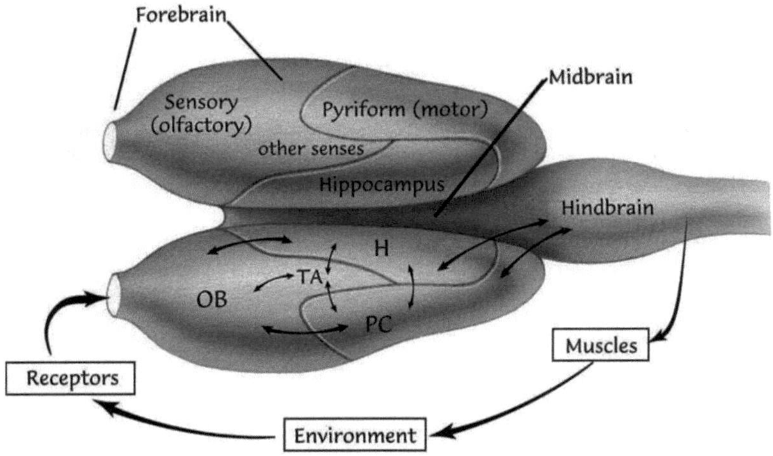

Fig. 8.1 The salamander brain is the most primitive among living vertebrates. The tissues of the head are too delicate for long-term electrophysiological observation, but the brain provides a simple model of the functions of the three main parts of each hemisphere (sensory, motor, associational) and above all their unity of organization in generating cognitive behavior that is implemented in the action-perception cycle (Merleau-Ponty 1942). The three parts are Olfactory Bulb (OB), Prepyriform Cortex (PC) and Hippocampus (H). They are elaborated in more advanced brains and become known as the limbic system from "limbus" meaning "border" or "belt" in Latin. The transitional area (TA) begins as an accessory to the olfactory system. It receives thalamic input from all other sensory systems. Its surface area expands in reptiles and birds into three-layered general cortex. In mammals, it is further elaborated to six-layered neocortex. From Freeman (2001)

Despite the small size and relative simplicity of the forebrain, the cooperative interactions among these three parts have enabled the survival and prevalence of the species. The organization of neurons into layers first appeared by evolution in these parts in the form of three-layered allocortex. Cortical processing in other modalities began with contributions of visual, auditory, and somatic information through the thalamus to a central transitional area (TA). In reptiles, the TA expanded in area into three-layered general cortex. The three primordial components formed a ring surrounding the general cortex of each hemisphere. That topology led to the nineteenth-century concept of the limbic system (Maclean 1969). In Fig. 8.1, we emphasize the anatomical basis for the unity of dynamics among the three parts in the execution and modeling of intentional behaviors (Bressler 1988; Kozma et al. 2003). The spatial images from the ECoG and EEG in mammalian brains reveal this functional unity in widespread spatial coherence (Chap.10).

The microscopic structure of cortex is a dense fibrous tissue containing axons, dendrites, capillaries, and the cell bodies of neurons and glia. It is called *neuropil* (Sect. 6.2.2). We find evidence that the distributions of the lengths of the axons and dendrites tend to be power-law (Miller 2002; Freeman and Breakspear 2007), which supports the scale-free cortical dynamics revealed by the ECoG (Freeman et al. 2008). It is likely that the most numerous connections with the shortest distances

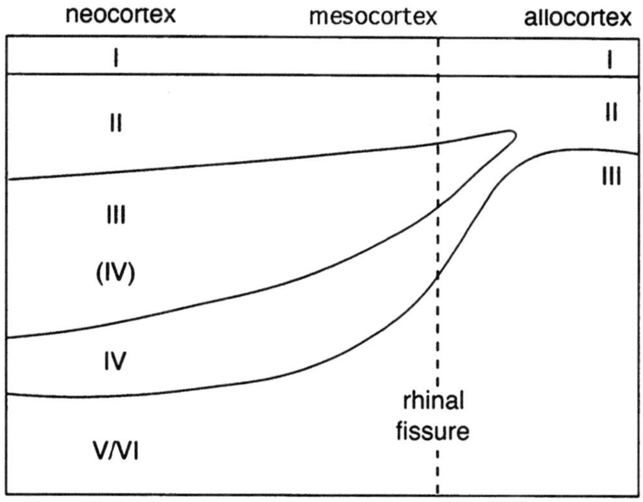

Fig. 8.2 Schematic summary of the transformation of allocortex to neocortex through mesocortex (including the entorhinal cortex) by the migration of cells from the floor of the lateral ventricle of each hemisphere into layer II of reptilian general cortex. In what became layers II, III, and IV of neocortex, they formed the specialized networks that perform the complex operations required for visual, auditory, and somatic information processing. The networks are embedded in the global sheets of interconnected neurons in layers I, V, and VI. The spatial images from ECoG and EEG (Sects. 10.3, 10.4) illustrate the unity of the scale-free dynamics that is sustained by these embedding layers in both allocortex and neocortex in all vertebrates from mouse to whale (Modified from Miller and Maitra (2002, p. 237) to emphasize the transition from 3 to 6 layers)

are both chemical and electrical (including gap junctions); longer connections are by chemical synapses, mainly axodendritic. Because the dendrites tend to align themselves in palisades oriented perpendicularly to the surface and the cell bodies tend to form layers, cortex is classified as *laminated neuropil*, in contrast to the *reticular* neuropil of the central neuraxis of the brain (seemingly random orientation of neurons in 3-D) and the *nuclear* neuropil of the basal ganglia (globular clusters of neurons in 3-D).

The upper layer I (Fig. 8.2) of allocortex contains input axons and the dendritic trees on which they synapse. The middle layer II has the cell bodies, often with triangular shapes giving the name *pyramidal* cells (*mitral* cells in the bulb). The lower layer III has interneurons (*stellate* cells in cortex, *internal granule* cells in the bulb) and the output axons with their recurrent side branches (axon *collaterals*) that synapse on interneurons but significantly on other pyramidal cells. Topologically, allocortex is a single synaptic layer containing three kinds of loops. Their complexity is simplified by defining and using K-sets (Chap. 5 in Freeman 1975), which provide a hierarchy of feedback models: excitatory cells (KI_E-sets with mutual excitation, which is positive feedback (Sect. 9.2)); inhibitory cells (KI_I-sets with mutual inhibition (Freeman 1986), which is also positive feedback (Freeman 1986)); and excitatory and inhibitory neurons (KII-sets with negative feedback, Sect. 9.7).

Higher-order K-sets form a hierarchy of models of cortical populations sufficing to replicate the properties of evoked potentials, ECoG and EEG (Kozma and Freeman 2001; Freeman and Erwin 2008).

The reptilian allocortex is upgraded in mammals by the massive invasion of neurons from the floor of the lateral ventricle that spread through the neuropil of layer II. That invasion creates three new layers giving the six-layered neocortex (Schüz and Miller 2002). Axons carrying sensory input from the thalamus organize the new layers II, III, and IV into local networks that are specialized to meet the unique requirements for extracting and processing sensory information in signal energies detected by each of the three neocortical distance sense modalities.[1] At the border between allocortex and neocortex is a transitional area often called mesocortex, of which the best known and most important is the entorhinal cortex, which serves as a major source of input to the hippocampus and is also its major path of output (Sect. 10.5).

The outer layer I and part of layer II combined with the inner layers V and VI in each hemisphere together retain the global structure and function of the allocortex, by which they provide a scale-free embedding sheet. This hemispheric sheet, roughly 2,000 cm^2 wide and 0.3 cm deep in humans, enabled an evolutionary increase in forebrain size with respect to body size by up to two orders of magnitude above the size of the rest of the brain. Comparison of mammalian brains with the relatively miniscule brains of dinosaurs testifies to the extraordinary increase in brain size that was made possible by the appearance of neocortex, owing to embedding of innumerable local circuits in the diffuse neuropil derived from allocortical neuropil. The dynamics of the mouse and whale brains is essentially the same despite the 10,000-fold difference in brain volume. The cortex is said to be *scale-free* (Sect. 9.4), meaning that it works the same way no matter how large or small it is. Despite enormous expansion in surface area, the images of ECoG and EEG (Sect. 10.5) illustrate the unity of global coordination of the diverse functions of the entire forebrain that is explicit in Fig. 8.1. The transcendent dynamic unity in diversity implies further that the upper limit of the aperture for imaging the EEG must be the entire head. A major aim of this and the next two chapters is to present evidence that can support that prediction.

The spatial imaging of allocortical ECoG is developed as follows. In Sect. 8.2, we use the temporal structure of allocortical ECoG to introduce the concept of the

[1] Braitenberg and Schüz (1998) wrote, "A recent hypothesis by Miller (2002), based on differences in both connectivity patterns and spontaneous activity between upper and lower layers, assigns to the upper layers the role of a neuronal 'library', storing most of the information encoded by assemblies ... while the lower layers are assumed to catalyze the process of assembly formation" (p. 150). Long cortico-cortical connections are not randomly distributed within the territory that they innervate. Instead they are distributed in patches with high local connection density, with intervening regions having few connections (Malach 1994). Kaas (1987) wrote, "Generalizing from cats and monkeys it appears that the evolutionary advance in brain organization is marked by increases in the numbers of unimodal sensory fields, not by increases in multimodal association cortex as traditionally thought" (p. 147).

neural frame as the carrier of macroscopic perceptual patterns.[2] The bulb creates them from microscopic sensory information that is carried by pulses from receptors in the nose with each inhalation; it transmits the patterns in the form of macroscopic clouds of pulses on axons in a divergent-convergent pathway that performs a spatial integral transform. In Sect. 8.3, we display the spatial images of the spatial patterns of bulbar impulse responses to electrical stimulation (averaged evoked potentials that conform to sums of damped cosine oscillations), which reflect mechanisms of microscopic sensory processing. Section 8.4 describes the spatial patterns of amplitude of gamma bursts that are induced by inhalation. Section 8.5 focuses on the spatial patterns of phase of gamma bursts that are induced by inhalation, which appear as macroscopic oscillations that increment with time rather than decaying. Section 8.6 describes the mechanism of readout of spatial images by a spatial integral transform and not by topographic mapping. Section 8.7 summarizes these points.

8.2 The Temporal Structure of the Bulbar and Prepyriform ECoG

Simultaneous recordings from the olfactory bulb (Fig. 8.3, OB) and prepyriform cortex (PC) confirm psychophysical experiments (Murphy et al. 1991) that olfaction operates by intermittent sampling and batch processing. Inhalation under limbic control brings an odorant sample to the nose from the environment with each sniff. The impact of the mix of chemicals in the inhaled air initiates a volley of action potentials that is delivered to the bulb by axons in the primary olfactory nerve (PON). The volley causes a surface-negative respiratory wave with each inhalation of the background air in control periods (Figs. 6.1, 8.3a) and the foreground conditioned stimuli in test periods. A brief burst of gamma oscillation begins on the rising phase of each cycle of the respiratory wave. The responses of the receiving neurons (mitral cells in the OB, pyramidal cells in the PC) reveal increased axonal firing rates in proportion to the amplitude of the dendritic wave (Sect. 6.6). The mixture of chemicals in the inhaled background air is unknown in complete detail, and whether a microscopic combinatorial code exists is unclear (Lettvin and Gesteland 1965; Firestein 2004). The pharmacological aspects of olfaction are not at issue here (Freeman 2001). What is important is that the gamma bursts only occur when the inhaled air contains a mixture with which the subjects are familiar, including odorants to which they are habituated (Fig. 8.3a). The bursts appear both in the prestimulus control period and in the conditioned stimulus (CS) test period but with differing spatial patterns of amplitude. But when a sufficiently strong novel stimulus

[2] There is substantial psychophysical evidence for batch processing and framing in all the cortical systems that support distance receptors, coupled with physiological evidence of central rhythms (Busch et al. 2009) and motor actions that assist in framing, including sniffs, whisks, saccades, and speech parsing (Atienza et al. 2002).

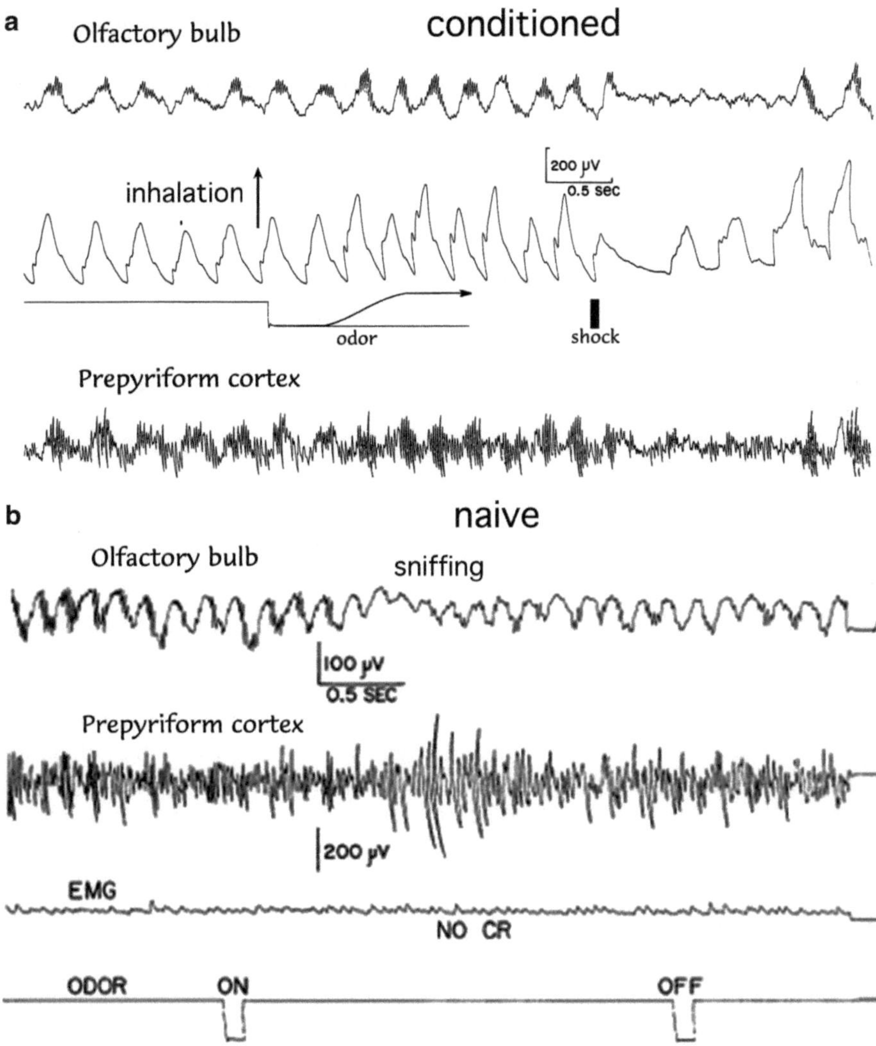

Fig. 8.3 (**a**) The *upper* and *lower* records respectively show the ECoG of the bulb and cortex. The *center* record shows the respiration from a pneumograph around the chest. Beneath is the warning odorant (CS) followed by a brief aversive shock to the cheek (an unconditioned stimulus, US). The rabbit responded to the CS by sniffing and to the US by breath holding. Gamma bursts were seen near the crests of the surface-negative theta waves; after training, the bursts had classifiable spatial patterns of amplitude (From Fig. 5 in Freeman and Schneider 1982). (**b**) The upper two records show the ECoG of the olfactory bulb and cortex when a novel odor was delivered to a rabbit, which elicited an orienting reflex (sniffing) and suppressed the gamma bursts. We see the chaotic activity as necessary for Hebbian learning (Sect. 8.4) to form a *new* assembly that is associated with a newly learned CS and avoid reinforcing an existing CS (From Fig. 11 in Freeman and Skarda 1985)

is given (B), the gamma bursts fail to appear, and the subjects give an orienting response (sniffing and searching with ears and eyes; Skarda and Freeman 1987). When the novel stimulus is given repeatedly without reinforcement, the burst suppression and orienting diminish and vanish after a few trials. If the novel odorant is paired with reinforcement, then gamma bursts reappear within two or three trials, as the subjects learn a new conditioned reflex (CR), and the bursts carry a new spatial pattern not seen before (Sect. 8.4).[3]

8.3 Spatiotemporal Images of Averaged Evoked Potentials

The temporal bandwidths of the ECoG waves are much too narrow to carry the information that is needed to support olfactory discrimination. Adrian (1950) postulated that coding was spatial, not temporal, and that the olfactory information would be spatially localized so that, for example, aliphatic compounds might excite one part of the bulb and aromatic, another. Therefore, we undertook a test of Adrian's postulate by spatial analysis based on multichannel ECoG recording from closely spaced electrodes fixed on the surfaces of the olfactory bulb and cortex (Fig. 8.4, OB). We began by determining whether focal electrical stimulation of the olfactory receptors in the nose or in the PON displayed topographic mapping from receptors to the bulb. We tested by imaging the impulse response (evoked potential, Sect. 1.4) to focal electrical stimulation of the axons in the primary olfactory nerve (PON) from the receptors (Fig. 8.4). In order to simulate normal receptor input at low density, the stimulus intensity of single shocks was kept close to threshold, so that the amplitude of the evoked potential (the impulse response to impulse input, $\delta(t)$) did not exceed the peak-to-peak range of the ongoing ECoG, and the frequency of the evoked oscillation was kept within the gamma range, which constrained the cortex to its linear dynamic range (Sect. 6.5). The impulse response was extracted by time ensemble averaging of sufficient single-shock responses (typically 100) to reduce the pre-stimulus fluctuations to 10% or less the peak amplitude of the peak range of the average evoked potential.

The array of 64 averaged evoked potentials demonstrated that surface recordings of dendritic potentials sufficed to resolve spatial pattern localization in the mm range. Shifting the focal stimulus site dorsoventrally caused the peak of the spatial impulse response to shift in the same way (from right to left in Fig. 8.4). Insertion

[3] The importance of the unpatterned "chaotic" fluctuations with novel stimuli (Fig. 8.3b) should be noted. For Hebbian learning to occur, the pair of neurons must fire action potentials coincidentally (Sect. 6.5), but the spatial patterns of firing must be novel, or else an existing AM pattern would be reinforced with no new learning (Uttley 1955). Therefore, the cortex must have a mode of function that is described as convergence into the basin of an "I-don't-know" chaotic attractor (Fig. 6.14, Sect. 6.9), which can provide the unstructured activity needed for trial-and-error associative learning (Skarda and Freeman 1987).

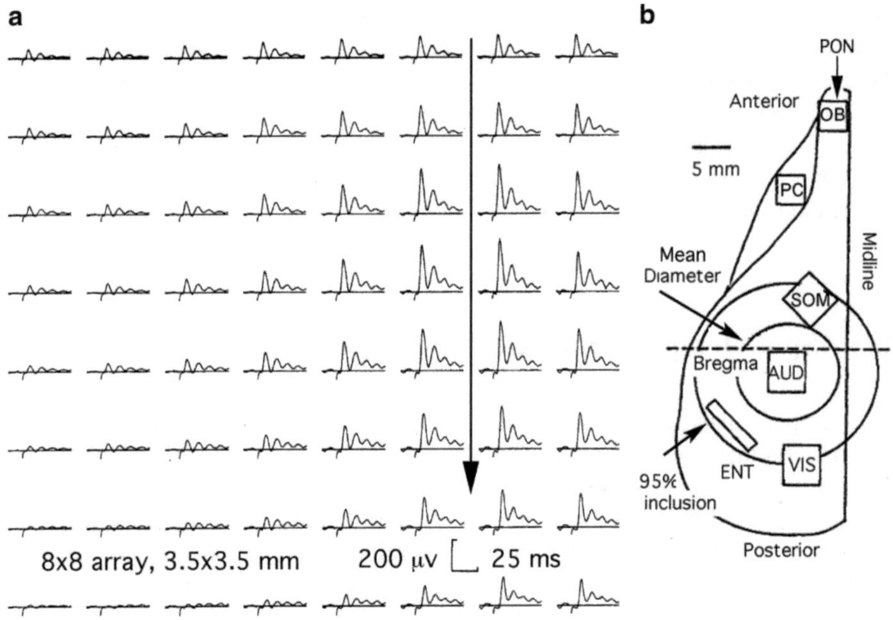

Fig. 8.4 (**a**) A set of 64 average evoked potentials were the bulbar response to single-shock electric stimulation of the primary olfactory nerve (PON) 2 mm anterior to the array (above the *arrow*) (number of shocks = 335). The *arrow* indicates the direction of propagation of the afferent volley of action potentials preceding the bulbar response. (**b**) The diagram shows the dorsal surface of the *left* hemisphere of the rabbit forebrain. The rectangles outline the size and location of an 8 × 8 recording array fixed on the olfactory bulb (OB), prepyriform cortex (PC), somatic cortex (SOM), auditory cortex (AUD), visual cortex (VIS), and (in the cat) entorhinal cortex (ENT) (see Sect. 10.3). The two circles show the mean and 95% inclusion diameters (15 mm and 28 mm) of coherent oscillations in the neocortical gamma ECoG (Fig. 9.5d, Sect. 9.4) (Evoked potentials from Freeman (1975, p. 221) drawing adapted from Barrie et al. 1996)

of the stimulating microelectrode in steps of 50 µm into the depth of the PON shifted the peak from posterior to anterior in the array (top to bottom). These correlations were in accord with the anatomical topographic organization of the PON on the bulbar surface (Freeman 1974a).

The grand average of the 64 averaged impulse responses was fitted with a damped cosine (Sect. 6.5) to measure the gamma response by fitting sums of linear basis functions (Freeman 1974b) using nonlinear regression.[4] From these data,

[4] A function that contains one exponential term can be fitted to a neural impulse response by linear regression, for example, the passive membrane decay potential on impulse input, by transforming to semilog coordinates. Contrariwise fitting a function containing cosines or the sum of two or more exponential terms requires nonlinear regression, which is a method of successive approximations that simultaneously adjusts all of the parameters for both signal and nonwhite noise also known as *clutter* (Freeman 1979b; Samar et al. 1999).

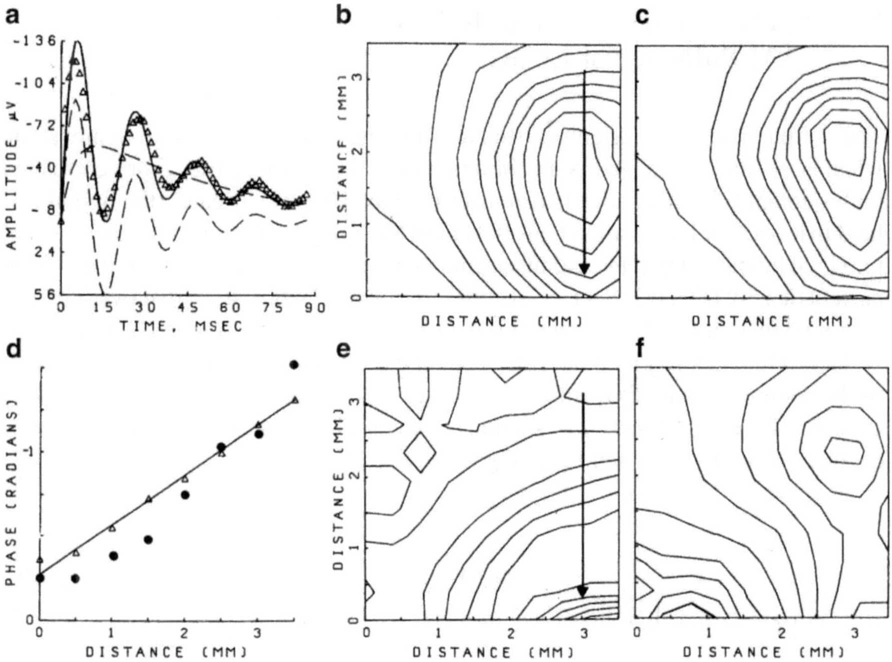

Fig. 8.5 (**a**) Grand average evoked potential was fitted by nonlinear regression with the sum of linear basis functions (*solid curve*): a damped cosine and the sum of three exponential terms (*dashed curves*). (**b**) Amplitudes of the 64 fitted cosines were displayed in a contour map. (**c**) A contour map showed the 64 amplitudes of the sum of exponential terms fitted to the baseline shift. (**d**) Phase lag was from anterior to posterior across the bulb (*arrow* as in Fig. 8.4). (**e**) Contours showed the phase of the cosine at the shared frequency. (**f**) Decay rates of the fitted damped cosines were proportional to peak amplitude as explained in Sect. 6.7 (Fig. 6.12c). (From Fig. 4.32, p. 230 in Freeman (1975))

a linear differential equation was constructed, for which the solution was a sum of exponential terms with which to simulate the impulse response (Fig. 8.5a), including the baseline shift corresponding to the respiratory wave (Fig. 8.3). Then the same equation was fitted to each of the 64 responses to get the amplitude of the gamma cosine (b) and the amplitude of the baseline shift (c). The frequency of the cosine was found to be everywhere nearly the same, so the 64 phase values were measured at the mean frequency, revealing an anteroposterior phase gradient (here 3.5 rad/mm) in the direction of conduction of PON axons (downward arrow). Division of the cosine frequency (here 45 Hz = 285 rad/s) by the gradient (rad/m) gave an estimate of the conduction velocity of the PON (here ~ 1 m/s). The results showed that the spatial amplitude and the phase gradient of the bulbar oscillation evoked by the afferent axons were both determined by the location of the input. They showed that the array for recording had the spatial and temporal resolution needed to test Adrian's hypothesis using the ECoG (Fig. 7.8, Sect. 7.5) that olfactory coding is spatial.

8.4 Spatial Images of Amplitude Modulation (AM Patterns) of ECoG

The definitive test of Adrian's postulate using the ECoG would be to demonstrate whether spatially localized peaks in amplitude of the bulbar ECoG are correlated with odorant CSs that subjects can discriminate. That was because the hypothesis held that odorant information conveyed in PON volleys determined the locations of maximal bulbar firing rates and the accompanying ECoG amplitudes from dendritic current density (Fig. 6.2, Sect. 6.2.1; Fig. 7.7, Sect. 7.5). In tests with naïve subjects, we found that during the pre-stimulus control state the carrier frequency of the burst with each inhalation (Fig. 8.3a) had a spatial pattern of *amplitude modulation* (AM) of the common carrier frequency (Fig. 8.6, *left*). This spatial AM pattern was like a signature for each subject; it was unique to each subject and yet never twice identical (set 1, "air") (cf. human EEG, Fig. 10.9, Sect. 10.4).

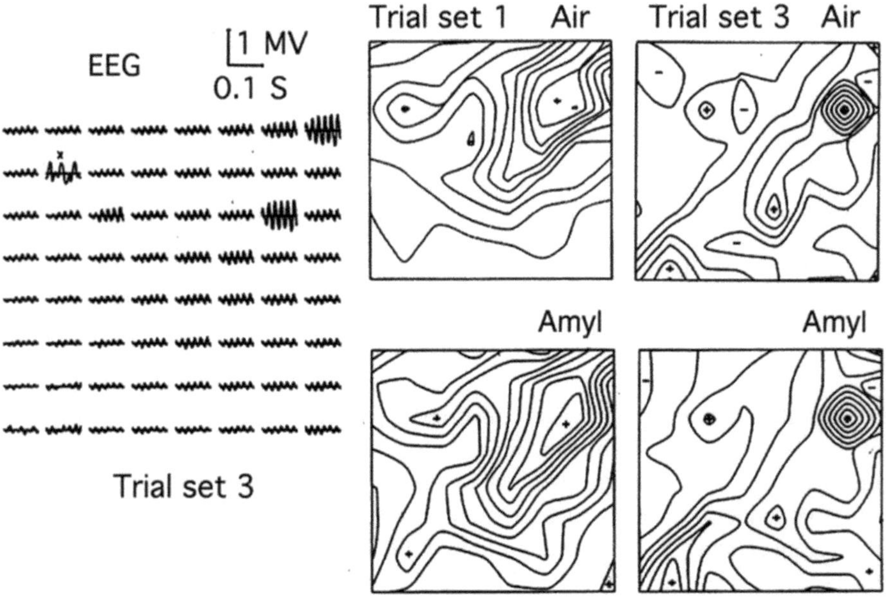

Fig. 8.6 The 64 ECoGs from a single burst show the shared frequency in the gamma range having 64 values of amplitude that form an AM pattern. The signal at channel 9 ("x") was recorded from the prepyriform cortex, replacing a broken channel. In statistical analysis, bad signals were replaced with averages of two neighboring signals. The middle frames show the contour plots of amplitudes comparing the mean AM patterns of ten trials after CR acquisition of pre-stimulus control bursts ("air") and teb bursts sniffing amyl acetate ("amyl"), a volatile chemical quickly clearing from the odorant delivery system. The frames at *right* show the changed AM patterns after 2 weeks of *consolidation* after reinforcement at weekly intervals. The conditioned response (CR) was sniffing, which was a fixed motor action that psychologists call an *autoshaped response* (Freeman and Viana Di Prisco 1986) (From Fig. 7 in Freeman and Schneider (1982))

Initial testing by delivering odorants at low concentrations to anesthetized or waking subjects had no effect on the ECoG. At easily detectable concentrations, the delivery of novel odorants to awake subjects suppressed the control AM pattern (Fig. 8.3b, Sect. 8.2) and elicited an orienting reflex response. When the stimulus was unreinforced (CS-), then within a few trials, the chaos and the orienting response abated, and the control AM pattern reappeared (CSo) (Emery and Freeman 1969). When the stimulus was reinforced (CS+), a new AM pattern emerged (Fig. 8.6, set 1, "amyl," here the odorant amyl acetate). We concluded that the ECoG responded to odorants only if the subjects were induced to pay attention by conditioning (Freeman and Schneider 1982).

In themselves, the AM patterns lacked any recognizable geometry. The mean amplitude and carrier frequency were unrelated to odorant type or intensity. The AM patterns were best displayed after = normalized to z-scores (zero mean and unit variance). Successive AM patterns in each subject were characterized by small deviations from the signature amplitude pattern on all channels, which accumulated over days, weeks, and months in drift. The spatial variations in shared carrier frequency and phase were so limited that 90–95% of the variance in principal components analysis was in the first component. The 64 amplitudes were derived equally well from the first component of PCA of the AM pattern of each burst, or the analytic amplitude from the Hilbert transform (Sect. 9.5), or the root mean square amplitude after band pass filtering in the beta or gamma range.

All three methods gave a 64x1 feature vector for each burst that specified a point in 64-space. Sets of AM patterns were represented graphically by projecting the feature vector tips into 2-space (Fig. 8.7a) by linear discriminant analysis (Freeman and Viana Di Prisco 1986). Similar AM patterns gave a cluster of points in a multivariate Gaussian distribution. Each cluster had a center of gravity (centroid) calculated as a central point in the graphic display with a circle around it of radius SD (Fig. 6.8, Sect. 6.4.4). Each cluster corresponded to a CS + or CS- or the control input (CSo). Classification of each burst was by calculating the Euclidean distance of its point to every centroid and finding the minimum distance. Significance of the level of classification was by calculating the binomial probability of linear separation having occurred by chance.

AM pattern categories held only within sessions. The Euclidean distances between centroids across sessions exceeded the distances within sessions. This reflected the irreversible drift of signature AM patterns, in some subjects up to four years of array recording. Sessions in which a new CS + was introduced were marked by significant jumps in intersession distances between centroids (Freeman and Grajski 1987, reproduced in Fig. XI, B, p. 268, Freeman 2000).

The AM pattern differences shown between "air" and "amyl" appeared at first to confirm the Adrian's postulate because the ECoG seemed to point to the areas in the bulb in which neurons were most strongly driven by the CS + or CS- and so could serve to direct the search with microelectrodes for the neural representations of the CSs in high rates of firing of selected neurons. However, the postulate of CS representation by ECoG amplitude failed in two respects. First, learning did not cease with reinforcement trials. It continued overnight and through the following week in

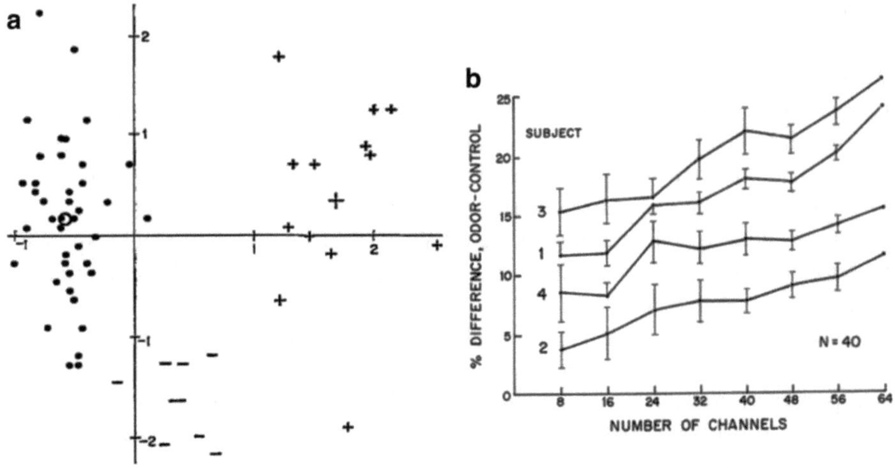

Fig. 8.7 (**a**) Each symbol shows the AM pattern of a burst plotted as a feature vector in 64-space and projected into 2-space. Similar patterns of multiple bursts gave a cluster of points around a center of gravity (centroid). Each CS+, CS-, or CS0 gave a separate cluster. Classification of bursts was accomplished by calculating the Euclidean distance to every center and finding the closest center. Goodness of classification was expressed as % correct by linear separation (From Freeman and Viana Di Prisco 1986). (**b**) The goodness of classification for 4 subjects was repeatedly calculated 40 times, each time with deletion of a different set of channels randomly selected in groups of 8 to 56. The classification efficacy increased with the number of channels. A running tally was kept for each channel of the scores when it was selected. The sum divided by the number of times used showed that the classificatory information was uniformly distributed; no channel had any more or less information than any other. Comparable results have been found for all of the sensory neocortices and for the EEG (Sect. 9.6; Figs. 10.3 and 10.8) (From Fig. 16 in Freeman and Baird (1987))

consolidation, so that the same stimuli gave different AM patterns in different contexts (Fig. 8.6, trial set 3) and with every change in the experience of the subjects, including reversals in reinforcement between CS + and CS-, adding new CSs, or switching to new CRs. Second, systematic testing by channel deletion showed that every channel contributed equally whether high, low, or volatile (Fig. 8.7b), so the density of the classificatory information was uniformly distributed over the entire array as in a hologram (Pribram 1991). In a metaphoric sense, both light and dark were required to make a pattern, and all levels of gray scales were equally informative.

The clustering of feature vectors implied the existence of a bulbar mechanism for generalization and abstraction. The olfactory system in common with other sensory modalities has massive convergence from receptors to cortical projection neurons, minimally 10^3 equivalent receptors for each mitral cell. Only a sample is selected on each sniff, but it differs on every sniff. The mechanism for generalization is inferred to be by formation of a *Hebbian assembly* during training with pair-wise strengthening of synapses between co-activated neurons (Fig. 8.8a). Alternating trials with a CS- were required with anti-Hebbian learning leading to habituation (Kozma and Freeman 2001). The reduction of gain with habituation is reversible; the strengthening of gain with association is not. Any subset excited thereafter ignites the entire

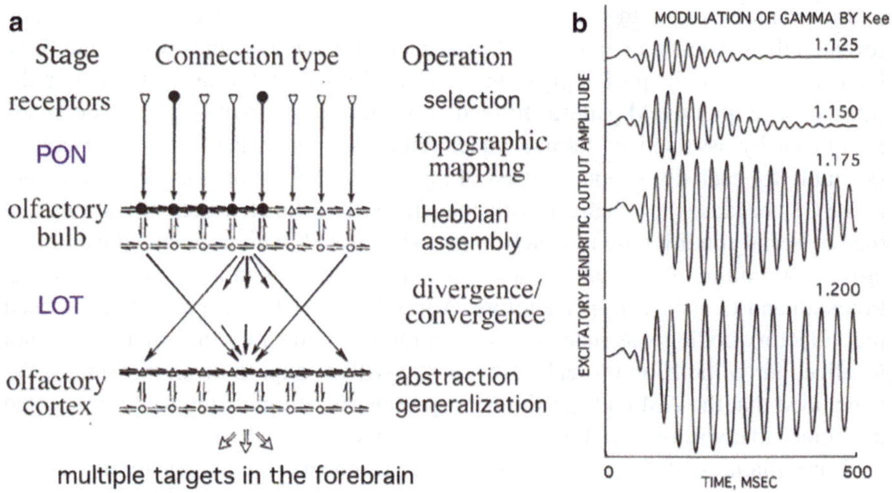

Fig. 8.8 (**a**) Sensation by a sniff begins with selection by a CS of some among equivalent receptors (two *black dots*), transduction to action potentials, and transmission to mitral cells in the bulb via topographic mapping. Generalization is by excitation (five *dots*) of a Hebbian assembly, which directs bulbar dynamics to a basin of attraction in a landscape of attractors that constitute a sample of olfactory memory (Fig. 6.14, Sect. 6.9). The resulting AM pattern is transmitted via a divergent-convergent projection that abstracts the perception by attenuating extraneous details of receptor-driven activity (Modified from Fig. 15 on p. 87 in Freeman 2001). (**b**) Simulation has shown that the increase in the strength of synapses that sustain mutual excitation greatly enhances the gamma oscillations from interactions of excitatory and inhibitory neuron populations (From Fig. 13 on p. 32 in Freeman (1979a))

assembly, which amplifies abstracts by deleting information on which neurons were excited on each trial, and generalizes to the category of the CS as defined by the cumulative sampling. Localization of the modified synapses to synapses between mitral cells was by measuring the averaged evoked potentials of the olfactory bulb (Figs. 8.4, 8.5) and cortex when the subjects were trained to use the evoking shock train as a CS (Emery and Freeman 1969). With CR acquisition, the duration of the initial negative peak, N1, of the evoked potential was prolonged, and the phase lag of the fitted cosine (Fig. 8.5a) increased. Confirmation of enhanced mutual excitation was by measuring the second peak of the PSTH during N1 (Freeman 1968; Nicoll 1971) and modeling its increase in density upon increasing the positive excitatory feedback gain, k_{ee} (Fig. 8.8b).

A significant secondary effect of the increase in positive feedback gain, k_{ee}, was to increase the negative feedback gain, k_n, between excitatory and inhibitory populations (Fig. 8.8b). Notably an increase in synaptic gain of 7% can support a 50-fold increase in gamma power. The simulation shows that the Hebbian synapse (Fig. 8.8a), which sustains positive feedback between the excitatory neurons (Amit 1995), is an extremely powerful amplifier of gamma oscillation, when the excitatory neurons are coupled with inhibitory neurons in negative feedback (a KII-set (Freeman and Erwin 2008). The Hebbian assembly is major component of the transition of cortex from sparse coding to high-density integration.

In summary, the high-resolution images of the allocortical ECoG in olfaction revealed the existence of spatial AM patterns in sequential frames, which might be likened to the systolic packaging of batches of information for dissemination on the Internet. The spatial AM patterns formed only after reception of CSs that were made significant by associative learning under reinforcement, which formed a Hebbian assembly for each discriminable class of input. They lacked invariance with respect to the CSs because they changed with consolidation in the aftermath of learning and then by serial conditioning in which all preexisting AM patterns changed incrementally with training to recognize each new stimulus (Freeman and Schneider 1982; Freeman and Grajski 1987). Therefore, Adrian's postulate was partially confirmed that olfactory coding was indeed spatial, but the classificatory information was not localized in the bulb, and it lacked invariance with respect to invariant stimuli. We concluded that the AM patterns did not represent the stimulus input. It represented a memory that was retrieved by the stimulus, which evolved with each recall.

The dynamics of the olfactory system is legendary for its sensitivity, acuity, speed of operation, and range of comprehension. When modeled by differential equations comprising a KIII set (Freeman and Erwin 2008), it has been developed and applied to pattern classification for a broad range of sensory modalities: olfactory (Fu et al. 2007; Yang et al. 2006), visual, mechanical (Kozma et al. 2007), auditory (Shimoide and Freeman 1995; Li et al. 2007), and chemical inputs (Hu et al. 2006). Its performance emulates its biological inspiration in learning from a few samples (5 to 20 examples of each category), rapid convergence (a single-phase transition instead of gradient descent), geometric rather than arithmetic increase in number of classes to be categorized with increase in number of nodes, and integration of multiple modalities of input by linear vector algebra (Kozma et al. 2005; Li et al. 2006). Its limitation is in the computational burden of solving hundreds of differential equations numerically. Facilitation is being sought in analog VLSI (Principe et al. 2001) and random graph theory (neuropercolation, Kozma et al. 2008).

8.5 Spatial Images of Phase Modulation (PM Patterns) of ECoG Bursts

The retrieval of the memory of an olfactory stimulus at the first synapse in the olfactory path is a microcosm for elicitation of the mental qualities of stimuli in all primary sensory cortices (Chap. 9). It is of paramount importance to understand how recall is done. Multichannel array recording of ECoG bursts on inhalation demonstrated that each burst had a narrowband carrier frequency that was shared over the entire bulb. The oscillation showed spatial modulation in amplitude and also in phase (Fig. 8.9). The phase modulation (PM) provided a basis for explaining the process of memory retrieval.

It was immediately apparent that the phase gradient of the carrier wave did not conform to the delay, direction, and velocity that were imposed by PON axons coming from the receptors to the bulb (*arrow* Fig. 8.5b, e). Instead, the gradient across

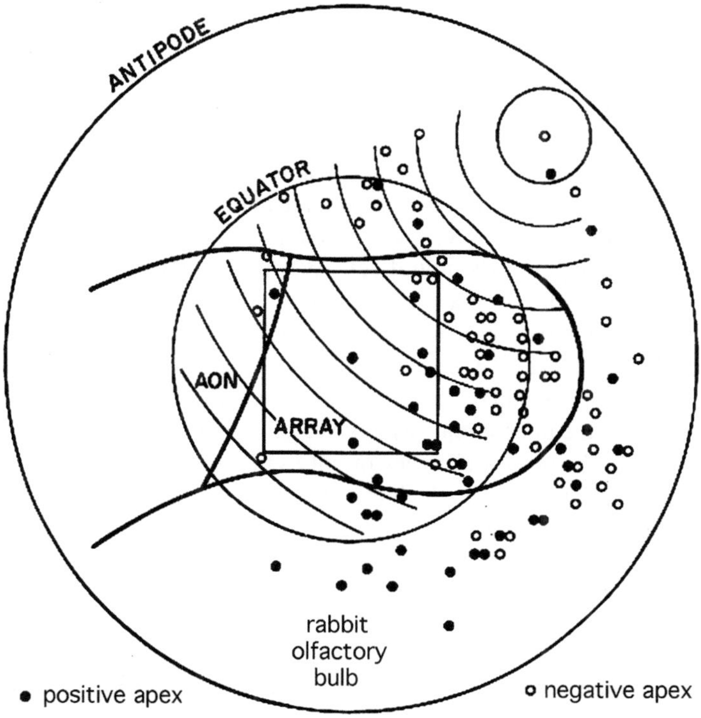

Fig. 8.9 The olfactory bulb of the rabbit is displayed as seen from the lateral view of the surface. The square outlines the 3.5×3.5-mm array. The roughly spherical bulbar surface is flattened into a plane, with equal distance from the north pole centered in the array to the south pole at the antipodes (*large circle*). Each 8 × 8-valued phase surface was constructed from the phase values at the frequency with maximal amplitude by the FFT of the 64 ECoG segments containing a burst. A conic surface was fitted to the 8 × 8 values in the phase surface by nonlinear regression with least squares residuals. Filled dots (•) showed maximal lead and open dots (o) showed maximal lag at the apices (From Freeman and Baird (1987))

the window of observation remained constant within each burst for its duration (Fig. 8.3a), but it varied randomly in slope and direction from each burst to the next. The 8 × 8 phase values on the bulbar surface could not be fitted with a plane or a bivariate Gaussian envelope (Freeman and Baird 1987), but it could be fitted with a right cone. An example of the cone is displayed as a set of concentric circles and arcs representing isophase contours at intervals of 0.1 rad (5.7°, Fig. 8.9). The phase modulation (PM) resembled the wave pattern of a stone dropped into a still pond at the apex (Fig. 9.11, Sect. 9.6.1). The location of the apex of the cone varied randomly with successive bursts over the bulbar surface and so did its sign. There was approximately equal incidence of maximal lead at the apex giving the extreme positive phase as at the point of an explosion or maximal lag giving the extreme negative phase as at the focus of an implosion.

The slope of the phase gradient in rad/m appeared to vary unpredictably from each oscillatory burst to the next, but it did so concomitantly with the ECoG frequency in rad/s. The ratio, carrier frequency/phase gradient, gave the phase velocity in m/s. The average, 1.89 ± 0.36 m/s, approximated the estimated conduction velocity of small myelinated axon collaterals of mitral cells radiating in all directions parallel to the surface and not the average conduction velocity of the unmyelinated PON axons, 0.42 ± 0.08 m/s, running in parallel anteroposteriorly across the bulbar surface. The maximal cumulative phase difference of a wave spreading across the whole bulb at the depth of the mitral cell layer in the rabbit was calculated to be $\pi/4$ rad (45°, cos = 0.707). At that distance, the oscillations began to go out of phase, so the strength of interaction that sustained spatial coherence was weakened. The reduced feedback gain (Fig. 6.12, Sect. 6.7; Fig. 9.11, Sect. 9.6.1) set a limit on the size of an interactive domain because the phase dispersion reduced the shared power. At some distance, the interaction was too weak to recruit more neurons into coherence. We adopted the phase difference of $\pi/4$ as the soft boundary condition of the size of the coherent domain, at which the shared power decreased to one half. That distance in the bulb corresponded to the circumference of the mitral layer of the rabbit bulb (~10 mm). The modal frequencies of bulbar carrier frequencies were found to vary inversely with bulbar size from 110 Hz in the mouse down to 25–30 Hz in the pig (Bressler and Freeman 1980). The negative correlation suggested that evolution adapted the carrier frequency to bulbar surface area so that every transmission of a gamma burst included all subpopulations of transmitting neurons in the bulb without degrading the integrated signal by excessive phase or frequency dispersion and the resulting interference (cancellation of excitatory and inhibitory oscillations by phase lag).

Every burst had a carrier frequency that was nearly constant for the duration of each burst and had a low but significant spatial variance (Sect. 6.3). The phase calculated at the carrier frequency with respect to the phase of the spatial ensemble average in the beta and gamma ranges was distributed in space, not zero lag as predicted from unit studies (Singer and Gray 1995; Roelfsema et al. 1997). Each burst was a standing wave, not a traveling wave. This property preserved the AM pattern for the duration of the burst, giving the same feature vector at each time step despite the rise and fall in mean power (Fig. 6.5, Sect. 6.3). The phase cone revealed distributions of the lead or lag at each recording point in onset of the burst. The spatial location and sign of the conic apex were fixed in each burst within the limit of resolution (the interelectrode distance), but varied randomly from each frame to the next. The frequency (rad/s) and phase gradient (rad/m) covaried randomly so that the ratio (the phase velocity in m/s) was relatively invariant in conformance with the conduction velocity of intrabulbar axons (Freeman and Baird 1987).

The difference between the input-dependent, the planar phase gradient of the stimulus-driven oscillatory impulse response, and the conic phase gradient of the endogenous bulbar response is profoundly significant. On the one hand, the artificial input causes a local subset of the bulbar negative feedback loops to oscillate like the ringing of a bell. The time of onset, location, dissemination, and amplitude are all determined by the shock input. The response decays exponentially from an early maximum. On the other hand, the burst is triggered, not driven, by inhalation of a CS. It begins after brief but random endogenous delay and radiates from or to an endogenously determined point at the conic apex, and it covers the entire bulb. The amplitude converges exponentially to a maximum and thereafter decays. These differences provide evidence that the bulb constructs a complex pattern when it receives a CS (Sect. 6.6). The pattern may contain far more information than does the stimulus that triggers it. Some of that information is in the spatial AM pattern. None is in the carrier frequency, cone location, or slope. The phase does provide information by revealing the location (Figs. 8.8; 9.6a), diameter (Fig. 9.6d, Sect. 9.4.1), and duration (Fig. 9.7, Sect. 9.4.2) of the bursts, both the bursts that are initiated by CSs (Fig. 9.7e) and the bursts found in the background activity of ECoG (Fig. 9.7a; Fig. 8.3a) and EEG (Fig. 10.6) (See *avalanches*, Sect. 6.9).

Based on these and related data, we postulate that the elicitation of a memory by a stimulus requires a change in cortex from low-density sparse firing of neurons at high rates to high-density firing of virtually all cortical neurons at low rates (Chap. 11). Confirmation has been undertaken by systematic correlation of the excitation and inhibition of pulse trains from single neurons (Freeman 1974b) with the spatial patterns of evoked potentials (Figs. 8.4, 8.5). The pulse firing evoked by electric shocks was sparse and occurred only within a small area of the bulbar surface, to which the stimulus site directed the PON volley. Spreading within that area was by small increments in that area with alternating excitation and inhibition in each quarter cycle. The evoked firing rates decremented with each time step (Fig. 4.48 on p. 259 in Freeman 1975). In contrast, the inhalation of a CS initiated increases in pulse density (Figs. 6.2, Sect. 6.2.2 and Fig. 6.11, Sect. 6.6) that spread at high velocity over the entire bulb within a quarter cycle of the carrier frequency. The form of spreading was called *anomalous dispersion* (Freeman 1990) because the macroscopic distance covered (10 mm) exceeded by an order of magnitude the average microscopic step size of serial synaptic transmission (0.5 mm, Sect. 4.5 in Freeman 1975).

We have adopted the term *phase transition* to describe the change (Kozma et al. 2005) as condensation from randomness to order (Fig. 6.5, Sect. 6.4; Sect. 9.4) followed by evaporation as bursts terminate (Fig. 6.6, Sect. 6.4.2; Fig. 9.12, Sect. 9.6.2; Chap. 11). It begins at a point on the 2-dimensional neural surface (the conic apex) and radiates as in the formation of a raindrop, or it begins afar and converges to a point at the apex. A possible explanation thereby emerges for the random incidence of phase lead vs. phase lag at the apex, which is based on anatomical evidence that distributions of connection distances among cortical neurons are power-law, not exponential (Freeman and Breakspear 2007; Freeman and Kozma 2010). That topological form of connectivity indicates that the dynamics is scale-free (Sect. 9.4), so that on reaching a threshold of intensity of interaction, the entire interconnected mass of neurons can undergo a phase transition, regardless of size (Wang and Chen 2003; Freeman and Vitiello 2006). If the randomly distributed connection weights momentarily favor the local connections, the phase transition is explosive. If they favor long connections, the phase transition is implosive (Kozma et al. 2005;

Freeman et al. 2009). Mathematical modeling of this phenomenon using chaos theory (Skarda and Freeman 1987) and many-body physics (Freeman et al. 2009; Freeman et al. 2012) is beyond the scope of this book.

8.6 Transmission of Macroscopic Patterns to Other Areas of Cortex

The next problem faced by the olfactory system is to send the distributed perceptual information to all parts of the brain that need it in a form that is compatible with the forms of all other sensory modalities. The solution to the problem is shown by the topological structure of the output path of the bulb, the lateral olfactory tract (LOT, Fig. 8.8a). Whereas the PON input path is organized topographically, the LOT output path is divergent-convergent. The axon of every transmitting mitral cell forms synapses on a wide distribution of target neurons, inside the bulb by axon collateral branches and outside the bulb into the basal forebrain by synapses *en passant* as the axons extend over the cortical surface. Each target neuron receives synapses from a wide distribution of transmitting neurons and sums its dendritic currents at its trigger zone. This one-to-many and many-to-one of the LOT performs a spatiotemporal integration on bulbar output, the Gabor transform (Fig. 8.8a, Sect. 8.4). The operation is analogous to that of a lens or a holograph (Pribram 1991). The bulb creates the AM pattern by convergence to an attractor with the narrowband carrier, which conforms to the properties of coherent light. The Gabor transform delocalizes the information expressed in the AM pattern. The operation enhances by coherence of the macroscopic perceptual activity that has everywhere the same frequency and phase, which the bulb has created in bursts, while it attenuates by smoothing the sensory-driven activity that lacks spatial coherence and synchrony.[5] However, cortex is not a holograph because it deletes massive quantities of information in categorizing and abstracting, and it has no inverse, whereas the holograph stores everything it receives and can retrieve it (Pribram 1991).

The prepyriform cortex and adjacent anterior olfactory nucleus (AON) have extensive connections recurrent to the bulb via the medial olfactory tract, forming t. Each of the three parts has its characteristic frequency, which is nonharmonic and incommensurate with the other frequencies (Kozma and Freeman 2001). The positive excitatory feedback among them continuously increases the activity, while the thresholds and refractory periods continuously curb it, with no convergence to any characteristic frequency, so the three-way interaction results in the broad spectrum

[5] It is noteworthy that this one-to-many and many-to-one topology can explain macroscopic percept transmission, but it is also compatible with transmission of any number of microscopic signals to local circuits, which could extend further olfactory sensory information processing deep into the forebrain. An example is provided by the functions of the accessory olfactory system in rodents devoted to reproductive function (Freeman 2001).

1/f background ECoG (Skarda and Freeman 1987).[6] When the prepyriform cortex is deprived of bulbar input by cutting or inactivating the LOT, the cortex goes silent, while bulbar activity becomes periodic, revealing the characteristic frequency of the bulbar limit cycle attractor. We infer that the prepyriform cortex is unlike the bulb, in that it lacks the capacity for self-organization and that its AM patterns are not emergent but driven by bulbar input. The LOT distributes the knowledge that is selected from the memories stored in the bulb over the prepyriform cortical surface. The dense neuropil of layers I and II (Fig. 8.2), which generates the prepyriform ECoG, displays it as classifiable AM patterns reflecting textures in clouds of macroscopic pulse densities. But the deep pyramidal cells in layer III, which are relatively sparse and have widely radiating basal dendrites, transmit the prepyriform output. We conceive that they down-sample the surface AM pattern by integrating over local patches, which spatially coarse-grains prepyriform activity and converts it from macroscopic pulse densities to microscopic pulse frequencies through many-to-one convergence. By this operation, the output of the olfactory system is at the same microscopic level as its input, but whereas the receptor pulse inputs signify sensory information, the prepyriform pulse outputs signify perceptual meaning. We conceive the deep pyramidal cells as possible category cells or precursors for *concept cells* (Quian Quiroga 2012), for which the cognitive correlates are not sensations, but abstract conceptions expressed in equivalent sensory representations associated with a concept (Sect. 11.5). The conversion and transmission of AM patterns in sparse form completes olfactory perception (Freeman 1991).

Because the density of the information in the ECoG that enables the categorization of CSs is spatially uniform, each part of the LOT can have the whole signal though at reduced resolution, just as a hologram can be broken into pieces with every piece retaining the whole. Every fraction of the LOT transmits the same message to multiple target areas in the basal forebrain, and each target can tune itself whether to accept or reject it in accord with its own state (Traub et al. 1996; Whittington et al. 2000) (Sect. 6.8). By this topology, the bulb interfaces between the olfactory receptors and the elements of the limbic system, which have differing constraints on embryological development of the brain. The integral transform by the LOT goes beyond feature binding of various facets of a stimulus. The bulbar output of pulse clouds carries also the memory of the stimulus. The bulbar operations of generalization by the Hebbian assemblies, phase transition to a limit cycle attractor, and abstraction by spatiotemporal integration comprise the first steps in the transition from the material sensory domain to the mental perceptual domain because the odorant substances are common to all in the world, but the Hebbian

[6] The interaction of an excitatory population (KI_E set) with an inhibitory population (KI_I-set, governed by point attractors) forms a KII-set governed by a limit cycle attractor (Freeman and Erwin 2008). The interaction of three KII sets (KII_{OB}, KII_{AON}, KII_{PPC}) forms a KIII set governed by a chaotic attractor (Fig. 6.14) (Skarda and Freeman 1987). The interaction of three KIII sets (olfactory, hippocampal, and septoamygdaloid) forms a KIV set that suffices to model the brain of the salamander (Fig. 8.1) (Kozma et al. 2003). KV is reserved for neocortex (Freeman et al. 2009).

assembly, the attractor, the pulse cloud vector field, the AM pattern in the ECoG, and the odor exist only in the brain and mind of the individual. Ultimately at the level of the motor systems giving the CR the lack of invariance is resolved downstream by analogous operations of selection by a perceptual signal of the basin of attraction of a response that is invariant with respect to category, but adapted in respect to the unique circumstances of responding by taking action.

In the primitive vertebrate forebrain (Fig. 8.1), the olfactory bulb and the transitional area interact directly with the hippocampus and the pyriform cortex.[7] In mammals, the bulbar transmission directly to the hippocampus is reduced to a rudiment dorsal to the corpus callosum. Bulbar transmission is mainly to the parts of the prepyriform cortex, with further transmission to the entorhinal cortex (Fig. 8.2). These are the AON anteriorly, which feeds back to the bulb in regulating bulbar output: the periamygdaloid cortex posteriorly, which forms the corticomedial part of the amygdaloid nucleus in service to the motor systems (LeDoux 2000); and between them the prepyriform cortex, which displays oscillatory bursts driven by the olfactory bulb (Fig. 8.3) and by central components of the limbic system including the entorhinal cortex (Bressler 1987; Kay et al. 1996; Kay and Freeman 1998). There are as yet no high-resolution array images of gamma ECoG from the hippocampus because the cortical layers are complexly folded and compressed. Therefore, little can be said at present about the patterns of its macroscopic fields. Recordings of action potentials combined with time series analysis of ECoG from selected sites have shown that the hippocampus maintains a *cognitive map* (O'Keefe and Nadel 1978) using *place cells*, which are closely related to animal navigation through behavioral space. Temporal recordings show the now familiar systolic packaging of activity in bursts that is described as theta-gamma linkage (Buzsaki 2006), but the spatial images of gamma phase and amplitude within the bursts are yet to be measured at high spatial resolution, and the search for phase gradients and discontinuities has not yet been undertaken.

The most significant bulbar and prepyriform outputs go to the entorhinal cortex, which is a transitional form often called *mesocortex* (Fig. 8.2) that receives input from all sensory systems and transmits it to the hippocampus. It is strategically located for the next step in perception, which is to integrate the macroscopic signal from olfaction with the signals from all other sensory modalities into a multisensory percept (a gestalt, Köhler 1940; Fig. 10.3, Sect. 10.3). The combined pattern is transmitted to the hippocampus, where the information about the behavioral time

[7] A hierarchical dynamic model of the primitive forebrain has been constructed from K-sets to simulate the performance of the action-perception cycle (the KIV model, Kozma et al. 2003) incorporating the sensory, motor, and associational lobes seen in Fig. 8.1. The model has been implanted as the controller of a prototype of NASA's Martian rovers, the SSR2K platform (Kozma et al. 2008). The autonomous robot generated repetitive AM patterns of gamma oscillations by phase transitions. The AM patterns enabled it to form multisensory gestalts in reinforcement learning by creating Hebbian assemblies from pairs of visual and kinesthetic inputs. Thereby, the robot learned to identify and avoid hazards in pursuit of its self-sustained goals.

and place of the subject at the moment of gestalt formation is incorporated. Of direct relevance here are recordings of spike trains from neurons in human medial temporal cortex, entorhinal cortex, and hippocampus (concept cells, Sect. 11.5), which manifest integration of cognitive information over multiple sensory modalities (Tanaka 2003; Quian Quiroga et al. 2009).

8.7 Summary

The case is made that the olfactory system provides the optimal model for perceptual dynamics in all sensory systems, culminating in hemisphere-wide coherent oscillations in the beta-gamma range.

Weak electrical shocks of sensory axons to the bulb give oscillatory impulse responses with narrowband frequencies in the gamma range. The averaged responses carry spatial patterns of amplitude and phase displayed as spatial AM and PM patterns. Evoked response amplitudes rise with short, fixed delay, and exponential decay from the initial maximum to the pre-stimulus background level. The spatial location of the peak and the rate and direction of its motion depend on the location of the stimulus site, so the evoked patterns represent the input.

Inhalation of conditioned stimuli (CSs) gives gamma bursts with amplitudes that increment after onset. Each burst spreads over the entire bulb within a quarter cycle of the carrier wave in a conic phase gradient. The directions and velocities conform to the conduction velocities of bulbar axon collaterals, not those of the input axons. The onset times and durations of bursts vary unpredictably ("jitter"). The spatial AM patterns are determined by modified synapses in the collateral connections, not by the locations of activated input axonal synapses. The AM patterns represent memories.

These differences in AM patterns and the conic PM patterns show that the bulbar ECoG images are self-organized by bulbar neural populations following reinforcement learning. The emergence of a burst with a classifiable AM pattern requires the formation of an attractor by reinforcement learning in two stages. During initial learning, the pair-wise excitation of bulbar excitatory neurons creates a Hebbian assembly, which forms cumulatively by strengthened synapses on reinforced (CS+) trials and weakened synapses on unreinforced (CS-) trials (Emery and Freeman 1969). Subsequently, during consolidation, a new attractor forms that integrates the assembly with contextual memory and provides a basin of attraction defined by the learning set.

In perception of a CS, the Hebbian synapse amplifies the gamma oscillation and assigns the CS to a category by generalization and abstraction; the attractor directs formation of an AM pattern that expresses the contextual memory. The bulb sends its output as a macroscopic pulse cloud through the LOT. The macroscopic AM pattern is enhanced by transmission through the LOT, which performs a Gabor spatiotemporal integral transformation by virtue of the coherent carrier frequency. The LOT also transmits the incoherent microscopic sensory-driven activity for further processing.

The main targets are the prepyriform cortex and AON. Feedback to the bulb through the medial olfactory tract supports the chaotic background activity and the coherent bursts of the three components. Neither target displays the capacity for self-organization of bursts through phase transitions. We infer that the main task of the prepyriform cortex is to spatially spread the knowledge retrieved and transmitted by the bulb, then to down-sample and spatially coarse-grain the display for transmission to forebrain targets by microscopic pulse trains. Whereas bulbar input consists of pulse trains that carry sensory information, prepyriform output consists of pulse trains that carry perceptual meaning.

References

Adrian ED (1950) The electrical activity of the mammalian olfactory bulb. Electroencephalogr Clin Neurophysiol 2:377–388

Amit DJ (1995) The Hebbian paradigm reintegrated: local reverberations as internal representations. Behav Brain Sci 18:617–657

Atienza M, Cantero JL, Dominguez-Marin E (2002) The time course of neural changes underlying auditory perceptual learning. Learn Mem 9(3):138–150

Barrie JM, Freeman WJ, Lenhart M (1996) Modulation by discriminative training of spatial patterns of gamma EEG amplitude and phase in neocortex of rabbits. J Neurophysiol 76:520–539

Braitenberg V, Schüz A (1998) Cortex: statistics and geometry of neuronal connectivity, 2nd edn. Springer, Berlin

Bressler SL (1987) Functional relation of olfactory bulb and cortex: I. Spatial variation of bulbo-cortical interdependence. Brain Res 409:285–293; II. Model for driving of cortex by bulb. Brain Res 409:294–301

Bressler SL (1988) Changes in electrical activity of rabbit olfactory bulb and cortex to conditioned odor stimulation. Behav Neurosci 102:740–747

Bressler SL, Freeman WJ (1980) Frequency analysis of olfactory system EEG in cat, rabbit and rat. Electroencephalogr Clin Neurophysiol 50:19–24

Busch NA, Dubois J, VanRullen R (2009) The phase of ongoing EEG oscillations predicts visual perception. J Neurosci 29(24):7869–7876. doi:doi: 10.1523/ JNEUROSCI.0113-09.2009

Buzsaki G (2006) Rhythms of the brain. Oxford University Press, Oxford UK

Emery JD, Freeman WJ (1969) Pattern analysis of cortical evoked potential parameters during attention changes. Physiol Behav 4:67–77. http://soma.berkeley.edu/archives/IID8/69.html

Firestein S (2004) A code in the nose. Sci STKE 2004:pe15 DOI: 10.1126/stke.2272004pe15

Freeman WJ (1968) Relations between unit activity and evoked potentials in prepyriform cortex of cats. J Neurophysiol 31:337–348

Freeman WJ (1974a) Topographic organization of primary olfactory nerve in cat and rabbit as shown by evoked potentials. Electroencephalogr Clin Neurophysiol 36:33–45

Freeman WJ (1974b) Average transmission distance from mitral tufted to granule cells in olfactory bulb. Electroencephalogr Clin Neurophysiol 36:609–618

Freeman WJ (1975) Mass action in the nervous system. Examination of the neurophysiological basis of adaptive behavior through the EEG. Academic Press, New York. Posted in e-formats (2004) http://sulcus.berkeley.edu/MANSWWW/ MANSWWW.html

Freeman WJ (1979a) Nonlinear dynamics of paleocortex manifested in the olfactory EEG. Biol Cybern 35:21–37

Freeman WJ (1979b) Measurement of cortical evoked potentials by decomposition of their wave forms. J Cybern Inform Sci 2–4:44–56

Freeman WJ (1986) Petit mal seizure spikes in olfactory bulb and cortex caused by runaway inhibition after exhaustion of excitation. Brain Res Rev 11:259–284

Freeman WJ (1990) On the problem of anomalous dispersion in chaoto-chaotic phase transitions of neural masses, and its significance for the management of perceptual information in brains. In: Haken H, Stadler M (eds) Synergetics of cognition, vol 45. Springer, Berlin, pp 126–143
Freeman WJ (1991) The physiology of perception. Sci Am 264:78–85
Freeman WJ (2000) Neurodynamics: an exploration of mesoscopic brain dynamics. Springer, London UK. http://soma.berkeley.edu/books/BD/MesoBrainDyn.html
Freeman WJ (2001) How brains make up their minds. Columbia University Press, New York
Freeman WJ, Baird B (1987) Relation of olfactory EEG to behavior: spatial analysis. Behav Neurosci 101:393–408
Freeman WJ, Breakspear M (2007) Scale-free neocortical dynamics. Scholarpedia 2(2):1357. http://www.scholarpedia.org/article/Scale-free_neocortical_dynamics
Freeman WJ, Erwin H (2008) Freeman K-set. Scholarpedia 3(2):3238. http://www. scholarpedia. org/article/Freeman_K-set
Freeman WJ, Grajski KA (1987) Relation of olfactory EEG to behavior: factor analysis. Behav Neurosci 101:766–777
Freeman WJ, Kozma R (2010) Freeman's mass action. Scholarpedia, 5(1):8040. http://www. scholarpedia.org/article/Freeman%27s_mass_action
Freeman WJ, Schneider W (1982) Changes in spatial patterns of rabbit olfactory EEG with conditioning to odors. Psychophysiology 19:44–56
Freeman WJ, Skarda CA (1985) Spatial EEG patterns, non–linear dynamics and perception: the neo–Sherringtonian view. Brain Res Rev 10:147–175
Freeman WJ, Viana Di Prisco G (1986) Relation of olfactory EEG to behavior: time series analysis. Behav Neurosci 100:753–763
Freeman WJ, Vitiello G (2006) Nonlinear brain dynamics as macroscopic manifestation of underlying many-body field dynamics. Phys Life Rev 3:93–118. http://dx.doi.org/10.1016/j. plrev.2006.02.001 http://repositories. cdlib.org/postprints/1515
Freeman WJ, O'Nuillain S, Rodriguez J (2008) Simulating cortical background activity at rest with filtered noise. J Integr Neurosci 7(3):337–344. http://repositories.cdlib.org/postprints/3373
Freeman WJ, Kozma R, Bollobás B, Riordan O (2009) Chapter 7. Scale-free cortical planar network. In: Bollobás B, Kozma R, Miklös D (eds) Handbook of large-scale random networks. Series: Bolyai mathematical studies. Springer, New York. Vol. 18, pp. 277–324. http://www. springer.com/math/numbers/book/978-3-540-69394-9
Freeman WJ, Livi R, Opinata M, Vitiello G (2012) Cortical phase transitions, non-equilibrium thermodynamics and the time-dependent Ginzburg-Landau equation. Int J Mod Phys B 26(6):1250035. doi:10.1142/S021797921250035X
Fu J, L G, Qin Y, Freeman WJ (2007) A pattern recognition method for electronic noses based on an olfactory neural network. Sensors Actuators B 125(2):489–497. http://escholarship.org/uc/ item/9tr82425
Herrick CJ (1948) The brain of the tiger salamander. Chicago University Press, Chicago IL
Hu M, Li J, Li G, Tang X, Freeman WJ (2006) Normal and hypoxia EEG recognition based on a chaotic olfactory model. Lecture Notes. Comp Sci 3973:554–559. http://escholarship.org/uc/ item/2fx253c1
Kaas JH (1987) The organization of neocortex in mammals: implications for theories of brain function. Ann Rev Psychol 38:129–152
Kay LM, Freeman WJ (1998) Bidirectional processing in the olfactory-limbic axis during olfactory behavior. Behav Neurosci 112:541–553
Kay LM, Freeman WJ, Lancaster LR Jr (1996) Simultaneous EEG recordings from olfactory and limbic brain structures: limbic markers during olfactory perception. In: Gath I, Inbar G (eds) Advances in Processing and Pattern Analysis of Biological Signals. Plenum, New York, pp 71–84
Köhler W (1940) Dynamics in psychology. Grove, New York
Kozma R, Freeman WJ (2001) Chaotic resonance: methods and applications for robust classification of noisy and variable patterns. Intern J Bifurc Chaos 10:2307–2322

Kozma R, Freeman WJ, Erdí P (2003) The KIV model – nonlinear spatio-temporal dynamics of the primordial vertebrate forebrain. Neurocomputing 52:819–826. http://repositories.cdlib.org/postprints/1049

Kozma R, Puljic M, Balister P, Bollobás B, Freeman WJ (2005) Phase transition in the neuropercolation model of neural populations with mixed local and non-local interactions. Biol Cybern 92(6):367–379. http://escholarship.org/uc/item/6rk1d0sj

Kozma R, Tunstel E, Aghazarian H, Huntsberger T, Freeman WJ (2007) Computational aspects of cognition and consciousness in intelligent devices. IEEE Computational Intelligence 2(3):53–64. http://escholarship.org/uc/item/0r0485fx

Kozma R, Huntsberger T, Aghazarian H, Tunstel E, Ilin R, Freeman WJ (2008) Intentional control for planetary rover SRR2k. Adv Robotics 21(8):1109–1127

LeDoux JE (2000) Emotion circuits in the brain. Ann Rev Neurosci 23:155–184

Lettvin JY, Gesteland RC (1965) Speculations on smell. Cold Spring Harb Symp Quant Biol 30:217–225

Li X, Li G, Wang L, Freeman WJ (2006) A study on a bionic pattern classifier based on olfactory neural system. Int J Bifurc and Chaos 16(8):2425–2434. http://escholarship.org/uc/item/06r9940s

Li G, Zhang J, Freeman WJ (2007) Application of novel chaotic neural network on Mandarin digital speech recognition. IEEE Conference Publications. http://escholarship.org/uc/item/9k25p3sg

Maclean PD (1969) The triune brain. Plenum, New York

Malach R (1994) Cortical columns as devices for maximizing neuronal diversity. TINS 17(101–104):129–152

Merleau-Ponty M (1942) The structure of behavior (trans: Fischer AL). Beacon Press, Boston, 1963.

Miller R (2002) Wheels within wheels: circuits for integration of neural assemblies on small and large scales. In: Schüz A, Miller R (eds) Cortical areas: unity and diversity. Taylor and Francis, New York, pp 423–458

Miller R, Maitra R (2002) Laminar continuity between neo- and meso-cortex: the hypothesis of the added laminae in neocortex. In: Schuz A, Miller R (eds) Cortical areas: unity and diversity, Ch. 11. Taylor and Francis, New York, pp 219–242

Murphy C, Cain WS, Gilmore MG, Skinner RB (1991) Sensory and semantic factors in recognition memory for odors and graphic stimuli: Elderly versus young persons. Am J Psychol 104(1):161–192. http://www.jstor.org/stable/1423153

Nicoll RA (1971) Recurrent excitation of secondary olfactory neurons: a possible mechanism for signal amplification. Science 171:824–825

O'Keefe J, Nadel L (1978) The hippocampus as a cognitive map. Clarendon, Oxford UK

Pribram KH (1991) Brain and perception: holonomy and structure in figural processing. Lawrence Erlbaum Assoc, Mahwah NJ

Principe JC, Tavares VG, Harris JG, Freeman WJ (2001) Design and implementation of a biologically realistic olfactory cortex in analog VLSI. Proceedings IEEE 89:1030–1051

Quian Quiroga R (2012) Concept cells: The building blocks of declarative memory functions. Nat Rev Neurosci 13:587–597. doi: 10.1038/nrn3251

Quian Quiroga R, Kraskov A, Koch C, Fried I (2009) Explicit encoding of multimodal percepts by single neurons in the human brain. Curr Bio 19:1308–1313

Roelfsema PR, Engel AK, König P, Singer W (1997) Visuomotor integration is associated with zero time-lag synchronization among cortical areas. Nature 385:157–161

Roth G (1987) Visual behavior in salamanders. Springer, Berlin

Samar VJ, Bopardikar A, Rao R, Swartz K (1999) Wavelet analysis of neuroelectric waveforms: A conceptual tutorial. Brain Language 68(1):7–60. http://dx.doi.org/10.1006/brln.1998.2024

Schüz A, Miller R (eds) (2002) Cortical measures: unity and diversity. Taylor and Francis, New York

Shimoide K, Freeman WJ (1995) Dynamic neural network derived from the olfactory system with examples of applications. IEICE Transaction Fundamentals E-78A:869–884

Singer W, Gray CM (1995) Visual feature integration and the temporal correlation hypothesis. Ann Rev Neurosci 18:555–586
Skarda CA, Freeman WJ (1987) How brains make chaos in order to make sense of the world. Behav Brain Sci 10:161–195
Tanaka K (2003) Columns for complex visual object features in the inferotemporal cortex: Clustering of cells with similar but slightly different stimulus selectivities. Cerebral Cortex 13:90–99
Traub RD, Whittington MA, Stanford IM, Jefferys JGR (1996) A mechanism for generation of long-range synchronous fast oscillations in the cortex. Nature 383:421–424
Uttley AM (1955) The probability of neural connections. Proc Roy Soc Ser B 144:229–240
Wang XF, Chen GR (2003) Complex networks: small-world, scale-free and beyond. IEEE Circuits and Systems 31:6–20
Whittington MA, Faulkner HJ, Doheny HC, Traub RD (2000) Neuronal fast oscillations as a target site for psychoactive drugs. Pharmacol Ther 86:171–190
Yang XL, Fu J, Lou ZG, Wang LY, Li G, Freeman WJ (2006) Tea classification based on artificial olfaction using bionic olfactory neural network. Lecture Notes in Computer Science 3972: 343–348. http://escholarship.org/uc/item/41t1c9g7
Zelano C, Mohanty A, Gottfried JA (2011) Olfactory predictive codes and stimulus templates in piriform cortex. Neuron 72:178–187. http://www.sciencedirect.com/science/article/pii/S0896627311007318

Chapter 9
Neocortical ECoG Images Formed by Learning

9.1 Introduction

The utility of the EEG and ECoG depends on finding methods of data recording, measurement, and analysis that provide reliable and robust neural correlates of ongoing cognitive behaviors (Freeman 1975; Basar 1998). The accessible temporal and spatial resolutions are in ms to s and mm to cm. The channel information capacity in the physiological bandwidth of one or a few cortical signals is too narrow to afford correlates for much more than binary-state changes such as wake-sleep, eyes open-shut, start-stop, and right-left and to type letters at an information rate of about 1 bit/s (Gao et al. 2003). The spatial domain in high-density array recording offers far greater channel capacity, provided the macroscopic neural formatting of images can be understood.

A viable hypothesis is that the sensory neocortices have the same sensory and perceptual formats as those of the olfactory system because the neocortex originated by elaborations of the allocortex that adapted them to the unique requirements of visual, auditory, and somatic sensory information processing (Fig. 8.2, Sect. 8.1). Despite the differences in inputs and local processing between modalities, their outputs must all have basically the same format in order for them to be integrated rapidly and efficiently (Sect. 10.3). This hypothesis serves to predict that spatial images exist in the neocortical ECoGs of all the sensory cortices, which are compatible with the olfactory spatial images. It follows that the entorhinal cortex, to which all sensory cortices contribute, enables the medial temporal cortex to create multimodal gestalts by receiving and integrating the feature vectors from all modalities (see Fig. 10.3, Sect. 10.3). It transmits its compound feature vector to the hippocampus, which incorporates information about where the perceiver was located in time and space when the perception took place, preparatory to inclusion of the percept into the lifetime memory of the subject (Squires 1992; Pribram 1991).

Specifically, the hypothesis requires that sensory neocortices be found to generate intermittent gamma bursts during acts of perception. Just as the olfactory system is a batch processor of sensory information, each sensory cortex should exhibit a

W.J. Freeman and R. Quian Quiroga, *Imaging Brain Function With EEG: Advanced Temporal and Spatial Analysis of Electroencephalographic Signals*, DOI 10.1007/978-1-4614-4984-3_9, © Springer Science+Business Media New York 2013

cinematic sequence of frames or 'microstates' (Pribram et al. 1998; Lehmann et al. 2009) comparable to those in olfaction. Each burst should have a narrowband carrier wave that is modulated in phase and amplitude. A significant correlation has to be shown between the spatial AM patterns and the stimulus categories. Testing the hypothesis requires finding the AM patterns in sufficient numbers and with sufficient reliability to perform the statistical tests. The preeminent requirement is for *markers* in the data by which to locate the gamma bursts in the ongoing ECoG (Sect. 6.3).

In the olfactory system, the markers are provided by the act of inhalation and the accompanying respiratory wave (Fig. 8.3, Sect. 8.2), which in small mammals is in the theta range. The difficulty in pursuing this hypothesis is that, unlike the olfactory and the hippocampal ECoGs, the neocortical ECoGs usually lack well-defined theta waves that are sufficiently reliable to serve as temporal markers (Fig. 9.1a). The coupling of theta waves and gamma bursts (*theta-gamma linkage*) (Buzsáki, 2006) is well documented in allocortical ECoG (Freeman 1975; Fell et al. 2003; Lisman 2005), neocortical ECoG (Chrobak and Buzsáki 1998; Freeman 2005b), and human scalp EEG (Schack et al. 2002; Freeman et al. 2003; Canolty et al 2006), but it is seldom strong enough to resolve repeated bursts single trials. It seems obvious that gating by motor controls of sensory input should occur, which is comparable to sniffing in other systems. These motor controls of sensory input include saccades in vision, whisking in somesthesis, and middle ear muscle contractions in audition. They might parse sensory inflows and leave their marks in ECoGs as sniffing does. However, gating in neocortex appears to be more dependent on internal factors (Sect. 9.7) than is gating of input by sniffing, and there appear to involve multiple overlapping bursts of narrowband oscillation that may reflect parallel processing in neocortex of multiple events in partially overlapping frequency bands. Therefore, testing the hypothesis requires finding alternatives to sniffing and theta waves as markers.[1] Section 9.2 describes proof of the existence of classifiable AM patterns by use of a window moved across multichannel ECoGs in fixed steps. Section 9.3 describes improved temporal resolution of AM patterns by use of the Hilbert transform. Section 9.4 describes phase patterns at high resolution. Section 9.5 focuses on the events between bursts that presage emergence of AM and PM patterns. Section 9.6 introduces cinematic display of AM and PM patterns. Section 9.7 discusses the neural mechanisms by which AM and PM patterns form. Section 9.8 summarizes the concepts.

9.2 Initial finding of Classifiable AM Patterns in Neocortical ECoGs

The search for neocortical AM patterns related to cognition was begun (Barrie et al. 1996) by fixing 8x8 high-density arrays of electrodes on the visual, auditory, or somatic cortices of rabbits (Fig. 9.1). The subjects were trained in the same

[1] Two examples of markers are described in other chapters. One uses recurring minima in spatial variance of phase (Fig. 6.5, Sect. 6.4.2; Fig. 10.6, Sect. 10.5), and the other uses the recurring maxima in burst power (Fig. 6.5c, Sect. 6.4.2; Fig. 8.3a, Sect. 8.2; Fig. 9.2, Sect. 9.2; Fig. 10.4, Sect. 10.4).

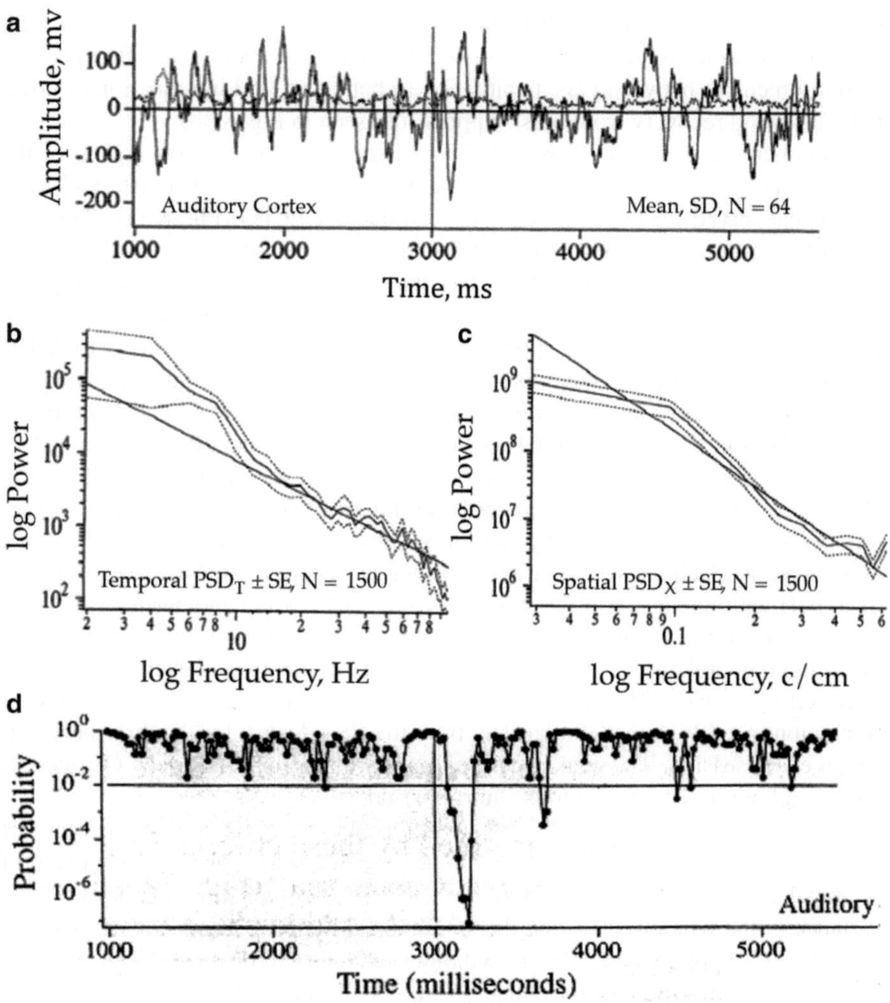

Fig. 9.1 (**a**) The example shows the spatial ensemble average (*high amplitude*) and SD_T (*low amplitude*, all positive) of the 64 neocortical ECoGs from an 8 × 8 array of electrodes (5.6 × 5.6-mm window, Fig. 8.4b, AUD) on the resting auditory cortex. The analog passband was 1–100 Hz. The digitizing step was 2 ms, giving 1500 time steps in each trial. (**b, c**) Temporal and spatial power spectral density after time or space ensemble averaging. (**d**) For working ECoG, a CS was delivered at 3 s in a 6-s recording trial. A 128-ms window was stepped at intervals along the band-pass-filtered ECoG. The graph is the probability of a difference having occurred by chance at each step between the root-mean-square amplitude patterns recorded from the auditory cortex during conditioned stimuli with reinforcement (CS+) and without (CS-) (Adapted from Figs. 2, 5 and 9 in Barrie et al. 1996)

paradigm as for the olfactory CS + and CS- (Sect. 8.5; Freeman and Grajski 1987) to discriminate reinforced conditioned stimuli (CS+) from unreinforced (CS-). The CSs were simple and easy to learn in a few trials: dim or brighter flash, 500-Hz or 5000-Hz tone burst, and puff of air to the cheek or rump. Each subject gave 20

CS + trials and 20 CS- trials with a total of 40 correct responses in each session, with up to nine sessions for each subject, including replications before and after reversal of reinforcement between CSs, habituation to both CSs, and introduction of novel CSs following recovery from burst suppression (seen in Fig. 8.3b).

Examination of the ECoG revealed broad-spectrum fluctuations typical of the three neocortices (Fig. 9.1a), showing neither of the prominent bursts and theta waves that served for locating bursts in the olfactory ECoG (Fig. 8.3a). The averages of long-duration (≥3 s) temporal and spatial power spectral densities (PSD_T, PSD_X, Sect. 7.2) were power law (1/f) with no reliably recurring local peaks (Fig. 9.1b, c). The same temporal passband (20–80 Hz) that sufficed for olfactory ECoG (Fig. 6.3, Sect. 6.3) was applied to each set of 64 signals in each 6-s trial. The search for classifiable AM patterns in the beta-gamma range was initiated by applying a 128-ms window that was shifted blindly by a fixed step of 64 ms along the set of 64 ECoGs in each trial. The 64 root-mean-square (rms) amplitudes were calculated so as to represent each frame by a 64x1 feature vector (Barrie et al. 1996). There were 78 time steps in each record after filtering. The 20 CS + and 20 CS- vectors were divided into training and test subsets of 10 each for multivariate classification. Each subset specified a cluster of 10 points in 64-space (Fig. 6.8, Sect. 6.4) with one center of gravity (centroid) for CS + and another for CS-, as in olfactory burst classification (Fig. 8.7, Sect. 8.4). At every time step, we calculated the binomial probability that linear separation of the two clusters could have occurred by chance. The result (Fig. 9.1d) showed multiple peaks of significant separation ($p < .01$) of the CS + clusters from the CS- clusters for the three neocortices in every session. The peaks occurred only in the interval between CS and CR onsets and not in the pre-stimulus control period.

The classification of feature vectors was repeated using three other measures of the feature vectors in place of rms amplitude: the factor loading first component of principal component analysis (PCA, Sect. 6.3), which contained 90–95% of the variance of the windowed ECoG; the amplitude at the peak frequency in the PSD_T from the FFT applied to the 128-ms window; and the analytic power from applying the Hilbert transform to the 64 signals (Sect. 9.5). The three measures of amplitude gave the same degree of classification. The result confirmed the hypothesis that the same form of cognitive information as that in the olfactory system would be found in the neocortical sensory systems, that is, classificatory information would be found in the spatial AM pattern of the carrier frequency of cinematic bursts of beta-gamma oscillation. The result laid the foundation for analyzing and simulating the mechanism by which multimodal percepts (gestalts) are synthesized in the limbic system (Sect. 10.3). In principle, the mechanism has been easily simulated with matrix algebra in autonomous robots (Kozma et al. 2003; Kozma et al. 2008). The multiple simulated feature vectors were concatenated by linear superposition (Sect. 6.5) and treated as a compound feature vector (Fig. 10.3). However, the result gave no information about the properties (locations, durations, diameters, frequencies, and textures) of the bursts on single trials to compare with the bursts in olfaction. For that information, a high-resolution marker was required, which was provided by the Hilbert transform for the close study of amplitude and frequency modulation.

9.3 High-Resolution of ECoG Amplitudes with the Hilbert Transform

9.3.1 Derivation of the Analytic Power, $A^2(t)$

The ECoG recording at each electrode and each time step gave the amplitude of the signal; the local FFT (Chap. 2) gave its rate of change. When the rate was constant, it was expressed as a frequency. However, ECoG and EEG signals typically showed modulation in amplitude (AM), frequency (FM), and phase (PM) in overlapping bursts at different center frequencies. Precise measurements of AM, FM, and PM was facilitated, when the signals were passed through a filter bank (Chap. 2; Fig. 7.12, Sect. 7.6), and the Hilbert transform (Sect. 6.4.1) was applied to each band, giving two time series for each band (Fig. 6.5). One was the analytic amplitude, $A(t)$, or its square, the analytic power, $A^2(t)$, which approximated the envelope of a burst dominating a band). The other was the analytic phase, $\phi(t)$, or its rate of change, the analytic frequency, $\omega(t)=\Delta\phi(t)/\Delta t$, in rad/s Fig. 6.5d, Sect. 6.3. The analytic frequency best displayed visually the segments in which cortical activity approached maximum order (Sect. 6.2.3). The results from the Hilbert transform and the FFT converged (Le Van Quyen et al. 2001; Quian Quiroga et al. 2002), when both transforms were applied to signals having the long durations needed for the lowest frequencies sought with the FFT (Chap. 3) or with wavelets (Chap. 4).[2] The clinical mode decomposition into customary, relatively narrow passbands (delta, theta, alpha, beta, gamma, epsilon, Sect. 2.10.1) gave high temporal resolution needed to follow rapid changes in phase and amplitude for components selected as candidates for cognitive correlation by classification.

What was at stake here is illustrated in Fig. 9.2. Parts a and b show the group classification that could be achieved by comparing CS+and CS- feature vectors in 128-ms windows locked in 64-ms steps fixed with respect to CS onset at t=0 s (Fig. 9.1d) while separating the passbands for beta and gamma oscillations. Parts c and d show the temporal locations and durations of bursts that were located by using the analytic amplitudes by after applying the Hilbert transform (Sect. 6.4.1) as markers. The single gamma peak of classification (Fig. 9.2a) was found only in the sensory cortex receiving the CS, whereas the two beta peaks were found in all cortices simultaneously (Fig. 10.3, Sect. 10.3). The significance of these three peaks is discussed in Sect. 11.5. Briefly, the increase in temporal resolution that was provided by the Hilbert transform made it worthwhile to learn how to deal with its difficulties and limitations.

[2] Empirical mode decomposition (Huang et al. 1998) has also been proposed for ECoG and EEG decomposition owing to its high spectral resolution. However, the spectral decomposition is not appropriate for tracking frequency modulation in ECoGs.

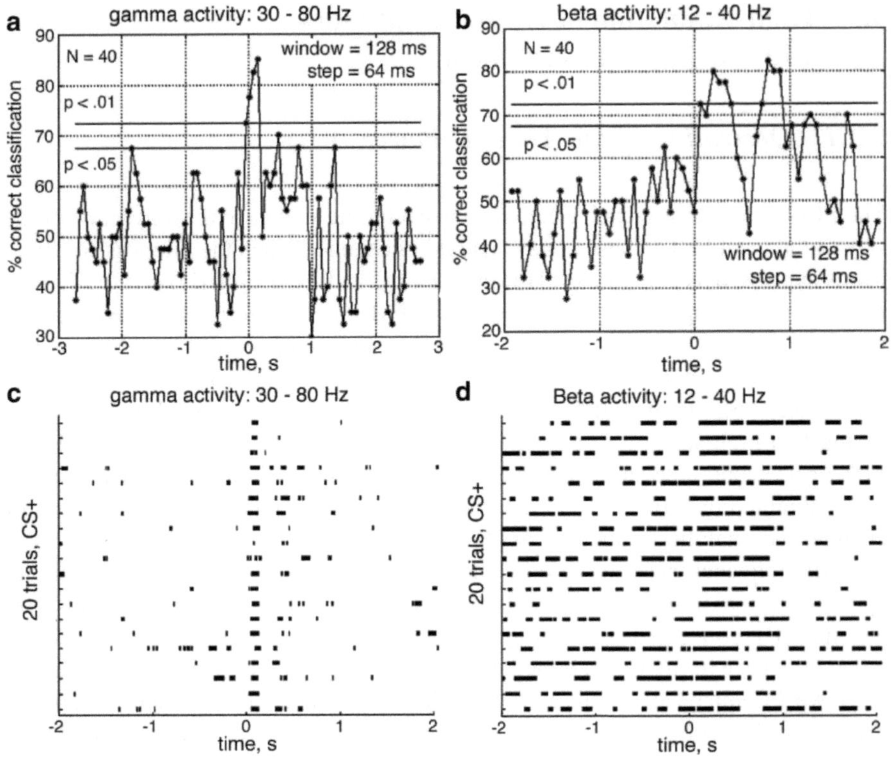

Fig. 9.2 (**a**) Significant AM pattern classification was found in the gamma range between the CS+and CS- stimuli but only during the first 200 ms after CS onset. (**b**) Significant AM pattern differences between CS+and CS- trials for bursts in the beta range in poststimulus time periods of 100–400 ms and 700–900 ms. The use of a locked-step window did not serve to locate the bursts on single trials. (**c**) The use of the index for pragmatic information, $H_e(t)$ (Fig. 6.7, Sect. 6.4.3), which by combining measures of AM pattern power and stability, optimally localized classifiable AM patterns on single trials. Gamma bursts much less often occurred in the pre-stimulus control period than in the test period. (**d**) The incidence of beta bursts in the control period was uniform than the incidence of gamma bursts. It increased during the first poststimulus second and declined in the next two seconds. (From Figs. 2 and 3 in Freeman 2006b)

9.3.2 *Use of $H_e(t)$ as a Scalar Index of the Order Parameter*

The construction of a marker with which to locate AM patterns by using the analytic power, $A_{ij}^2(t)$, was in seven steps (Fig. 9.3). Step 1 (a): As in all previous assays, the ECoG amplitudes were normalized to z-scores by subtracting the global channel means and dividing by the global SD. Step 2: A spatial low-pass filter was applied to each frame (Fig. 1 in Freeman and Barrie 2000). Step 3: (b) A band-pass temporal filter was applied to every signal. Step 4: The Hilbert transform was used to calculate the analytic power of each channel, $A_{ij}(t)$, at each time step. The 64 values were

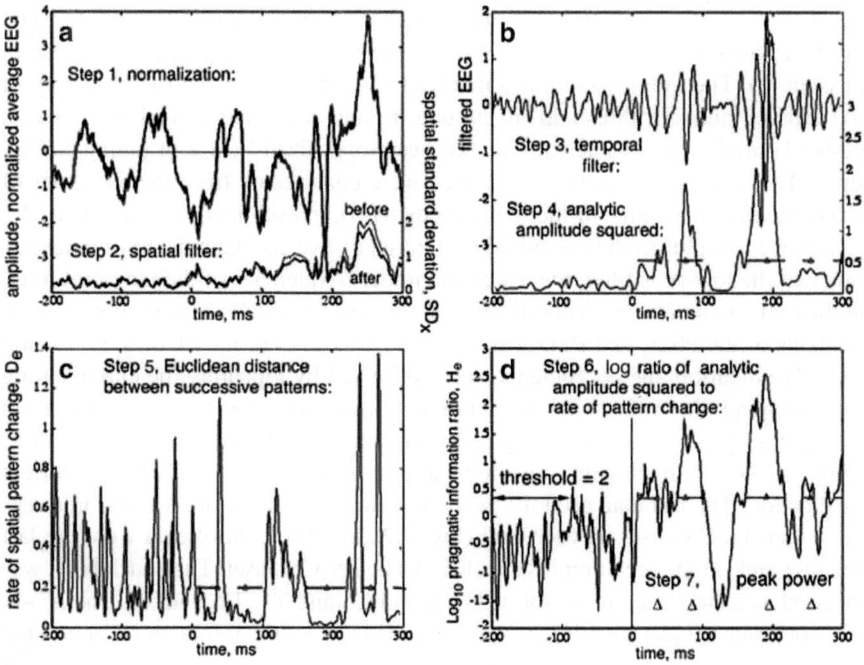

Fig. 9.3 The algorithms are illustrated that were used to locate each stable spatial frame in which to calculate its feature vector for classification. (**a**) Average (*upper curves*) and SD (*lower curves*) of 64 EEGs from one trial in a 500-ms segment extending across CS arrival at 0 ms; Step 1, after channel demeaning and amplitude normalization; Step 2, before and after low-pass spatial filtering at 0.3 c/mm. (**b**) Upper curve: Step 3, average EEG after temporal band-pass filtering at 20–80 Hz. Lower curve: Step 4, spatial average, $\underline{A}^2(t)$, of the analytic amplitude squared, $A_j^2(t)$, averaged over channels, $j = 1, 64$, at each digitizing step, t. The increase in $\underline{A}^2(t)$ was not entirely due to an increase in synchrony, which was shown to increase to a sustained level before $\underline{A}^2(t)$ began to rise. (**c**) Step 5: The 64 values of $A_j^2(t)$ gave a 64 × 1 feature vector, $\mathbf{A}^2(t)$, that specified an AM pattern and a point in 64-space. The Euclidean distance, D_e, between successive points, $\mathbf{A}^2(t) - \mathbf{A}^2(t-1)$, gave the rate of change in the AM pattern. Successive low values indicated pattern stability. (**d**) Step 6: The pragmatic information was given by the ratio $H_e = \underline{A}^2(t)/D_e(t)$. Qualifying segments were identified by the criteria that H_e remained above a threshold, here $t_e = 2$, longer than a minimal duration, here $m_e = 10$ ms, in segments shown by the bars across the curve representing $\log_{10} H_e$. Step 7: In each qualifying segment, the maximum of $\underline{A}^2(t)$ and its time of occurrence, t_{max}, were calculated (triangles). The 64 × 1 feature vectors used for classification of the AM patterns were given by the 64 values of peak instantaneous $H_e(t)$ (From Fig. 1 in Freeman 2004a)

divided by the mean power, $\underline{A}^2$(t), which gave a normalized 64x1 feature vector, $\mathbf{A}^2$(t), and a point in 64-space at each time step. Step 5: The succession of points gave the Euclidean distance, $D_e(t) = \mathbf{A}^2(t) - \mathbf{A}^2(t-1)$, taken in 64-space by each digitizing step of the trajectory, Δt (Fig. 6.6, Sect. 6.4.2) and similarly for phase (Fig. 9.6a). The tip of the feature vector tended to dwell in a small region of 64-space during peaks of A(t) and transit widely in the troughs between peaks. Step 6: (d) The mean rate of change in energy dissipation, $\underline{A}^2$(t), at each time step was divided by the rate of change in location of the feature vector, D_e(t), giving an index

(Fig. 6.7, Sect. 6.4.3) defined as the *pragmatic information,* $H_e(t)=\underline{A}^2(t)/D_e(t)$ (Fig. 6.7a, Sect. 6.4.3).[3] The distribution of the values of $\log_{10} H_e(t)$ was close to Gaussian (c). The index served as a marker to locate bursts when the trajectory of $H_e(t)$ crossed above a threshold magnitude, t_c, and remained there longer than a threshold duration, m_c. Values of t_c and m_c were optimized by use of tuning curves. Step 7: The feature vectors to be classified were constructed from the 64 values of $A_{ij}(t)$ at the maximal value of the scalar index, $H_e(t)$, in each identified segment.

The seven-step procedure required evaluation of five coefficients: the spatial frequency of the low-pass filter, the lower and upper frequencies of the temporal filter, and the two thresholds. Each coefficient was evaluated initially in accord with measurements of the PSD_X and PSD_T (Sect. 7.3; Fig. 9.1b, c) of the ECoGs. Following initial classification, the coefficients were optimized by repeating the classification while stepping the values across a range to construct tuning curves for each coefficient (e.g., threshold, t_e, in Fig. 9.4a; spectral and temporal window durations in Fig. 10.3a, b) in classifier-directed parameter optimization of the % correct classification. The test data used for the optimization consisted of the first three segments after the CS onset on each trial (Fig. 9.2c, d). The control data were the last three segments in the pre-stimulus epoch 1–3 s before CS onset. Then each burst was expressed as a 64x1 feature vector: C3, C2, and C1 and T1, T2, and T3. The 37–40 trials from each subject gave 12 clusters of points in 64-space, 6 for control bursts and 3 each for CS+and CS- bursts. For optimization, the three 64x1 feature vectors for T1, T2, and T3 bursts were concatenated into two 192x1 feature vectors, one for CS+and one for CS-. The optimized results were displayed by time and trial in a raster (Fig. 9.2c, d) showing their temporal locations and durations and as clusters in192-space (Fig. 9.4b). Then the goodness of classification was evaluated for the three pairs of 1x64 feature vectors in the control pre-stimulus period (c) and the test period (d). The results were simulated with artificial AM patterns (Freeman 2006a).

The qualifying bursts (Fig 9.2c, d) were found in the time intervals having high probability of the incidence of classifiable AM patterns with the locked-step method (a, b). Bursts in the gamma range (c) were found mainly 40–200 ms following CS onset. Bursts in the beta range (d) increased in numbers after 200 ms and declined in numbers thereafter. Conditioned responses (the CR was the onset of sniffing) occurred with start latencies of about 700±500 ms (Barrie et al. 1996). The precision of measurement was not sufficient to correlate the latencies with bursts. In any case, the focus of this study was on proof of the existence of AM patterns that were classifiable with respect to learned CS+and CS-. The CR was used to verify the capacity of the subjects for discrimination of the CS.

Analysis of variance of the blocks of AM patterns in each frequency range showed that only two sets of test bursts emerged as differing significantly from the control bursts in every subject and cortex (for statistics, see Table 1 in Freeman

---

[3] Atmanspacher and Scheingraber (1990) described the concept of the ratio of the rate of energy dissipation to the rate of order formation as a "fundamental extension of Shannonian information" (pp. 731–732). Its use as a measure of the knowledge created from information by cortex as it forms a textured burst is described in Sect. 11.5.

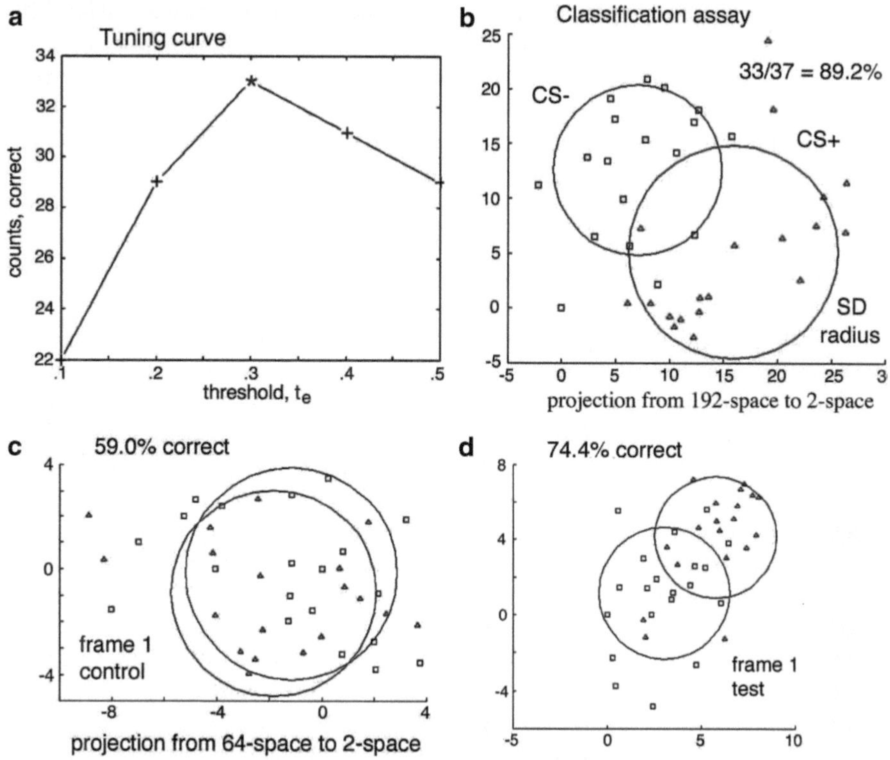

Fig. 9.4 (**a**) An example shows the tuning curve used to optimize the threshold, t_e, by systematically stepping the coefficient and repeating the classification test to find the optimal separation of points. (**b**) The multidimensional scaling technique of nonlinear mapping (Sammon 1969) projected clusters from 192-space into 2-space, optimizing the separation of clusters while preserving the relative distances between all of the data points. The circles show the standard deviations (SD) of the clusters calculated in the display plane. (**c**) The 64 × 1 feature vectors in the 1st frame of the control state showed no separation. (**d**) The clusters from the 1st frame of the test period were the best separated, as predicted from the results such as in Fig. 9.1d (From Figs. 2 and 4 in Freeman 2005a)

2005a). One set was the first set of gamma bursts, T1, which started 66 ± 16 ms poststimulus and lasted 57.2 ± 5.7 ms (Fig. 9.2a). The other was the third set of beta bursts, T3, which began 490 ± 47 ms poststimulus and lasted 64.7 ± 3.3 ms, ending before the mean onset of CR near 700 ms (Fig. 9.2d). The AM patterns in these two sets of bursts differed from each other (CS+ vs. CS-) as well as from the control bursts, C1–C3, and from the other test bursts. The results confirmed predictions of the long distances and durations of spatial correlations (Freeman and Vitiello 2006; Kitzbichler et al. 2009) and prior estimates of the exceptionally long durations (Fig. 9.7e, f) and large diameters (Fig. 8.4b) of AM patterns related to behaviors. The second burst, T2 in the beta range, was notable for its latency (268 ± 19 ms near the human P300, Sect. 1.5.1), its high power (0.417 ± 0.091 SD), and its lack of classification with respect to CSs.

A further test was required to determine whether the information in the data that served for classification might be concentrated in any smaller number than the whole set of 64 channels instead of being spatially uniform. Initial assays by repeating the classification assay after removing channels with the highest amplitude or least amplitude or greatest variance or greatest change from control to CS showed only assay reductions that depended on the number of channels removed by whatever criterion. A global test was conducted by deleting randomly selected channels in groups of $N = 8, 16, \ldots, 56$ and for each group deletion and repeating the assay 40 times (See Figs. 8.7 and 10.8 for examples), each time with an independent selection. In a separate tally for every channel, the sum of 40x7 values of goodness of classification was divided by the number of times the channel was used. The rationale was that if a channel contained classificatory information, it could be assigned a value of 1 and if not then 0. Any test not having that channel would give a value of 0, and any test that included that channel would give a value of $1/N > 0$, so likewise with two channels and so on. By this test, channels having high and low amplitudes had equal classification efficacy to within 0.1% of the variance of the assay. In other words, pattern classifications required both high and low activity corresponding to light and dark patches. The best classification rate was achieved by using all available channels. The result corroborated similar findings of non-locality with the olfactory ECoG (Fig. 8.7b, Sect. 8.4 from Freeman and Baird 1987), the neocortical ECoG (Fig. 13 from Freeman and Barrie 2000), the auditory ECoG (Sect. 10.2 from Ohl et al. 2003), multiple sensory cortices, the entorhinal cortex (Fig. 10.3c, from Freeman and Burke 2003) and the scalp EEG (Fig. 10.8, Sect. 10.5 from Ruiz et al. 2010). The hypothesis was strengthened that the perceptual information in neocortical activity (as distinct from sensory information in pulse trains) was delocalized, most likely by a divergent-convergent topology (one-to-many, many-to-one) of connections resembling that in the olfactory system (Fig. 8.8, Sect. 8.4) and in holography using the Gabor transform (Sects. 8.6, 10.3).

9.4 High-Resolution Images of Neocortical Phase Patterns

9.4.1 Derivation of the Phase Cones of Classifiable Bursts

The short-term FFT gave the PSD_T of the spatial ensemble average of each filtered burst. Each frequency component had values of amplitude and phase. The maximal amplitude gave the mean carrier frequency across the frame of measurement. The FFT of the 64 signals gave the 64 values at peak power of each channel, which was displayed as an AM pattern and represented by a feature vector. The 64 values of phase defined an 8×8 phase surface that usually displayed a spatial pattern of phase modulation (PM) (Fig. 9.5c), which as in the ECoG of the olfactory bulb could be fitted with a right cone (Fig. 9.5d). The AM pattern of every classifiable burst was accompanied by a phase cone, which was characterized by the location and sign of the apex (maximal lead or lag) and the slope of the cone (Fig. 8 in Freeman and Barrie 2000).

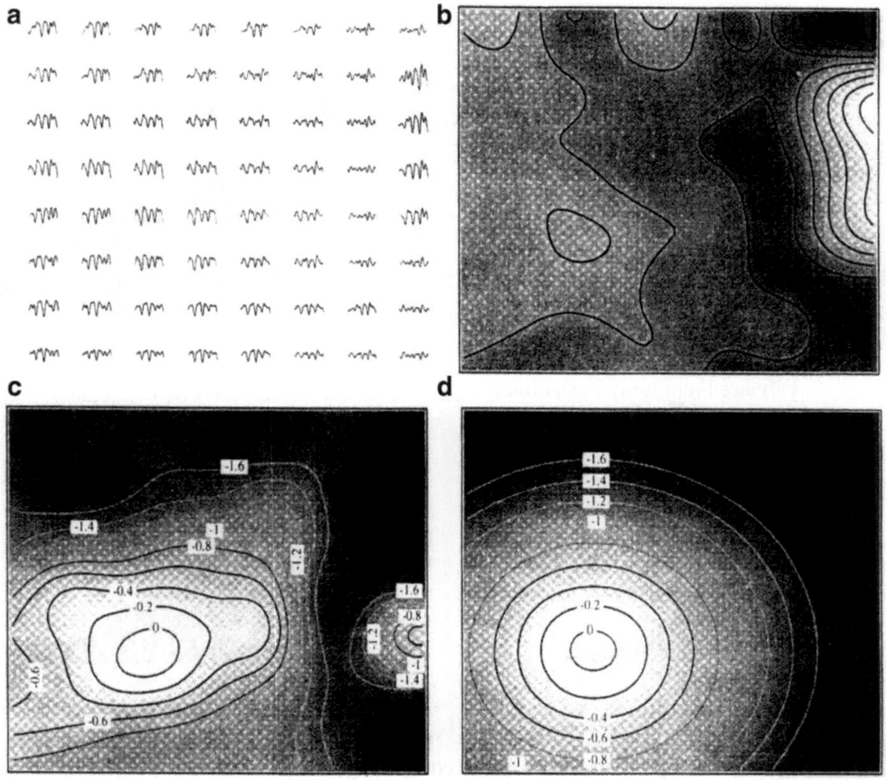

Fig. 9.5 (**a**) Representative example of one 128-ms EEG segment from visual cortex (Fig. 8.4b, VIS) illustrating amplitude modulation (AM) of the common wave form on 64 channels, temporally filtered (20–80-Hz band pass). (**b**) Pattern of root-mean-square (RMS) amplitude from (**a**). (**c**) Spatially filtered (0.03–0.5 cycles/mm band pass) phase distribution at 22 Hz from the FFT of (**a**) referenced to the phase of the spatial ensemble average. (**d**) Nonlinear regression of a conic surface onto the phase distribution in (**c**) indicating a phase gradient of 0.55 rad/mm, a phase velocity of 0.25 m/s, and a half-power diameter of 2.9 mm. The isophase contours in (**c, d**) were at intervals of 0.2 rad (11.4°) (From Fig. 3 in Freeman and Barrie 2000)

The 64 phase values of the PM pattern carrier wave were re-referenced to the phase at the frequency of the spatial ensemble average waveform. The phase surface from the 8x8 ECoG signals (Fig. 9.5c) was fitted with a phase cone by nonlinear regression (Sect. 8.4), which gave the location and sign of the apex and the slope of the cone (Fig. 9.5d). Then each burst could be conceived as manifesting a vector field (Sect. 6.2.2), in which the density of the activity at each of 64 points in time and space could be expressed as a complex number given by the analytic amplitude and phase, or more informatively the ECoG amplitude at each point and frequency and its rates of change in time (rad/ms) and space (rad/mm) and the velocity (mm/ms). The complex number gave the direction and velocity either away from (explosion •) or toward (implosion o) the apex (Fig. 8.9, Sect. 8.5) (Freeman and Barrie 2000).

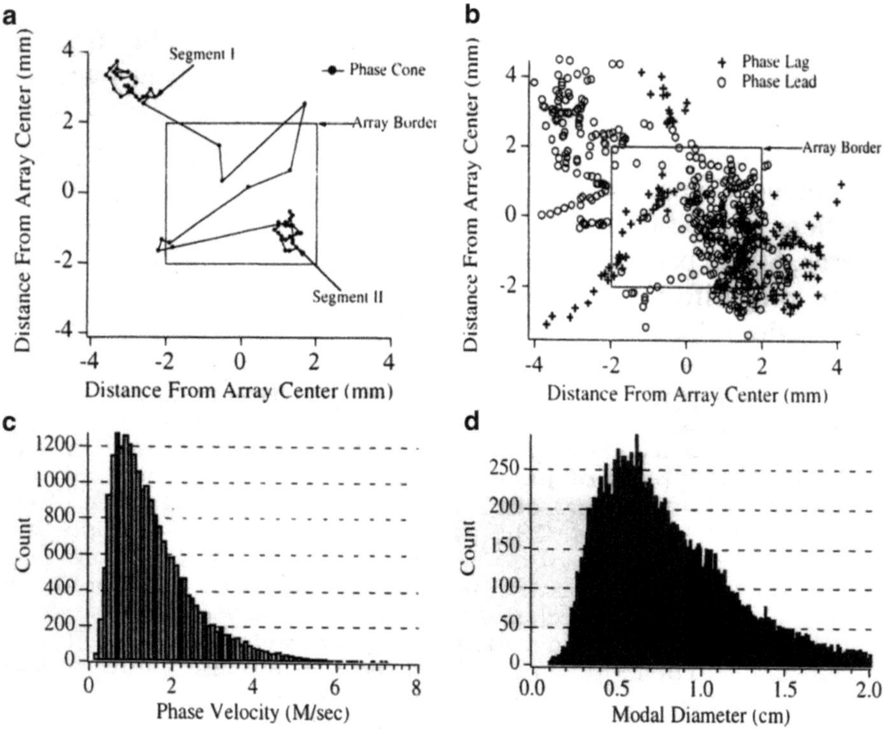

Fig. 9.6 (**a**) Calculating the successive phase cones at 2-ms steps revealed the trajectory of the neocortical dynamics by the location and sign of the apex. The rectangle shows the outline of the array on the cortical surface (Fig. 8.4b, Sect. 8.3). Clusters were revealed when the next apex, 2 ms later, was < 0.79 mm from the preceding apex and < 1.2 mm from the site of initial detection with no change in the sign (*lead* or *lag*). A transition between two clusters was labeled "Segment I" and "Segment II." (**b**) Locations of phase-cone apices were projected on the visual neocortical surface with the 64-channel recording array (*square*). Each point represented an apex from one phase cone at the peak frequency in a 128-ms moving window. (**c**) The phase velocities in m/s were calculated by dividing the carrier frequency in radians/s by the phase gradient in radians/m. (**d**) The half-power radii were calculated from the product of the phase velocity in m/s and 1/8 of the duration of the carrier wave length in s ($\pm\pi/4 = \pm 45°$ with $\cos = 0.707$) (From Fig. 5 in Freeman and Barrie 2000)

The duration of each burst was assessed by shifting the 128-ms window in 2-ms steps forward and backward from peak power along the 64 signals and fitting a phase cone to the phase surface at each digitizing step (Fig. 9.6a). The successive apices occurred in tight clusters, during which the location shifted by less than the interelectrode distance (0.79 mm, the limit of the spatial resolution). Between clusters, the apex shifted randomly over the surface (Fig. 9.6c). The duration of a phase cone was defined by the number points falling within a cluster multiplied by $\Delta t = 2$ ms. The ratio of carrier frequency (in rad/s) divided by the phase gradients (in m/rad) gave the phase velocity (in m/s, (Fig. 9.6c). As in the olfactory bulbar ECoG, the distributions of phase velocities were correlated with the conduction velocities of axons running parallel to the surface (Freeman and Barrie 2000).

The cone diameter in mm was estimated by the distance from the apex to the isophase contour of $\pi/8$, at which the cumulative phase dispersion reduced the power, $A^2(t)$, to one-half ($\cos \Phi(t) = 0.707$; $\cos^2 \Phi(t) = 0.500$) as in the bulb (Fig. 8.4b. VIS, Sect. 8.3). The results showed that the phase values from the FFT gave the burst parameters of carrier frequency, temporal and spatial location, duration, and diameter, and that these parameters did not contribute significantly to the burst classification with respect to CSs.

The burst amplitude waxed and waned in temporal modulation as already seen in allocortical bursts. The rate of change in spatial AM pattern during the temporal modulation was evaluated by calculating the Euclidean distance in 64-space, $D_e(t)$, between successive steps of the normalized pattern (zero mean, unit SD, Fig. 6.6, Sect. 6.4.2) and dividing the change in distance by the step duration, $D_e(t) = D_e(t)/\Delta t$. Low rates of change showed that the spatial AM patterns were invariant within the limits of experimental error over the burst durations despite the wide temporal modulation of amplitude (Fig. 6.5, Sect. 6.4.2). In other words, the ECoG had already converged to a stable AM pattern by the time following the discontinuity of a phase transition (Sect. 6.4.3) that the amplitude had risen sufficiently high for the pattern to be measured and classified. This observation is important for an explanation of the stubborn fact that the sign of the phase apex of the fitted cone varied randomly. A cortical pacemaker could explain phase lead at the apex but not phase lag. The implication is that the spatial AM pattern is selected at the singular moment of transition of the cortex from one basin of attraction to another (Fig. 6.14, Sect. 6.9) and the AM pattern can unfold either from the center or the periphery. An explanatory model has been devised using quantum field theory (Freeman and Vitiello 2006, 2008); it is beyond the scope of this book.

9.4.2 Phase Structure of the Background Activity

Well before the termination of each burst and phase cone, another overlapping burst emerged with differing frequency, conic apex, and amplitude distribution, which was not classifiable. When displayed in pairs of overlapping segments, the phase cones in sets of 40 trials appeared to occur at random, although spectral analysis showed that cones tended to recur preferentially at frame rates in the theta range (Fig. 7 in Freeman and Barrie 2000). More generally, every frequency component from the FFT had a phase surface, many but not all of which could be fitted with a cone. We attempted systematically to extract and chart more than two overlapping cones by decreasing the stepped window duration and thereby calculate the mean duration and diameter. The number of cones increased, and the means and SDs of the durations and diameters decreased in proportion to window duration (Fig. 9.7). The distributions with shorter windows were power law, and the means and SDs changed in proportion to window size, which implied that the phase cones were scale-free and perhaps fractal in the sense that the mean values varied in proportion to the window of measurement (Fig. 9.7). They manifested the origin of the

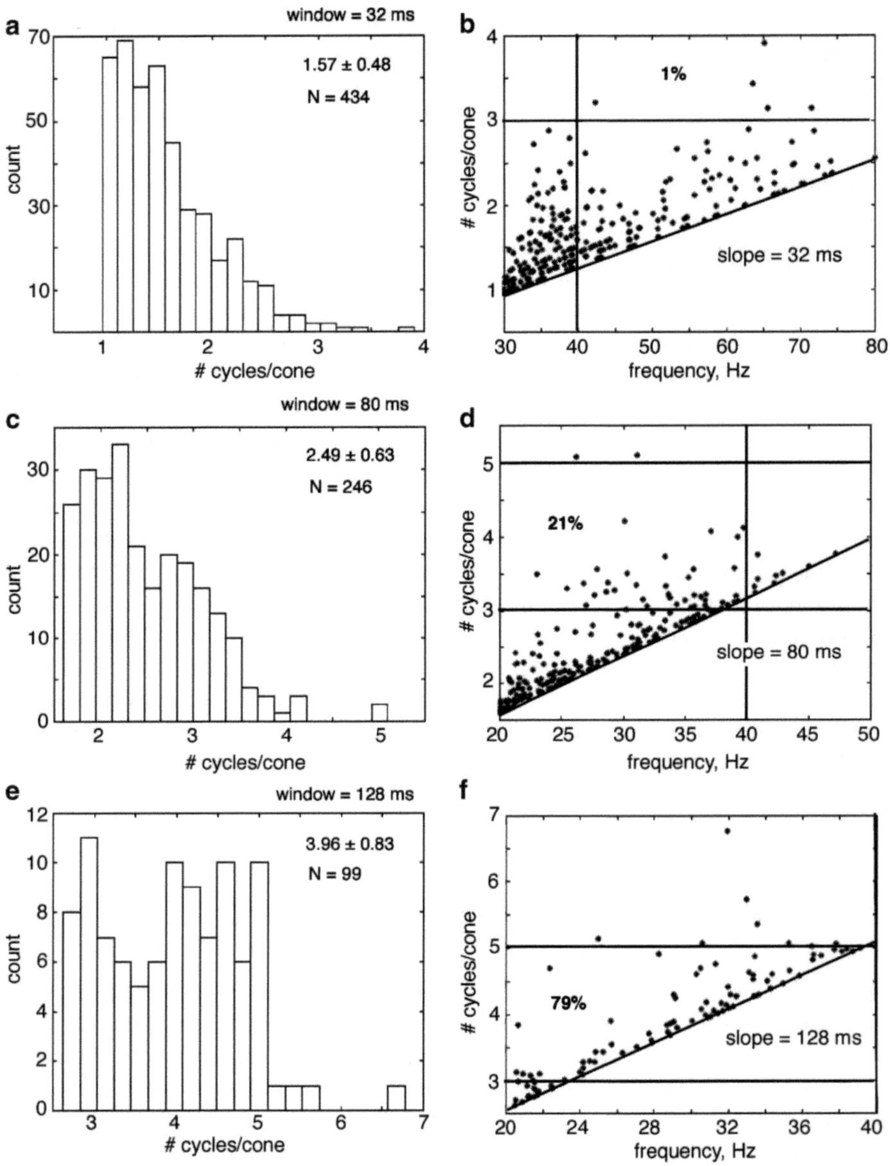

Fig. 9.7 (**a, c, e**) Histograms showed the number of cycles in each qualifying cone at the frequency determined by the peak power FFT of ECoG segments. The means and variances of the distributions varied with the window duration, which is typical of fractals. The distribution for short windows was power law, whereas that for long windows was not. The cones accompanying classifiable AM patterns had significantly larger durations (Fig. 8.4, Sect. 8.3). The short bursts were interpreted to be neural avalanches by which the cortex maintained itself at criticality. (**b, d, f**) Scatter plots showed the relation between the number of cycles and the frequency. Segments were excluded that had durations shorter than the sloping line, for which the units were c/c/s = the window duration. The % denoted the proportion of phase cones that had ≥ 3 cycles/cone. A preponderance of events exceeded the power-law distribution, showing that they were nonrandom and manifested broken symmetry (From Fig. 2.6 in Freeman 2004b)

background ECoG in mutual excitation (Sect. 6.7). In other words, the phase cones resembled the bubbles in a pan of boiling water holding itself at a critical temperature (Fig. 11.2, Sect. 11.4) or the avalanches (Plenz and Thiagaran 2007) of a sand pile holding itself at a critical angle in pseudo-criticality (Sect. 6.9). The bursts showed self-similarity across the beta-gamma frequency range, meaning that the selection of the carrier frequency was unconstrained by the CSs. The most significant variable was the number of cycles per cone at whatever carrier frequency. A window (32 ms) gave a power-law distribution with only 1% having more than three cycles (Fig. 9.7a). A long window (128 ms) gave relatively few bursts, but 79% of the bursts had 3–5 cycles per cone, which was comparable to the range of number of cycles found for classifiable AM patterns in allocortical ECoGs.

The search for fine phase structure was extended to the human ECoG from a patient undergoing presurgical evaluation for treatment of intractable seizures (Freeman et al. 2006a, b). The electrode array location was the right inferior temporal lobe (Fig. 10.4, Sect. 10.4). The size of the 1x1-cm array of 64 electrodes was optimized by spatial spectral analysis to determine the optimal Nyquist frequency (Fig. 7.1, Sect. 7.2), giving an interelectrode interval of 1.25 mm (vs. 0.79 mm in rabbit). The digitizing step was 5 ms (vs. 2 ms). Measurements of carrier frequencies and phase surfaces revealed phase cones both in resting and in working ECoG that had the same properties as those in the rabbit ECoG. The only significant differences were those in mean duration and diameter, which were correlated with the differences in sampling intervals of time (the digitizing rate) and space (the interelectrode spacing). We regarded them as evidence of the scale-free properties of the ECoG. In the working ECoG, the beta-gamma bursts carried spatial AM patterns that were correlated with higher cognitive function, described in Sect. 10.4. Hence, the comparison across species supported the concept that the neocortex in an expectant state holds itself in criticality (Sect. 6.9), giving self-organized, scale-free, self-similar background activity having myriads of microscopic phase cones. Upon initiation of an appropriate task to perform, the cortex springs into action through a phase transition (Kozma and Freeman 2002), by which it reorganizes itself from disorder to a coherent, ordered, oscillatory pattern, which we conceive as the retrieval of a macroscopic memory, which is initiated by the microscopic representation of a sensory stimulus that is delivered afferent action potentials that were evoked by the stimulus.

9.5 High-Resolution Images of Analytic Phase Between Bursts

Beginning with the respiratory cycle in olfaction (Fig. 8.3, Sect. 8.2), we have emphasized the cinematic sequences of bursts in frames at theta rates, by which the several sensory systems receive and process batches of sensory information. Each new burst has its carrier frequency and its AM and PM patterns, all usually though not invariably differing from those preceding. Each new frame forms by a macroscopic change in ECoG pattern that within a few milliseconds involves a large area, ranging from a few cm^2 up to and including the entire scalp (Sect. 10.5). The jump

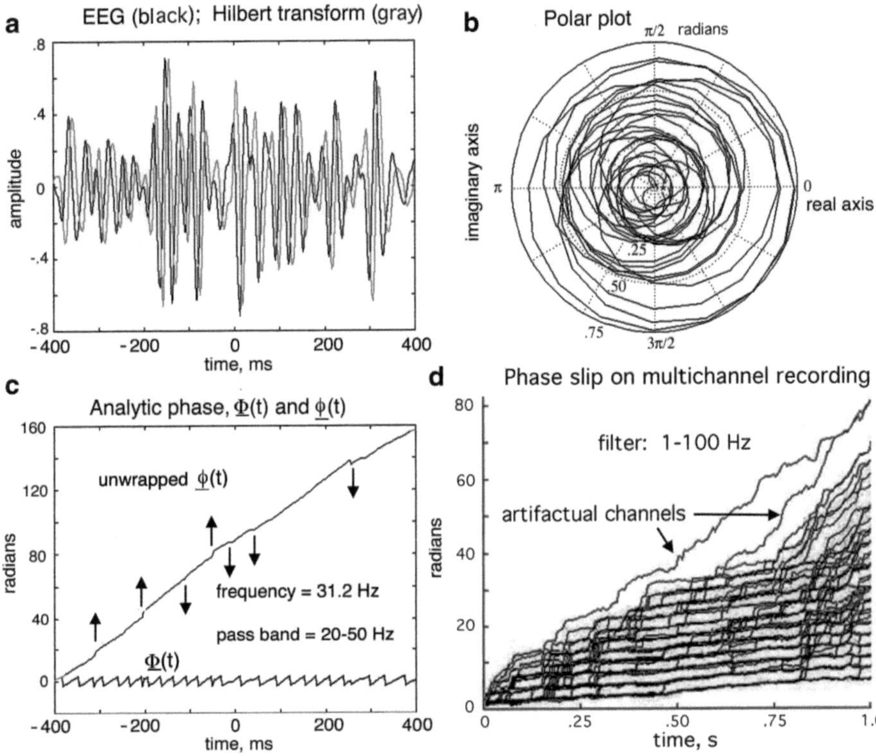

Fig. 9.8 (**a**) The spatial ensemble average of the filtered ECoG (real part, *black*) and its transform (imaginary part, *gray*) form a complex number which is a vector specifying the analytic signal (Chap. 4). The two components are in quadrature (Sect. 6.8). (**b**) The polar plot is shown of the real and imaginary parts. Time is implicit in counterclockwise rotation of the tip of the vector. (**c**) Sawtooth shows the analytic phase, $\phi(t)$, given by the arctangent of the angle of the vector with the real axis. The ramp shows the phase with reference to the arbitrary value at the start of the calculation 400 ms before the onset of a CS at time = 0. Arrows indicate temporal discontinuities in phase. (**d**) The problem of cumulative phase slip in multichannel recording is illustrated for very broadband filtering (From Fig. A1.3 in Freeman 2004a)

cannot be detected by recording action potentials, yet it takes place on the time scale of microscopic events. The Hilbert transform (Sect. 6.4.1; Freeman 2007) is used to describe frequency modulation (FM), so we adapted it to the study of ECoGs between bursts. We focus on the abrupt emergence of a narrowband burst of oscillation in what usually appears as a switch from one center frequency to another. The Hilbert transform can display that jump as a *phase slip* (Pikovsky et al. 2001), that is, a discontinuity in the analytic phase, $\varphi(t)$, that appears as a spike in the analytic frequency, $\omega(t)$ (Fig. 6.5d, Sect. 6.4).

An example of repetitive phase slips is shown in Figs. 9.8c and 9.9. The 64 visual cortical ECoGs in a 6-s segment were band pass filtered, and the spatial ensemble average was computed, giving the real part (Fig. 9.8a, *black curve*). The Hilbert

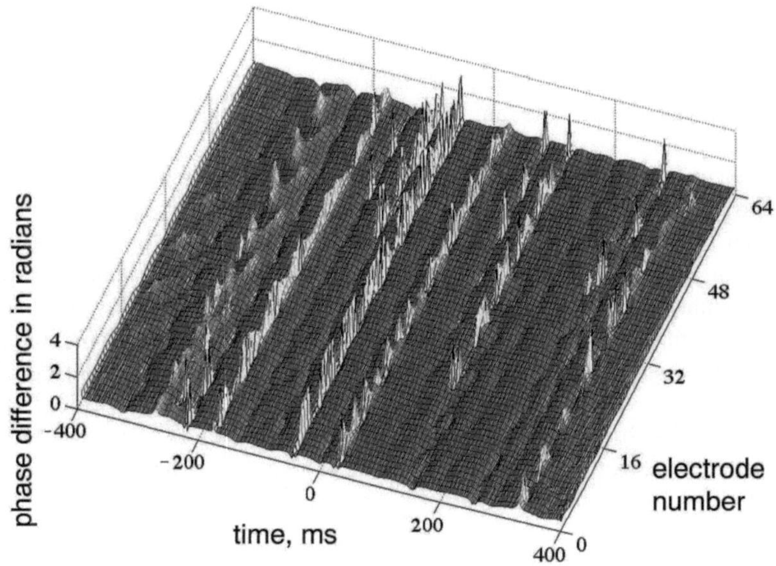

Fig. 9.9 The raster plot from rabbit neocortical ECoG showed the coordination of phase slip in the unwrapped analytic phase, Δ $\phi_j(t)$, changing with time (*left abscissa*) and channel (*right abscissa*). The phase differences from 8 columns of 8 rows were aligned to show near coincidence of sudden jumps and dips given by fast-forward and backward rotation of the vector tip displayed as the trajectory in Fig. 9.8b. Phase slip tended to coincide across the array at minima in the amplitudes of the filtered ECoG, except when a phase cone was swept across the array (−300 ms to −250 ms). Similar patterns were found in human scalp EEG. (Freeman et al. 2003) (From Fig. A1.4 in Freeman 2004a)

transform gave the imaginary part (a, *gray curve*) in quadrature. The two time series were displayed in polar coordinates (b) as the trajectory of a rotating vector. The length of the vector gave the analytic amplitude, $A(t)$, from the square root of the sum of squares of the two parts. The angle between the vector and the real axis gave the analytic phase, $\varphi(t)$, from the arctangent of the ratio of the imaginary to the real part. The raw phase was sawtooth (c). Each small ramp represented the rotation of the vector a half cycle counterclockwise about the origin of the polar plot. The breaks showed where the vector crossed the imaginary axis. The raw phase values were concatenated (*unwrapped*) into an ascending ramp by adding π radians at each break, when the phase dropped from $+\pi/2$ to $-\pi/2$, implementing the typical procedure of adding π rad whenever the phase dropped to 0 rad.

The slope of the ramp was equal to the peak frequency in rad/s from the FFT of the filtered signal of the same segment duration. The overall ramp mainly depended on the center frequency of the filter passband. An upward or downward jump of the ramp indicated either artifactual *phase slip* or a *phase transition*. The challenge was to distinguish them. A common cause of artifactual phase slip was leakage of power from adjacent passbands (Sect. 2.7), which caused interference between multiple overlapping signals. It tended to occur when an expected zero crossing failed to occur, which prevented a break and dropped the ramp from that time on by π rad (a low-frequency bias) or when jitter caused an extra zero crossing, which abruptly raised or lowered the ramp from that time on by π rad (high-frequency noise). The artifacts were enhanced by broadband filtering of an array of ECoGs (Fig. 9.8d). With the passage of time, the ramp splayed into a comb with the teeth separated by π radians.

The splay was minimized by narrowband filtering but that was only suitable when the passband was closely adapted to that of a specific carrier frequency. Since the carrier frequency jumped unpredictably between bursts, the passband had to be broad enough to encompass the expected range of variation (Freeman 2004a), so splay was unavoidable. Band-pass filtering by itself induced splay, whether the filtering was imposed by neural negative feedback interactions (Sect. 6.8) or by off-line data processing, regardless of passband width, because filtering imposed beats (Fig. 9.8a) at intervals 0.641 times the width of the passband (Rice 1950; Freeman 2009; Ruiz et al. 2010). During the minima in amplitude, the analytic phase was undefined and subject to errors of measurement (Fig. 9.9). The spatial variance of the analytic phase was also heightened by phase cones because the distribution of phase values was swept across each branch point in succession (Fig. 9.9: −300 ms to −250 ms). In order to construct accurate images of analytic phase surfaces and track long-term trajectories of multichannel analytic phase, the artifacts had to be removed by calculating the analytic phase reference of the spatial ensemble average of the multichannel EEG or ECoG and thereafter adding or subtracting π rad (Freeman 2007) to every slope value that lay outside the phase-cone distribution ($\pi/4$ rad) and likewise any jump that deviated from the mean ramp by $a \pm \pi$ rad displacement.

9.6 Cinematic Display of Temporal Dynamics of AM and PM Patterns

9.6.1 Vortices

The advent of high-rate digitizing, high-density cortical electrode arrays and high-speed data processing opened an exciting new frontier for ECoG and EEG imaging: cinematic display of fine textures (Kozma and Freeman 2008; Freeman and Vitiello 2009). The recent improvements in the use of filter banks and the Hilbert transform that minimize the artifacts and optimize feature extraction revealed novel spatiotemporal structures that were predicted by brain theory (Sect. 6.2.3 in Freeman 1975; Freeman and Vitiello 2010). Preliminary explorations using movies gave images of the filtered ECoG, which in perspective resembled the flapping of a flag in the wind. Close examination of successive images of *contour plots* of electric potential revealed hills (*white*) and valleys (*black*, Fig. 9.10), which showed propagating waves that had been indirectly manifested in phase cones. The sequence of the symbols ("+" for phase lead, "–" for phase lag) near the upper edge of the array showed a pulsating oscillatory focus that resembled the action of a jellyfish. The sequence

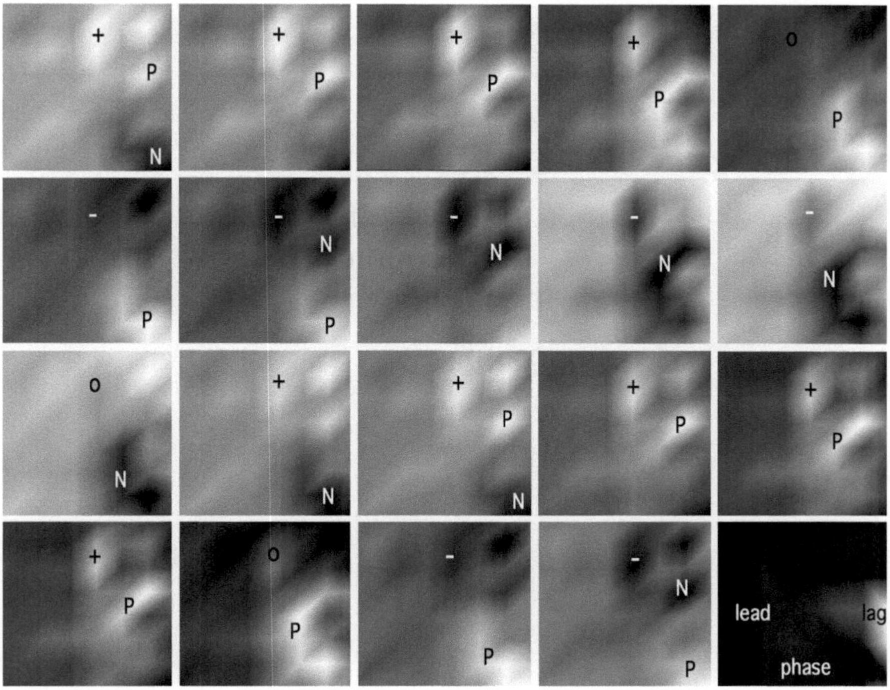

Fig. 9.10 Frames from a 5.6×5.6-mm (8×8) electrode array are shown in time steps of 4 ms of ECoG filtered in a passband 20–25 Hz. A stationary focus (+ o – o) oscillated in place with cycle duration near 48 ms. Another stationary focus rotated counterclockwise with cycle duration near 46 ms (P N P N). Each pattern persisted for several cycles, then terminated. Several independent phase-locked ECoG patterns commonly overlapped, giving the appearance of a pan of boiling water. For cinematic displays, see Freeman (2011) (From Fig. 4 in Freeman and Vitiello 2009)

of symbols near the right edge ("P" for positive = phase lead. "N" for negative = phase lag) showed a vortex with its center of rotation near the apex of a phase cone (lower right frame). The two events partially overlapped at different frequencies in the beta range. The frame intervals (4 ms, every other digitizing step) encompassed approximately 1.5 cycles from bursts lasting 2 to 3 times that long. The varieties of dynamic spatial patterns observed are schematized in Fig. 9.11. The patterns include rotation either clockwise or counterclockwise and pulsation either inwardly or outwardly without detectable rotation (Fig. 8.9 in Sect. 8.4).

On the one hand, the appearance of vortices might manifest optical illusions resembling interference patterns (moiré) produced by overlapping phase cones with different carrier frequencies and locations of their apices. On the other hand, vortices are predicted to accompany phase transitions in interactive media (Freeman and Vitiello 2010), so these volatile patterns might offer a fresh clue to the locally dense dynamics of neuropil mediated by ephapsis, gap junctions and chemical synapses. Vortices tend to form stable islets of energy that are embedded in turbulence. That property in cortex may help to explain the prolongation of classifiable bursts

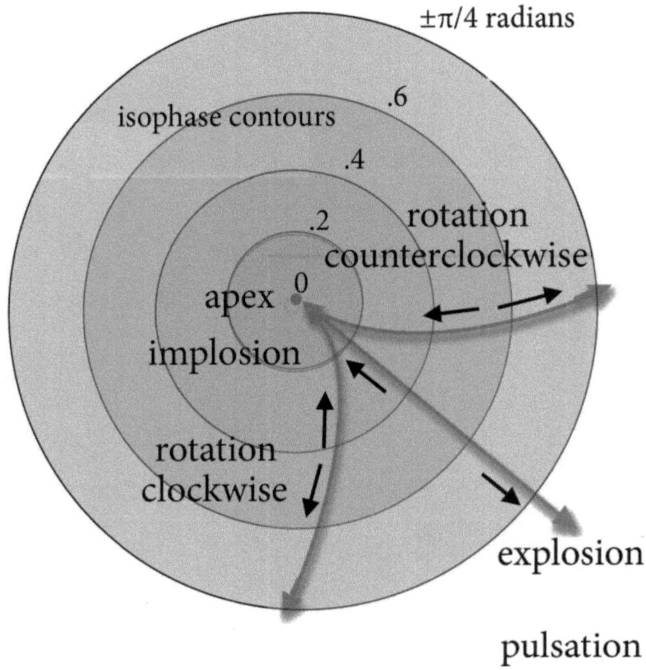

Fig. 9.11 The conic phase gradients were either inward (*implosion*) or outward (*explosion*) (Fig. 8.9, Sect. 8.4). They rotated clockwise or counterclockwise or pulsed without rotation. (From Fig. 5 in Freeman and Vitiello 2009)

(Fig. 9.7e, f). Both explanations (and possibly others) should be explored further by digitizing at much higher spatial and temporal rates in order to locate and track the centers of rotation as they drift across the cortex, followed by decomposition using temporal wavelets (Brockmeier et al. 2012) and arrays of spatial cones (Sect. 8.4) as templates.

9.6.2 Null Spikes

Cinematic displays of analytic amplitudes readily showed the smooth waxing and waning of the mean amplitudes of fixed AM patterns of bursts. Between bursts in the filtered ECoG, the analytic power on all channels appeared to approach zero in *beats* (Fig. 6.5c; Sect. 7.3). The modal values of beats occurred at intervals (Fig. 7.6, Sect. 7.4) that depended solely on the width of the passband (Freeman 2009; Ruiz et al. 2012), independently of the center frequency as predicted by Rice (1950). The analytic power did not go to zero but could go to extremely small values (Fig. 7.5b, d, Sect. 7.4). Brief decreases in analytic power called *null spikes* (Freeman 2009) appeared in the intervals between beats (Fig. 9.12c), either in clusters or singly (Fig. 7.6d). The optimal procedures for locating them were to search the

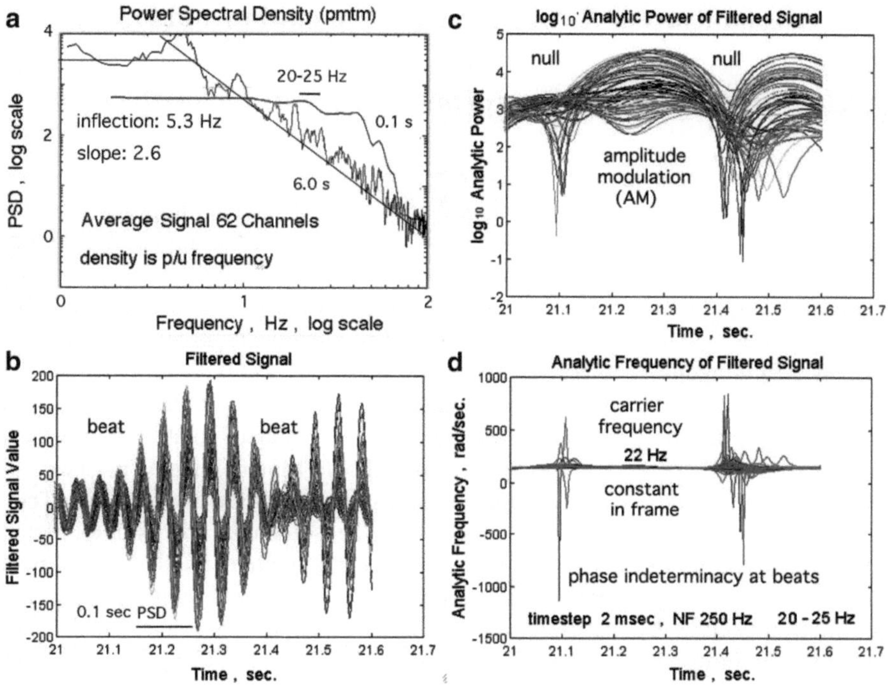

Fig. 9.12 (**a**) The long-term PSD_T (6 s) is compared with the short-term PSD_T (0.1 s) using the multitaper method (pmtm). (**b**) The 64 ECoGs were band-pass filtered (20–25 Hz). (**c**) The $\log_{10}$ of analytic power from the 64 signals revealed a brief downward dip in surface potential that was localized to a point in the array (Fig. 7.8d, Sect. 7.5). (**d**) The analytic frequency was indeterminate at minima in analytic amplitude, leading to the high variance accompanying null spikes. (From Fig. 3 in Freeman and Kozma 2010)

short-term FFT of the spatial ensemble average of an array of recordings, using a moving window 0.1 s in duration stepped at 0.05-s intervals in search of spectral peaks in the beta-gamma range (Fig. 9.12a), coupled with a peak in the theta range near the concave downward inflection of the PSD_T (Sect. 7.2) suggesting theta-gamma linkage (Sect. 8.6).

Initial guesses for the center frequency of a band-pass filter were set at a selected peak in the spectrum, and the bandwidth was set by dividing the accompanying theta peak frequency by 0.641 (Rice 1950). Initial guesses were improved by use of tuning curves (Fig. 9.4, Sect. 9.3a; Fig. 10.3a, b, Sect. 10.3). The widths of the isolated single spikes conformed to the point spread function (Fig. 7.7c) of a dendritic column in the neocortical ECoG, which indicated that the shape of the spikes was determined by volume conduction of a point source-sink in the cortex. The magnitude of the downward spikes varied widely, ranging downwardly from the local surface by as much as 10^{-6}. The null spikes were accompanied by localized discontinuities in the analytic phase, which were obscured by the loss of amplitude resolution. The extremely low values of amplitude of null spikes were predicted by the Rician distribution of analytic power (Rice 1950; Fig. 7.5b, d, Sect. 7.3).

9.7 Mechanisms of AM/PM Pattern Formation and Dissolution

9.7.1 Piecewise Linear Approximations for Attractor Dynamics

Examination of null spikes has been exploratory, because they occurred at, near, or beyond the limits of spatial and temporal resolution of the data imposed by sampling. The temporal and spatial location of null spikes varied randomly, seldom coinciding precisely with the location of an electrode or a digitizing step, so the variance was due to hit-or-miss search. The inference was drawn that the null spike might manifest a singularity in cortical dynamics. There were three other pieces of evidence for singularity in cortical dynamics besides the null spike. One was the discontinuity in analytic phase seen as phase slip (Fig. 9.8c). Another was the apex of the phase cone (Fig. 9.6a). Yet another was the center of rotation of a vortex (Fig. 9.10). In theory, an emergent phase cone should have collocated its apex and center of rotation at the site of a preceding null spike and phase discontinuity. They seldom coincided. A salient reason was that the measurements are distorted by the overlap of multiple events in the volume conductor. The grand challenge now is to bring these strands together by improvements in the spatial and temporal sampling rates and especially by the design of new basis functions and wave functions as templates for informed decomposition of the raw ECoG and EEG e.g., (Brockmeier et al. 2012). Decomposition using sophisticated wave functions and spatial templates based on learned feature vectors for AM patterns and phase cones for PM patterns may suffice to test this prediction.

What makes singularity important is the possibility of a role in the phase transition. In a physical model of dissipative cortical dynamics, the phase transition has the form of spontaneous breakdown of symmetry (SBS, Freeman and Vitiello 2006). The symmetry appears in sensory cortices in the unstructured, random background activity of unit and ECoG activity. SBS appears as a change in the cortical response to impulse input from decay (to the pre-stimulus background, Fig. 6.12, Sect. 6.7) to diverging oscillation simulated in 8.8b, Sect. 8.4). In an engineering model (KII, Sect. 8.1) using linear feedback control theory, the SBS is by transfer of cortical self-regulation from a chaotic (approximated by a point) attractor to a limit cycle attractor (Fig. 9.13).[4] Modeling the process of SBS with attractor dynamics required the construction and solution of a set of differential equations. The simplest model (KII set) that sufficed to represent the neurodynamics of SBS was a laminated population of excitatory neurons interacting with inhibitory neurons. It was devised to model allocortical dynamics (Chap. 6 in Freeman 1975; Kozma and Freeman 2001) and may now serve to model neocortical population dynamics as well.

[4] The stability of the 1/f background noise set by the nonlinear feedback gain of interactions among excitatory neurons (Fig. 6.10, Sect. 6.6) is governed by a non-convergent attractor (Principe et al. 2001) sometimes called a strange or chaotic attractor (Skarda and Freeman 1987). The representation is by use of stochastic differential equations (Kozma and Freeman 2001). In a piecewise linear model, the random fluctuations are removed by averaging, and modeling is by ordinary differential equations. The non-convergent attractor is replaced by a point attractor.

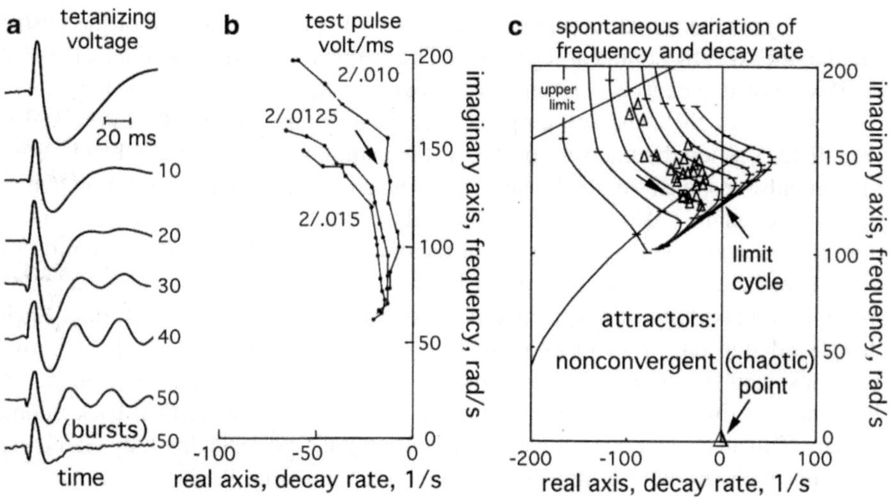

Fig. 9.13 This figure is a condensed representation of cortical piecewise linear dynamics. The purpose is graphic display of the point and limit cycle attractors that govern the phase transition in perception. Full exposition of KII dynamics is in Freeman (1975) (**a**) Artificially increasing the cortical background activity by tetanizing the LOT (200 pulses/s, 10–50 v, .01 ms pulse duration) increased the frequency and decay rate (ω, α) of the impulse response (averaged evoked potential (AEP)) to a fixed test pulse. Tetanizing replicates the sensitizing effect of increased background activity with arousal as expressed in *Qm* (Fig. 6.10, Sect. 6.5). (**b**) The root loci are shown in the upper half of the complex plane, s = a ± jw for three test inputs. (**c**) The symbols (Δ) show the spontaneous variation of the AEP with fixed stimulus intensity. The symbols are superimposed on curves showing a family of root loci from solutions of the set of differential equations that simulate the relation between ω and α with changes in signal-to-noise ratio. The arrow shows the direction of decreased negative feedback gain, k_n, with increased amplitude and/or decreased background noise. The tick marks show the effects of differing values of forward excitatory or inhibitory loop gains, k_{ee} and k_{ii}. Displacement of curves to the right with increased arousal (Q_m in Fig. 6.10, Sect. 6.6) and/or with Hebbian learning (Fig. 8.8, Sect. 8.4) forced the root loci into the right half of the complex plane with $\alpha>0$. The root loci imply that it is the null spike (Fig. 7.8d, Sect. 7.5) that reverses the sign of the decay rate of the impulse response and precipitates a phase transition. (From Fig. 6.21a in Chap. 6, Freeman 1975)

Cortical dynamics was viewed as intrinsically nonlinear owing to the thresholds and refractory periods of the axons (Fig. 6.10, Sect. 6.6), but in the background state, the cortex was conceived as holding itself in a small-signal, near-linear range (Sect. 6.5). The range was revealed by the amplitude distribution of the ECoG (Fig. 6.4, Sect. 6.3). In that range, the solutions of the equations for impulse input (Fig. 6.9) were obtained by piecewise linear approximation. The impulse responses to electrical stimulation (Fig. 9.13a) were ensemble averages of evoked potentials collected by repeated stimulation (typically $N=100$) sufficient to flatten the pre-stimulus baseline in the average. The averaged evoked potentials (AEP) were fitted with a sum of linear basis functions that included a damped cosine, which evaluated the frequency of oscillation, ω, and the exponential rate of change in the envelope, α (Fig. 6.13, Sect. 6.7):

$$v(t) = v_0 \cos(\omega t + \phi) \exp(\alpha t), \tag{9.1}$$

where v_0 gave the initial amplitude and φ gave the phase in radians of the cosine fitted to the AEP by nonlinear regression. The frequencies and decay rates of sets of AEP were displayed as experimental root loci (Fig. 9.13b).

In pursuit of SBS, the key property was the spontaneous variation in ω and α (Fig. 9.13c), when the intensity of the evoking stimulus was fixed at a low level to give impulse responses in the linear range. The cluster of ω and α values plotted in the complex plane ("Δ") appeared to be random, but factor analysis of the ω and α matrix (Freeman 1964) gave two main factors that were displayed in a family of bow-shaped curves. One factor was the level of arousal, which was expressed by the asymptotic maximum of normalized pulse density, Q_m (Fig. 6.10, Sect. 6.6). Each downward step on a root locus represented the effect of decreasing the negative feedback gain, k_n, with reduced ω and α, hence tending toward destabilization. Evidence for this interpretation was obtained by reducing the cortical background activity with an anesthetic (intravenous pentobarbital) and replacing it by tetanizing the input pathway with a high-frequency shock train (Fig. 9.13a). Increasing the artificial background activity replicated the form of the bow-shaped root loci for three amplitudes of test shock. The second factor was fluctuations in the signal-to-noise ratio from the beats (Sect. 7.3) in the ECoG caused by the distribution of characteristic frequencies about the center carrier frequency (Figs. 6.5d; 9.12d).

The rightward shift was replicated by increasing the strength of the Hebbian synapse between excitatory neurons that formed a Hebbian assembly. Evidence for this interpretation was obtained by measuring the changes in the impulse response when the subjects were trained to respond to the test pulse as a conditioned stimulus (Emery and Freeman 1969). The training strengthened the re-excitation of coupled neurons and sensitized the cortex selectively to the reinforced CS + (here the electric shock). By this analysis, a phase transition could occur only if the subject was aroused, had a preformed Hebbian assembly, actually received the appropriate CS, and was primed by vanishing of the order parameter in a downbeat. These factors can be thought of as safeguards that preclude phase transitions precipitated by noise.

9.7.2 A Possible Role for Singularity in Perception

A role for singularity in a phase transition was predicted by extrapolation of the family of root loci into the right half of the complex plane, which was inaccessible to observation using evoked potentials. Each evoked potential that decayed to the background level designated an operating point in the left half of the complex plane controlled by the point attractor at the origin of the complex plane. Any event with its operating point in the right half failed to converge to the point attractor. It diverges into an expanding oscillation (Fig. 8.8b, Sect. 8.4), after the sign of the exponent governing the envelope of the oscillatory impulse response changed from negative (decaying) to positive (regenerative). Calculation of the KII root loci predicted that the operating point would converge to the imaginary axis, in other words, to a limit cycle attractor. Approach of the operating point to the attractor would give infinite amplitude, that is, singularity. However, the amplitude was bounded by the refractory periods (Fig. 6.10, Sect. 6.6). The conjecture is proposed instead that it is the density that increases geometrically (Sect. 6.2.2) resembling condensation of a gas to a liquid. The increase in density would engage all neurons in an active focus non-exclusively in the spectral range of the carrier. At maximal density, the entire body of knowledge embedded in the synaptic web of the neuropil might be brought to the interactions. In this scenario, it is the high density of activity in the primary sensory cortices that can support the rich context that comes with recall and recognition of a memory evoked by a stimulus.

What is still unresolved is the neural mechanism by which the transmitted neocortical information is further processed. It was proposed in Sect. 8.6 that the macroscopic activity pattern of the recalled memory was transmitted from the bulb to the prepyriform cortex, where it was down-sampled to the microscopic level and transmitted by divergent pathways to diverse targets in the cerebral hemispheres. That interpretation was invoked because the cortical ECoG phase gradient conformed to the properties of the input path (LOT) with no evidence for self-organization. The olfactory precedent suggested that primary sensory neocortices may transmit perceptual information in nonlocal form by massive clouds of action potentials to neighboring cortical or striatal regions, where it might similarly be locally integrated for sampling and coarse-graining and returned to the microscopic level of single neuron firing. Whether it occurs in unimodal neocortical processing, and if so where, remains to be determined. Evidence is presented in Chap. 10 that it does occur in multimodal neocortical processing in the entorhinal cortex. The coding might be detected in the firing of *concept cells* (Quian Quiroga 2012) (Sects. 8.6, 10.3, 11.5). Testing this will require measurements of ECoG AM and PM patterns in conjunction with unit recording, especially the beta and gamma phase gradients.

To recapitulate, a perceptual action begins with an act of observation accompanied by activation of an attractor landscape that serves to predict the sensory consequences of the action in the form of Bayesian priors. Reception of a CS ignites the relevant Hebbian assembly, which provides the energy needed to increase cortical response amplitude and shift the operating point to a positive exponential. The reversal of sign shifts control from a point attractor to a limit cycle attractor. In a linear model, the point attractor at the center of a non-convergent "chaotic" attractor (Footnote 4, Sect. 9.7.1) is symbolized by a pole at the origin of the complex plane ($\omega = 0$, $\alpha = 0$), and the limit cycle attractor is represented by a complex conjugate pair of poles on the imaginary axis (Fig. 9.13c). The assembly directs the cortex to the basin of an attractor, which imposes the AM pattern on the carrier frequency for transmission.

Thereafter, the termination of the AM pattern requires an exit strategy because each current pattern must evaporate in order to make way for the next. The ECoG data suggest a simple mechanism: the null spike (Fig. 9.12b). Simulations show that after a downbeat in the sum of outputs of a collection of *independent* oscillators, the oscillation resumes with no change in AM and PM patterns. However, we postulate that if the oscillation is sustained by interactions among a collection of *coupled* oscillators, then the null spike might initiate irreversible erasure of the AM and PM patterns. In other words, if as already proposed (Sects. 6.4.3 and 9.3.2) the vectorial order parameter He(t) and the analytic power, $\mathbf{A}^2(t)$, both go to zero, then during a downbeat the coupling also goes to zero. The AM pattern conceivably evaporates, leaving a learned synaptic residue of change and the neurons free for capture in the next perceptual event. The termination of every burst is intrinsic by virtue of the dispersion of the carrier frequencies within the population (Fig. 6.5d, Sect. 6.4.1) that create the downbeats by interference. A possible role worthy of exploration is that it the null spike accesses a singularity, a nucleus of condensation perhaps comparable to a grain of dust in the formation of a raindrop. As already mentioned, the supporting evidence would require showing whether the site of the apex of a phase cone collocates with the preceding null spike.

9.8 Summary

As in allocortex, the ECoG evidence shows that spatiotemporal structures emerge in the neocortical activity by spontaneous breakdown of symmetry of the random background activity sustained by mutual excitation among pyramidal cells. The ECoG spectrum has the form of black noise, characterized by a power-law relation of log power to log frequency that is steeper than expected for random (brown) noise, owing to the mechanism of stabilization of the activity by refractory periods.

The steady state ECoG contains multiple overlapping bursts of oscillation, each with a spatial pattern of phase in the form of a cone. The distributions of cone durations are power law and fractal, with the exception of the largest and longest lasting bursts. These prolonged bursts carry AM patterns that are classifiable with respect to CSs. They serve to distinguish the resting ECoG from the working ECoG. We interpret the background bursts as evidence that neocortex maintains itself in a symmetric state of self-organized criticality. The myriad small phase cones manifest neural avalanches, which we interpret as responses to background noise input to neocortex that reflect governance by a point attractor.

The analytic phase from the Hilbert transform of broadband ECoG confirms the stationarity of the carrier frequency and the conic phase pattern carried by bursts. The analytic amplitude from the Hilbert transform of narrowband ECoG confirms the stationarity of the classifiable AM patterns expressed as feature vectors, $\underline{A}^2(t)$. The feature vector is adopted as an order parameter. The ratio of mean analytic power divided by the rate of change in spatial pattern, $H_e(t) = \underline{A}^2(t)/D_e(t)$, provides a scalar index, the pragmatic information, with which to evaluate optimal AM pattern classification. The analytic frequency phase, $\varphi(t)$, and frequency, $\omega(t)$, serve to define the spatial location, size, time of onset, and duration of classifiable AM patterns.

What is most remarkable is that the fine-grain search for neocortical events came to focus on two bursts in the CS-CR interval: a gamma burst closely following CS onset and a beta burst preceding the CR. The finding suggests that in well-trained animals a single gamma burst suffices for CS recognition, and a single beta burst suffices for CR selection. The rarity of classifiable bursts, the variation in their carrier frequency, and their embedding in neural avalanches explain the difficulty of finding them.

Refinements in calculating the analytic signal of multiple channels made possible the cinematic display of AM and PM patterns. The method opened investigation of the spatiotemporal dynamics by which the patterns form and dissipate. Of special interest was the appearance of vortices that suggested field effects and null spikes that suggested singularities in phase transitions. Definitive study of these events will require up to tenfold increases in spatial and temporal sampling rates of ECoG and EEG without sacrifice of aperture size.

References

Atmanspacher H, Scheingraber H (1990) Pragmatic information and dynamical instabilities in a multimode continuous-wave dye laser. Can J Phys 68:728–737

Barrie JM, Freeman WJ, Lenhart M (1996) Modulation by discriminative training of spatial patterns of gamma EEG amplitude and phase in neocortex of rabbits. J Neurophysiol 76:520–539

Basar E (ed) (1998) Brain function and oscillations. Vol. 1: Principles and approaches. Volume II: Integrative brain function. Neurophysiology and cognitive processes. Springer Series in Synergetics, Berlin

Brockmeier AJ, Hazrati MK, Freeman WJ, Principe J (2012) Locating spatial patterns of waveforms during sensory perception in scalp EEG. Proc IEEE Conf EMBS #2098, San Diego, 28 Aug–01 Sept. http://cnel.ufl.edu/files/1340860427.pdf

Buzsaki G (2006) Rhythms of the brain. Oxford University Press, Oxford UK

Canolty RT, Edwards E, Dalal SS, Soltani M et al (2006) High gamma power is phase-locked to theta oscillations in human neocortex. Science 313:1626–1628

Chrobak JJ, Buzsáki G (1998) Gamma oscillations in the entorhinal cortex of the freely behaving rat. J Neurosci 18(1):388–399

Emery JD, Freeman WJ (1969) Pattern analysis of cortical evoked potential parameters during attention changes. Physiol Behav 4:67–77. http://soma.berkeley.edu/archives/IID8/69.html

Fell J, Klaver P, Elfadil H, Schaller C, Elger CE, Fernández G (2003) Rhinal–hippocampal theta coherence during declarative memory formation: interaction with gamma synchronization? Eur J Neurosci 17(5):1082–1088

Freeman WJ (1964) Correlation of goal-directed work with sensory cortical excitability. Recent Adv Biol Psychiatry 7:243–250. http://soma.berkeley.edu/archives/IIC1/64.html

Freeman WJ (1975) Mass action in the nervous system. Examination of the neurophysiological basis of adaptive behavior through the EEG. Academic Press, New York. Posted in e-formats (2004) http://sulcus.berkeley.edu/MANSWWW/MANSWWW.html

Freeman WJ (2004a) Origin, structure, and role of background EEG activity. Part 1. Analytic amplitude. Clin Neurophysiol 115:2077–2088. http://repositories.cdlib.org/postprints/1006

Freeman WJ (2004b) Origin, structure, and role of background EEG activity. Part 2. Analytic phase. Clin Neurophysiol 115:2089–2107. http://repositories.cdlib.org/postprints/1486

Freeman WJ (2005a) Origin, structure, and role of background EEG activity. Part 3. Neural frame classification. Clin Neurophysiol 116(5):1118–1129

Freeman WJ (2005b) NDN, volume transmission, and self-organization in brain dynamics. J Integr Neurosci 4(4):407–421

Freeman WJ (2006a) Origin, structure, and role of background EEG activity. Part 4. Neural frame simulation. Clin Neurophysiol 117(3):572–589

Freeman WJ (2006b) A cinematographic hypothesis of cortical dynamics in perception. In: Karakis S, Basar E (eds.) Intern J Psychophysiol 60(2):149–161

Freeman WJ (2007) Hilbert transform for brain waves. Scholarpedia 2(1):1338. http://www.scholarpedia.org/article/Hilbert_transform_for brain_waves

Freeman WJ (2009) Deep analysis of perception through dynamic structures that emerge in cortical activity from self-regulated noise. Cognitive Neurodynamics 3(1):105–116. doi:10.1007/s11571-009-9075-3

Freeman (2011) http://soma.berkeley.edu/videos/?video=6

Freeman WJ, Baird B (1987) Relation of olfactory EEG to behavior: Spatial analysis. Behav Neurosci 101:393–408 http://soma.berkeley.edu/articles/IIIF8.BehavNeuro87SpatialAnal-OriginalScanBitmap.pdf

Freeman WJ, Barrie JM (2000) Analysis of spatial patterns of phase in neocortical gamma EEGs in rabbit. J Neurophysiol 84:1266–1278. http://soma.berkeley.edu/articles/BC.Phase.Cones.pdf

Freeman WJ, Burke BC (2003) A neurobiological theory of meaning in perception. Part 4. Multicortical patterns of amplitude modulation in gamma EEG. Int J Bifurc Chaos 13: 2857–2866. http://repositories.cdlib.org/postprints/3345

Freeman WJ, Burke BC, Holmes MD (2003) Aperiodic phase re-setting in scalp EEG of beta-gamma oscillations by state transitions at alpha-theta rates. Hum Brain Mapp 19(4):248–272. http://repositories.cdlib.org/postprints/3347

Freeman WJ, Grajski KA (1987) Relation of olfactory EEG to behavior: factor analysis. Behav Neurosci 101:766–777

Freeman WJ, Holmes MD, West GA, Vanhatalo S (2006a) Fine spatiotemporal structure of phase in human intracranial EEG. Clin. Neurophysiol. 117(6): 1228–1243. http://soma.berkeley.edu/articles/EJ%20CLINPH02-28-06%20txt-fig.pdf

Freeman WJ, Holmes MD, West GA, Vanhatalo S (2006b) Dynamics of human neocortex that optimizes its stability and flexibility. Intern J Intelligent Syst 21:1–21. http://repositories.cdlib.org/postprints/2385

Freeman WJ, Kozma R (2010) Freeman's mass action. Scholarpedia 5(1):8040. http://www.scholarpedia.org/article/Freeman%27s_mass_action http://soma.berkeley.edu/videos/?video=6

Freeman WJ, Vitiello G (2006) Nonlinear brain dynamics as macroscopic manifestation of underlying many-body field dynamics. Phys Life Rev 3:93–118. http://repositories.cdlib.org/postprints/1515

Freeman WJ, Vitiello G (2008) Dissipation and spontaneous symmetry breaking in brain dynamics. J Phys A: Math Theor 41:304042. doi:10.1088/1751-8113/41/30/304042. http://repositories.cdlib.org/postprints/3385

Freeman WJ, Vitiello G (2009) Dissipative neurodynamics in perception forms cortical patterns that are stabilized by vortices. J Physics Conf Series **174**(2009)012011. http://repositories.cdlib.org/postprints/3379

Freeman WJ, Vitiello G (2010) Vortices in brain waves. Intern J Modern Physics B 24(17): 3269–3295. http://dx.doi.org/10.1142/S0217979210056025

Gao X, Xu D, Cheng M, Gao S (2003) A BCI-based environmental controller for the motion-disabled. IEEE Trans Rehab Eng 11(2):137–140

Huang NE, Shen Z, Long SR, Wu MC, Shih HH, Zheng Q, Yen N-C, Tung CC, Liu HH (1998) The empirical mode decomposition and the Hilbert spectrum for nonlinear and non-stationary time series analysis. Proc R Soc Lond 454:903–995

Kitzbichler MG, Smith ML, Christensen SR, Bullmore E (2009) Broadband criticality of human brain network synchronization. PLoS Comput Biol 5(3):e1000314. doi:10.1371/journal.pcbi.1000314

Kozma R, Freeman WJ (2001) Chaotic resonance: methods and applications for robust classification of noisy and variable patterns. Intern J Bifurc Chaos 10:2307–2322

Kozma R, Freeman WJ (2002) Classification of EEG patterns using nonlinear dynamics and identifying chaotic phase transitions. Neurocomputing 44:1107–1112

Kozma R, Freeman WJ (2008) Intermittent spatio-temporal de-synchronization and sequenced synchrony in ECoG signals. Special Issue: Synchronization in Complex Networks, Suykens J, Osipov G [eds]. Chaos 18, 037131.

Kozma R, Freeman WJ, Erdí P (2003) The KIV model – nonlinear spatio-temporal dynamics of the primordial vertebrate forebrain. Neurocomputing 52:819–826. http://repositories.cdlib.org/postprints/1049

Kozma R, Huntsberger T, Aghazarian H, Tunstel E, Ilin R, Freeman WJ (2008) Intentional Control for Planetary Rover SRR2k. Advanced Robotics 21(8):1109–1127

Lehmann D, Pascual-Marqui RD, Michel M (2009) EEG Microstates. Scholarpedia 4(3):7632. doi:10.4249/scholarpedia.7632

Le Van Quyen M, Foucher J, Lachaux J-P, Rodriguez E, Lutz A, Martinerie J, Varela F (2001) Comparison of Hilbert transform and wavelet methods for the analysis of neuronal synchrony. J Neurosci Meth 111:83–98

Lisman J (2005) The theta/gamma discrete phase code occurring during the hippocampal phase precession may be a more general brain coding scheme. Hippocampus 15(7):913–922

Ohl FW, Deliano M, Scheich H, Freeman WJ (2003) Early and late patterns of stimulus-related activity in auditory cortex of trained animals. Biol Cybernetics. doi:10.1007/s00422-002-0389-z, 6 pages

Pikovsky A, Rosenblum M, Kurths J (2001) Synchronization — A universal concept in non-linear sciences. Cambridge University Press, Cambridge

Plenz D, Thiagaran TC (2007) The organizing principles of neural avalanches: cell assemblies in the cortex. Trends Neurosci 30:101–110

Pribram KH (1991) Brain and perception: holonomy and structure in figural processing. Lawrence Erlbaum Assoc, Mahwah NJ

Pribram KH, King JS, Pierce TW, Warren A (1998) Some methods for dynamic analysis of the scalp recorded EEG. Brain Topography 8(4):367–378

Principe JC, Tavares VG, Harris JG, Freeman WJ (2001) Design and implementation of a biologically realistic olfactory cortex in analog VLSI. Proceedings IEEE 89:1030–1051

Quian Quiroga R, Kraskov A, Kreuz T, Grassberger P (2002) Performance of different synchronization measures in real data: a case study on electroencephalographic signals. Phys Rev 6504:U645-U6 58 – art. no. 041903

Quian Quiroga R (2012) Concept cells: The building blocks of declarative memory functions. Nature Reviews Neuroscience 13:587–597. doi: 10.1038/nrn3251

Rice SO (1950) Mathematical Analysis of Random Noise - and Appendixes – Technical Publications Monograph B-1589. Bell Telephone Labs Inc., New York

Ruiz Y, Pockett S, Freeman WJ, Gonzales E, Guang L (2010) A method to study global spatial patterns related to sensory perception in scalp EEG. J Neurosci Methods 191:110–118. doi:10.1016/j.jneumeth.2010.05.021

Ruiz Y, Li G, Gonzalez E, Freeman WJ (2012) A new approach to detect and study spatial–temporal intracranial EEG frames. Digital Signal Processing 22:133–139. doi:10.1016/j.dsp.2011.09

Sammon JW (1969) A nonlinear mapping for data structure analysis. IEEE Trans Comput C-18:401–409

Schack B, Vath N, Petsche H, Geissler H-G, Möller E (2002) Phase-coupling of theta–gamma EEG rhythms during short-term memory processing. Int J Psychophysiol 44(2):143–163

Schroeder M (1991) Fractals, chaos, power laws. Minutes from an infinite paradise. WH Freeman, San Francisco

Skarda CA, Freeman WJ (1987) How brains make chaos in order to make sense of the world. Behavioral & Brain Sci 10:161–195

Squires LR (1992) Memory and the hippocampus: a synthesis from findings with rats, monkeys, and humans. Psychol Rev 99(2):195–231. doi:10.1037/0033-295X.99.2.195

Chapter 10
ECoG and EEG Images in Higher Cognition

10.1 Introduction

In the preceding two chapters, the challenge has been to understand the transition from sensation to perception. The problem and the solution could be remarkably similar in all sense modalities. The stimuli are microscopic forms of diverse energies in the form of molecules, photons, vibrations, and phonons, which are captured by molecular structures embedded in cilia of neurons in the eye, ear, nose, and skin. Cilia are macromolecular threads that extend from or through the individual receptor neurons. The cilia selectively transduce and amplify the incident energies by supplying their own metabolic energies and expressing them in ionic currents. The receptor cells convert the currents to pulse trains and transmit the microscopic information by pulse frequency modulation in proportion to transmembrane current density.

At the behavioral level, macroscopically the brain samples the fields of energy surrounding the body by repetitive actions: the sniff, saccade, whisk, and harkening, thereby intermittently offering sensory information to the receptors, which send it to the sensory cortices. The macroscopic searching actions parse the sensory input but do not provide the spatial structure of the resulting percepts. In this respect, the sensory systems resemble the digestive system. Foodstuffs of all kinds are broken down into their molecular and ionic components before absorption and then reconstructed in forms that are at once immunologically unique to each person and yet common to other persons in physiological function, hence our metaphors for thinking: chew, digest, and ruminate. The microscopic sensory receptors act as point processors that disintegrate ("shatter" according to Skarda 1999) the sensory input into bits. When transmitted to and received by cortex, the bits may ignite a Hebbian assembly (Fig. 8.8) that directs the cortical trajectory into a basin of attraction (Fig. 6.14). Convergence to the attractor results in construction of an AM pattern that is created by neural interactions through modified synapses, and that expresses the memory elicited by the stimulus information. This basic physiological process is obvious in olfaction, in which the desired result of sampling the environment is an odor, that is, the knowledge about an odorant and not its molecular features. Is it safe

W.J. Freeman and R. Quian Quiroga, *Imaging Brain Function With EEG: Advanced Temporal and Spatial Analysis of Electroencephalographic Signals*,
DOI 10.1007/978-1-4614-4984-3_10, © Springer Science+Business Media New York 2013

to eat? Does it signal danger? The immediacy and simplicity of this operation of convergence to an attractor is good reason why the ECoG in olfaction more readily provides an insightful model of human perception than do the other sense modalities (Ackerman 1990).

Spatial images of the correlates of cortical dendritic activity show that the cortices hold themselves in a domain of self-organized criticality (Sect. 6.9), which is manifested in their spontaneous background activity. We posit that the critical state contains an organized sensitivity that is expressed by a landscape of limit cycle attractors (Fig. 6.14, Sect. 6.9), which has been selected by messages sent to the cortices including the olfactory bulb from the limbic system (Sect. 8.1) by preafference (Kay and Freeman 1998) and which serves to predict the form of the information desired (Zelano et al. 2011). The impact of the receptor action potentials that contains the predicted information initiates a phase transition from a gas-like receiving phase to a liquid-like transmitting phase (Sect. 6.4.3) through a substantial increase in the intensity of synaptic interactions that resembles that of a condensation. This provides binding on a grand scale, because every neuron transmits its output directly or indirectly to many if not all other neurons in the transmitting cortex and receives from many if not all other neurons for many tens of ms in a burst of oscillation. The intracortical matrix of synaptic connections that has been modified during consolidation (Sect. 8.5) determines the AM pattern of the carrier, so the population output is not a representation of the material stimulus. It is the recollection of a memory of the stimulus, which is a percept.

The core process of realizing perceptual information from sensory information is similar in all sense modalities. The similarities between allocortical and neocortical dynamics have been obscured by the greater complexity of the neocortical ECoG and by the intricate preprocessing of visual, auditory, and somatic information that is provided by nuclei in the brain stem and thalamus and by local circuits in the middle layers of the neocortex.

In this chapter, the challenge is to begin to understand the higher-order phase transition from perception to conception. Four experiments are described that illuminate different facets of the process. Section 10.2 describes the macroscopic patterns of dendritic potentials from neurons in the auditory cortex that have been thoroughly studied in regard to their microscopic processing of sensory information. The authors introduce the distinction between generalization by classical and instrumental learning and categorization by insight learning. The creation of a perceptual category in dynamical terms requires formation of a new attractor in an existing landscape, which maintains a generalization gradient expressed in its basin of attraction. Section 10.3 addresses the problem of multisensory convergence and synthesis of a Gestalt by combining feature detectors from multiple sensory cortices. Section 10.4 extends the study of AM patterns in the ECoG to a small area in the temporal lobe that is remote from sensory cortices, for which the cognitive correlates of both ECoG patterns are familiar forms of human conceptual activities. Section 10.5 sets the upper limit of the scale-free dynamics of neocortex at the whole of both cerebral hemispheres by demonstration of the classifiability of AM patterns of carrier frequencies in the beta range of the EEG. Section 10.6 summarizes the Chapter.

10.2 Categorization Versus Generalization in Concept Formation

When called upon to learn a complex task, subjects commonly show a sudden increase in performance from chance levels to reliable choice, which is often called *insight learning* through an "Aha!" experience. Ohl et al. (2001) trained four gerbils in a shuttle box by instrumental avoidance conditioning to auditory CSs and recorded the ECoG from an electrode array on the auditory cortex (Fig. 10.1a). The gerbils readily learned to move in response to a CS^+, for example, to a 500-Hz tone and to stay in response to a CS^-, for example, a 5,000-Hz tone. Thereafter, they easily learned to discriminate a 5,000-Hz tone that was frequency modulated (FM) either *rising* or *falling* at different rates of FM in what the authors described as generalization.

Then the task was made complex by randomly changing the center frequency of the FM tone on successive trials across a wide spectral range. The gerbils were at first slow to learn the task, but when they did so, they abruptly increased their % correct response rate (Fig. 10.1b). The authors described the process as *category learning*, by which the subjects learned the abstract concepts of "rising" and "falling." The characteristic psychometric functions were replaced by new categories. They concluded (Ohl et al. 2005): "In a nutshell, generalization is a general feature of learned stimulus-cued behaviors reflecting the converse of stimulus specificity, while categorization is a cognitive process based on the parcellation of the represented world into equivalence classes of meaning, valid for an individual in a particular context and in a particular time."

Before training the gerbils, the authors fixed a 3×6 electrode array on the right auditory cortex covering the area in which single-unit responses and averaged local field potentials were previously shown to provide a sensory map of the microscopic sensory input (Ohl et al. 2000). The gerbils performed the task despite the inconvenience of the cable attached for recording the ECoG during learning. The same recording equipment and software used by Barrie et al. (1996) for the 64 channels of rabbit ECoGs (Fig. 8.4b) were used to record and analyze the 18 channels of gerbil ECoGs. The signals were recorded in 6 s trials and filtered in the 20–80 Hz range. A 120-ms window was stepped at 20-ms intervals along the ECoGs from each trial. The 18 root mean square amplitudes at the steps gave 300 18×1 feature vectors that revealed the dynamic trajectory of the AM pattern, A(t), over each trial (Fig. 10.1c).

The mean of a string of feature vectors, **A**(t), from the control ECoG ($t<0$) had a center of gravity at time T in midburst, A(T), and SD radius (Figs. 7.7a, and 8.9d) that were comparable to those from rabbit ECoG (Fig. 6.8, Sect. 6.4.4). The authors reexpressed the AM pattern trajectory as a *dissimilarity function* of time by calculating the Euclidean distance of the tip of the feature vector from the global center of gravity in 18-space. The dissimilarity function had essentially the same rationale as the distance measures, $D_e(t)$ and $H_e(t)$, that were used to demonstrate convergence to a stable AM pattern within each burst and for classification of single spatial AM patterns with respect to CSs (Fig. 9.2c, d, Sect. 9.3.1). The dissimilarity function as

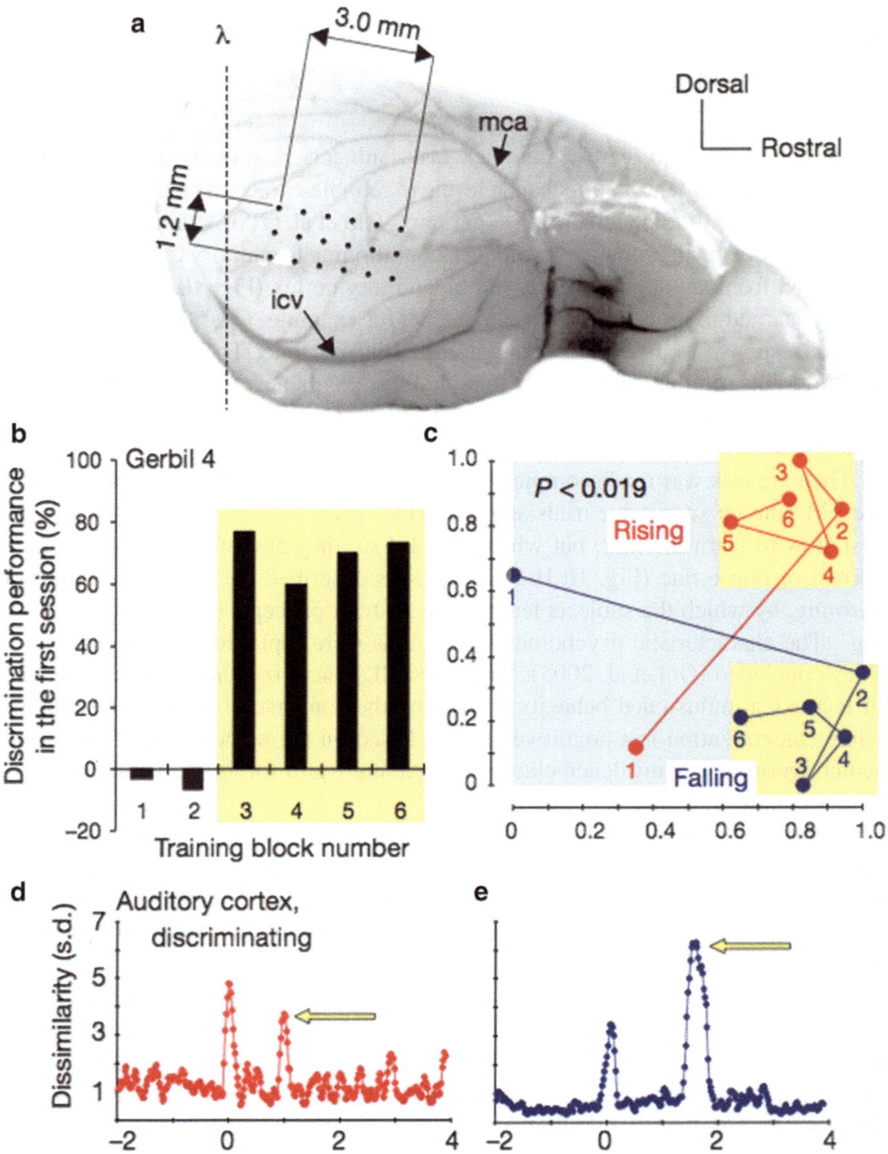

Fig. 10.1 (**a**) Conditioning to discriminate auditory FM tones was conducted with a 3×6 electrode array fixed on the auditory cortex (*icv* inferior cerebral vein, *mca* middle cerebral artery). (**b**) The behavioral transition to categorization was marked by an abrupt increase in % correct CR anywhere between the 2nd and 5th block of trials on different subjects. (**c**) Concomitantly with the increase in % correct CR the feature vectors from the spatial AM patterns formed two clusters, one for each CS (*yellow areas*, Wilcoxon's test, $p < 0.05$), shown in 2-space by projection from 18-space (as in Fig. 9.4d and 10.5b). (**d, e**) Sample time series of the index of dissimilarity revealed two peaks (*arrows*) after the behavioral transition (as in Fig. 9.2b) for both CSs but not in the naïve state and not in the visual cortex (From Ohl et al. 2001)

predicted peaked at stimulus onset but, with the development of learning, showed additional peaks in the 4-s interval between the tone presentation (CS) and the conditioned response (CR). These additional peaks marked epochs of the ongoing ECoG during which cortical states emerged with category learning, called "marked states" by the authors.

In accord with prior findings (Fig. 8.7b, Sect. 8.4), the information contained in the marked states was uniformly distributed over the AM patterns, regardless of local amplitude and variance (Ohl et al. 2003). The microscopic tonotopic information, inherent in the activity state marked by the first (trivially expected) peak of the dissimilarity function, was lost in the marked states associated with the later peaks in the dissimilarity function (arrows Fig. 10.1d, e), so that two average spatial AM patterns expressed the categories of "rising" and "falling" FM tones. The patterns expressed the information required for decision-making and suppressed the irrelevant, sensory-specific information that had no value at the level of complexity built into the experimental design. In dynamical terms the three centers of gravity of the feature vectors (control, "rising," "falling") expressed attractors as members of an attractor landscape (Sect. 6.9), in which the basins of attraction expressed the generalization gradients that were formed and maintained by repeated sensory sampling during reinforcement learning. In psychometric terms, after a new stimulus pair was introduced in a new learning block, learning did not begin with a gradual learning curve with zero discrimination performance in the initial training session (Fig. 10.1b, blocks 1 and 2 in one representative animal) but with full discrimination performance already in the initial session, that is, for the novel stimuli (blocks 3 and following). Upon testing the psychometric function in this state, a sigmoid function (the so-called curve of categorical perception) was found to have replaced the generalization gradient. The transition from the "discrimination phase" of learning to the "categorization phase" of learning occurred abruptly, at individually different points in time in different animals and led to a stable state until the experimental conditions were changed (Fig. 10.1b). The transition can thus be considered a cognitive state transition underlying an "Aha!" experience. While in the early patterns maximum information about the stimulus class was contributed by a central patch co-localized with the dominant thalamocortical input (topographic representation principle), in the later patterns maximum information about the behaviorally relevant category was spatially distributed over the entire recording area (holographic representation principle) (Ohl et al. 2003). Both types of spatial organization coexisted in the same brain area separated in time.

10.3 Convergence of Percepts into Multisensory Gestalts

The ECoG evidence thus far supported the hypothesis that the olfactory, visual, auditory, and somatic systems transmitted their perceptual information by spatial amplitude modulation of intermittent clouds of oscillatory pulse density waves. The spatial uniformity of the classificatory information density (Fig. 8.7b, Sect. 8.4) and the large size of the domains of coherent oscillation (Fig. 8.4, circles in b) implied that each cortex broadcasted its unique perceptual information in the same form from all its subareas to every neural population it targeted (Fig. 8.8a, Sect. 8.3). We inferred from the AM patterns that the transmitted pulse clouds carried both the stimulus features and the memories of the category of stimulus (Sect. 8.7). We inferred further that the Gabor transform took the transmitted signals out of the spatial domains specific to the olfactory, visual, auditory, and somatic input and placed them into a spatial frequency domain, such that the feature vectors of two or more cortices could be combined by dendritic integration (simulated with matrix concatenation), wherever the transmitted pulse clouds overlapped. We knew that every sensory cortex sent part of its output by synaptic relays to the entorhinal cortex, which emerged early in mammalian evolution from the transitional area of the amphibian brain (Fig. 8.1, TA). From this background, we predicted that the activity transmitted from a stimulated sensory cortex during an action–perception cycle would be found to be integrated in both time and space with activity in the entorhinal cortex, during the extended durations of classifiable bursts (Fig. 9.7e, Sect. 9.4). More specifically, we predicted that bursts would appear in the CS-CR interval, in which sensory and entorhinal cortices would be found to share phase-locked carrier frequencies with high temporal correlations. What we found surprised us. All of the observed cortices shared the prevailing carrier frequency (Freeman and Rogers 2003) and its classifiable AM patterns (Freeman and Burke 2003), not the initial gamma burst (Fig. 9.2a, Sect. 9.3.1) but the following beta bursts (Fig. 9.2b).

To test the prediction, we fixed small arrays on the visual (16 electrodes), auditory (16), somatic (14), olfactory (2), and entorhinal cortices (16) totaling 64 electrodes. The entorhinal cortex of the rabbit was found to be surgically inaccessible, whereas the entorhinal cortex of the cat lay on the flat surface of the bony tentorium. This anatomical feature made it possible to insert a 2×8 array between the bone and the cortical surface (Fig. 8.4b, ENT). The 4×4 arrays were fixed on the surfaces of the visual, auditory, and somatic cortices. Two electrodes sampled the olfactory bulb. An example of the simultaneously recorded ECoGs (Fig. 10.2) shows the similarity of the broad-spectrum neocortical ECoGs, in contrast to the prominence of gamma activity in the cat olfactory ECoGs (OB). Four cats thus surgically prepared were trained to press a bar for water in response to visual or auditory CS+ and to refrain with a CS– in a classical appetitive instrumental conditioning paradigm (Freeman et al. 2003c).

The prediction was tested by searching for episodic AM patterns of oscillations in the range of 20–80 Hz over the entire array. For each subject and session, 20 CS+ and 20 CS– trials with correct responding were collected. The ECoGs were processed in the same manner as in experiments illustrated in Figs. 9.2, 9.7 and 10.1. A window of fixed duration was stepped along the 63 ECoGs after band-pass filtering. A feature vector was constructed from the normalized root mean square amplitudes at each step. Classification of each feature vector was by finding the nearest center of gravity and the number correctly classified. An initial survey showed that significant numbers were correctly classified in the test period but not in the control period (Fig. 10.3a).

An empirical criterion was devised for rapid assessment of the goodness of classification. The number of correctly classified bursts at each step was divided

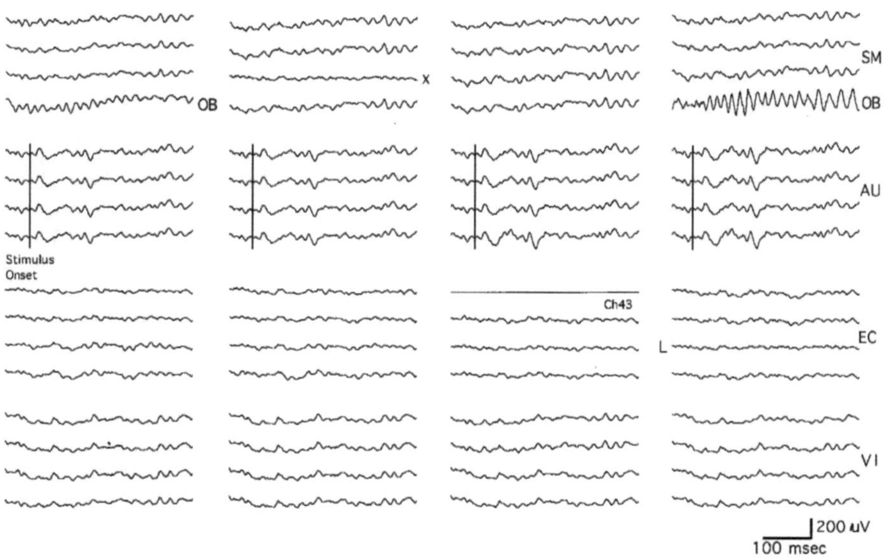

Fig. 10.2 A sample of 64 ECoGs is shown from a single trial: 16 from the visual (*VI*), auditory (*AU*), and entorhinal (*EC*) cortices; 14 from the somatic cortex (*SM*); and 2 from the olfactory bulb (*OB*) of a cat trained to respond to a tone burst (*vertical lines*). One bad channel (*Ch43*) was used to record the CR (bar press for water). The "x" indicates a bad channel. The "L" indicates the channel with lowest amplitude, by which to judge the maximal fluctuations that could be attributed to the monopolar reference electrode (From Freeman et al. 2003c)

by the SD of the numbers in the control periods for all subjects and trials in each session. This converted the number correct to a deviation in units of SD. Since the numbers in the control period were normally distributed, the ratio could be interpreted as a *t*-value for assessment of significance.[1] The test/control (t:c) ratio served for classifier-directed optimization of the search parameters. The optimal window duration was 164–256 ms (A). The step size was half the window length.

[1] The question whether episodic synchronization occurred among channels was addressed with an index of synchrony between pairs of signals that was developed by Pikovsky et al. (2001) and applied by Tass et al. (1999) to evaluate coupling between a magnetoencephalographic (MEG) signal and an electromyographic (EMG) signal in subjects with Parkinsonian tremor. The index was based on normalized Shannon entropy and was modified to give zero for a uniform distribution of phase differences in a moving window and unity for global phase locking. Here the index was generalized to the nearly 2,000 channel pairs by combining them into a t-value at each step of the window (Freeman and Rogers 2002). The validity of the algorithm was first tested on 64-channel data by replicating the intermittent synchrony of bursts in the several rabbit neocortices. Then it was applied to the data from the multicortical ECoGs of cats (Freeman and Rogers 2003). All values were significantly above chance levels, but the salient features were four spikes in synchrony adjacent to the times of the peaks in correct classification (Fig. 10.3d, arrows in c).

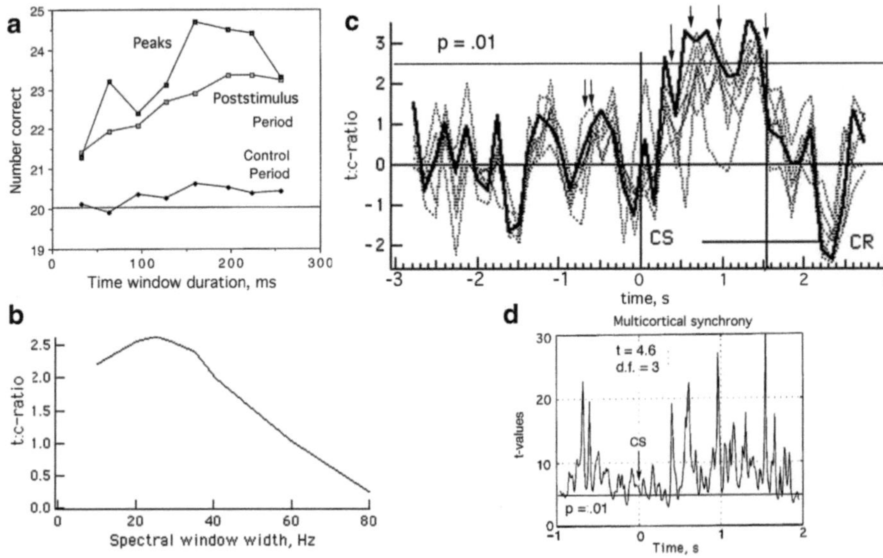

Fig. 10.3 (**a**) Examples are shown of tuning curves using the classification measure to optimize the parameters used in preprocessing. The grand means of the set of feature vectors counted as correct were tallied across sessions and subjects and summed in three groups: the mean in the control interval (1.6–2.6 s), the test interval (3.3–4.4 s), and the highest number in the test interval. The separation was optimized by varying the duration of the stepped window. The optimal step size was half the window duration. (**b**) The choice of center frequency of the pass band was optimized. (**c**) The optimized parameters gave three or four peaks in the interval between the onsets of CS and CR onsets (*solid curve*) but none in the first 250 ms after CS onset. *Dotted curves* show reduction of classification by deletion of channels from the five cortices (From Freeman and Burke 2003). (**d**) The degree of synchrony over all 63 signals was estimated by calculating an index based on normalized Shannon entropy adapted from Tass et al. (1999), revealing a series of spikes in synchrony in the CS-CR interval (*arrows* in C) (From Freeman and Rogers 2003)

The optimal center frequency of the pass band was 44 Hz (B). The optimal width of the pass band was 24 Hz (C). From previous experience (Barrie et al. 1996), the times of onset of AM patterns were expected to vary across trials between the onsets of the CS and CR. Yet the classification procedure searched across trials in a window that was stepped across all trials at the same time. The variation in latency of evoked potentials was termed "jitter" by Tallon-Baudry et al. (1998) and Quian Quiroga (2000), and was dealt with by wavelet denoising (Sect. 5.4). For Fig. 10.3a, the variation in burst latency was dealt with by searching at every step for an optimally classifiable AM pattern, which might occur in either of two steps before or after the current window. The classification test was repeated five times, and the trajectories of the six truncated feature vectors were superimposed. The data pooled across cats gave three peaks of significant classification ($p < .01$) in the CS-CR interval (Fig. 10.3c), which corresponded to the three peaks in rabbit ECoG of correct classification in the same interval see also (Fig. 9.2, Sect. 9.3.1).

Deletion of the EEG data from each cortical area reduced the classification assay in the test period. It had no significant effect in the control period. The classification test was repeated five times, each time removing the channels from one of the five cortical areas. In each instance, the deletions reduced the goodness of classification (dotted curves). The strongest effect was by removal of the two olfactory ECoGs, while the least effect was by deletion of the entorhinal ECoGs. The mean test/control (t:c) ratios from the control period (1.6–2.4 s) and the test period (3.6–4.4 s), calculated after the deletions, were as follows: none, 0.34 vs. 2.71; EC, –0.01 vs. 2.36; VC, 0.01 vs. 2.17; SM, 0.00 vs. 2.04; AC, 0.01 vs. 1.60; and OB, 0.07 vs. 0.74. We interpreted the findings of intermittent long-range spatial coherence across cortices (Fig. 8.4b) as revealing underlying phase transition in the time period in the action–perception cycle at which percepts were assembled into Gestalts.[2] Questions of where, how, and exactly when the integration took place are considered in Sect. 11.5.

10.4 Demonstration of Classifiable AM Patterns in Human ECoG

Additional support for the hypothesis of macroscopic perceptual coding comes from a study by Panagiotides et al. (2010) of ECoGs in a neurosurgical patient, who was undergoing treatment for intractable epilepsy. In addition to conventional scalp and intracranial neurosurgical electrode arrays (Fig. 10.4), a 10- × 10-mm square array of 64 electrodes, with informed consent of the patient, was inserted through a temporal burr hole and fixed on the anterior surface of the right inferior temporal gyrus for 8 days of continuous recording. The location and duration of recording were determined solely by the diagnostic needs of the neurosurgeons (Fig. 10.5). The 64 ECoGs were analog filtered (0.5–120 Hz), digitized (200 Hz), and low pass filtered (55 Hz). The electrode array (Fig. 7.1, Sect. 7.2) was designed in accordance with an evaluation of the ECoG PSD_X (Sect. 7.5), which set the optimal interelectrode interval at 1.25 mm and the spatial Nyquist frequency at 0.8 c/mm (Ramon et al. 2009).

The radial phase gradients (phase cones) found in animal ECoGs were fully documented in ECoGs from the human subject (Freeman et al. 2006a), which indicated that the spatial resolution afforded by the grid (Fig. 7.8, Sect. 7.5) would be adequate

[2] The dynamics of Gestalt formation has been modeled using K-sets (Sect. 8.2; Freeman and Erwin 2008) to implement the action–perception cycle in intentional robots (Kozma et al. 2003, 2008). The modeling was simplified by the linearity of the four operations: concatenation of macroscopic input feature vectors, spatiotemporal integration following a phase transition, transmission with the Gabor transform, down-sampling by local integration over the global AM pattern, and partitioning of the output to multiple targets. The operations were simulated with matrix algebra. Owing to linearity, the operations were commutative. The weights of the connections were expressed in matrices and adapted by learning in updating the memory bank of limit cycle attractors (Fig. 6.14).

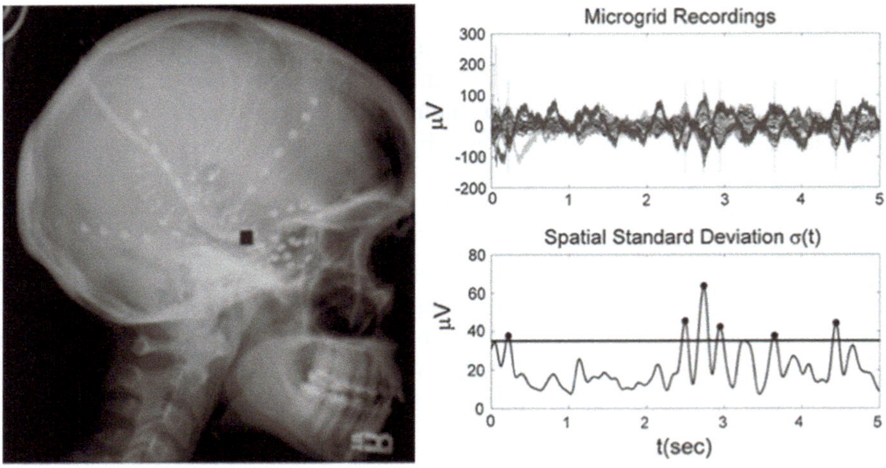

Fig. 10.4 The radiograph shows where the 1- × 1-cm array was placed on the anterior surface of the right inferior temporal gyrus. The ground and reference were Cz and Pz, respectively. A 5-s segment of 64 ECoGs is shown exemplifying 8 days of continuous recording. The marker with which to locate AM patterns was the spatial standard deviation, σ(t), of the 64 amplitudes at each time step. A threshold of 35 mV was used to identify time regions of high spatial variance. Only feature vectors at the time points of local maxima in suprathreshold segments of σ(t) were stored for further analysis (points marked with *dots*) (From Panagiotides et al. 2010)

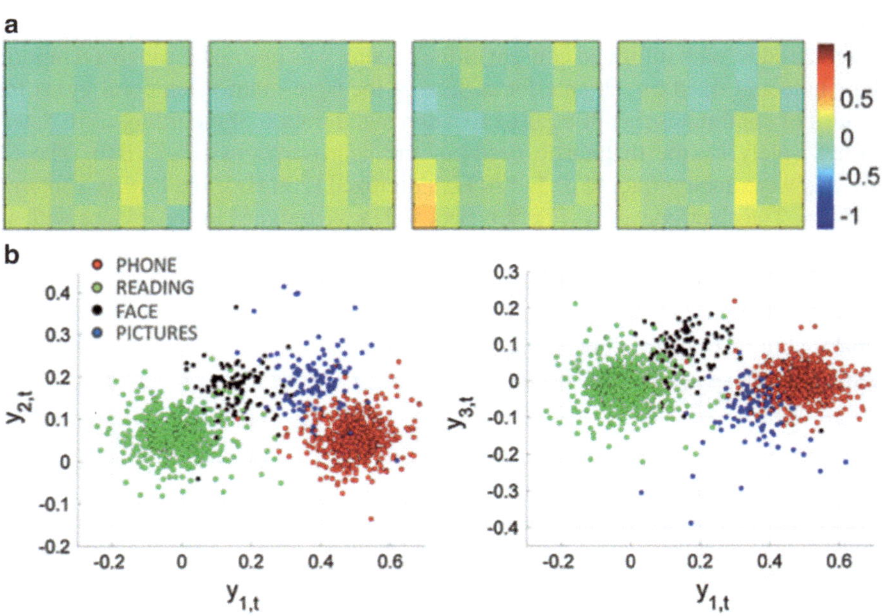

Fig. 10.5 (**a**) Spatial patterns of the means of the feature vectors show the AM patterns in the four conditions, which are deviations from the characteristic signature pattern of the human subject. (**b**) Scatter plots show the first two linear discriminant coefficients, y_{kt}, from voltage measurements. All four conditions are clearly distinguished. (From Figs. 6 and 7 in Panagiotides et al. 2010)

for classification of AM patterns. The means and SDs of the durations and diameters were larger in proportion to the digitizing step size (5 vs. 2 ms) and the interelectrode interval (1.25 vs. 0.79 mm) (Freeman et al. 2006b), which was attributed (Freeman and Breakspear 2007) to the power-law distributions of ECoG parameters (Fig. 9.7, Sect. 9.4), in which the mean and SD vary with the size of the measuring window.

The main problem to be solved was the definition of an appropriate set of behaviors. The animal studies (Freeman et al. 2003c) were designed to acquire ECoGs from sensory and entorhinal cortices. The selections of behavioral correlates and the time and space frames in which to search were based on the CSs, the intervals between CS and CR, and the cortices to which the CS information was directed. On reviewing and citing diverse evidence for involvement of the temporal lobe in human behaviors, the authors (Panagiotides et al. 2010) concluded: "The temporal pole has also been implicated in tasks involving theory of mind (inferences about the intensions, desires and beliefs of others). Functional neuroimaging studies have revealed activation during theory of mind tasks such as thinking of other people's emotions, making moral decisions or judging intentionality in viewing geometric shapes that move around" (p. 57). This encouraged us to search for classifiable AM patterns in ECoGs from the temporal lobe. The prior demonstration of the independence of classification from ECoG phase and frequency led to adoption of the magnitude of the spatial variance of amplitude as the marker of choice for selection of potentially classifiable ECoG segments (Fig. 10.4c).

The authors reviewed the entire set of videos of the patient during the 8 days of recording. Behaviors involving sustained visual and auditory interactions were sought and identified. Four recurring behaviors were selected, and the AM feature vectors during those behaviors were extracted: (a) face to face interaction with another person (5 min, 199 vectors), (b) looking at pictures (27 min, 254 vectors), (c) reading a magazine or a book (10 min, 1,347 vectors), and (d) speaking with family on the telephone (14 min, 1,138 vectors). The plots of the AM patterns from the four conditions (Fig 10.5a) showed small deviations widely scattered from the signature pattern. Scatter plots of the first three factors in multiple discriminant analysis of the principal components (Sect. 8.5) showed that the feature vectors from the four conditions formed four clusters that were linearly separable. The demonstration confirmed the hypothesis of the existence in temporal lobe ECoG of macroscopic AM pattern correlates of higher cognitive functions, which had been advanced by Quiroga et al. (2009) in the basis of recording the microscopic pulse trains of concept cells (Sect. 11.5).

However, the finding of classifiable AM patterns in the ECoG did not show whether they were locally constructed as in the olfactory bulb or imposed from another area as in the prepyriform cortex. For the two olfactory areas, the distinction was made on the basis of the phase gradients, those in the bulb being conic (Fig. 8.9) and those in the cortex conforming to the properties of the LOT Ch. 4 in (Freeman 1975). Conic phase gradients were found in human ECoG (Freeman et al. 2006a, b) comparable to those found in neocortical areas in animals (Fig. 9.5). It is reasonable to conjecture that entorhinal ECoG reflected macroscopic coding in its AM patterns, but the patterns could have been generated locally or imposed from other areas nearby. Measurements of phase in search of conic vs. input-dependent gradients would help to resolve the issue.[3]

10.5 Demonstration of Classifiable AM Patterns in Human Scalp EEG

We predicted that we would find significant AM patterns in scalp EEG because the accumulating evidence suggested that the diameters of domains of coherent beta and gamma oscillations might be sufficiently large to overcome the barriers posed by the intervening tissues. The classifiable AM patterns in ECoGs spanned the entorhinal and sensory cortices following training of the subjects to discriminate visual or auditory CSs (Sect. 10.3). The power-law distributions of multiple ECoG parameters (Sects. 6.9, 8.1, 8.5, 9.2 and 9.4) gave evidence for scale-free dynamics arising in criticality (Sect. 6.9), which carries the implication of the existence of exceedingly long correlation distances (Freeman and Vitiello 2009). Numerous authors have reported high correlations among EEG signals filtered in various frequency ranges (Sect. 7.3). Of particular relevance are the extensive cortical "microstates" recurring at rates in the theta range, with the suggestion that they serve as "building blocks" in thought processes (Lehmann et al. 2009), the "widespread phase synchrony at all frequencies" reported by Pockett et al. (2009) and the "large-scale networks" in cognition described by Bressler and Menon (2010).

We sought AM patterns with 64 electrodes in a standard 10–20 montage (Pockett et al. 2009). The recordings were referential to a balanced lead in the left occipital area. Six normal volunteers learned to press a button in response to a visual CS (red or blue light flash) or an auditory CS (faint or soft tone), which were presented at random intervals, either alone or in pairs. Their reinforcement was the pleasure of success on being notified of correct responses or the frustration of making a mistake. Once trained, the subjects were instructed that a new CS would be given, which they had to learn by trial and error. The new CS used the same visual and auditory CSs, but it combined them into a multimodal CS, so as to force transcortical integration of the sensory information. For example, a response to red-faint would be rewarded (CS+) if the button was pressed, whereas a response to blue-soft (CS–) would be penalized. One set of blocks of 40 artifact-free trials was acquired without yet

[3] Measuring spatial phase gradients is an arduous task, requiring identification of an ECoG segment with a prominent spectral peak, band-pass filtering, calculation of a phase surface with respect to the frequency of the ensemble average, and fitting a conic surface by nonlinear regression (Freeman and Barrie 2000). A simpler assay could facilitate preliminary explorations (Ruiz et al. 2009). The analytic phase difference in rad was calculated between each pair of signals and grouped in accord with the distance between them in mm for the duration of an epoch of stable carrier frequency. The group averages in rad were plotted with distance in mm and fitted with a straight line, giving the gradient in rad/mm. This with the carrier frequency gave the phase velocity and half-power radius. The presence of a cone might be detected by grouping the phase differences with direction as well as distance.

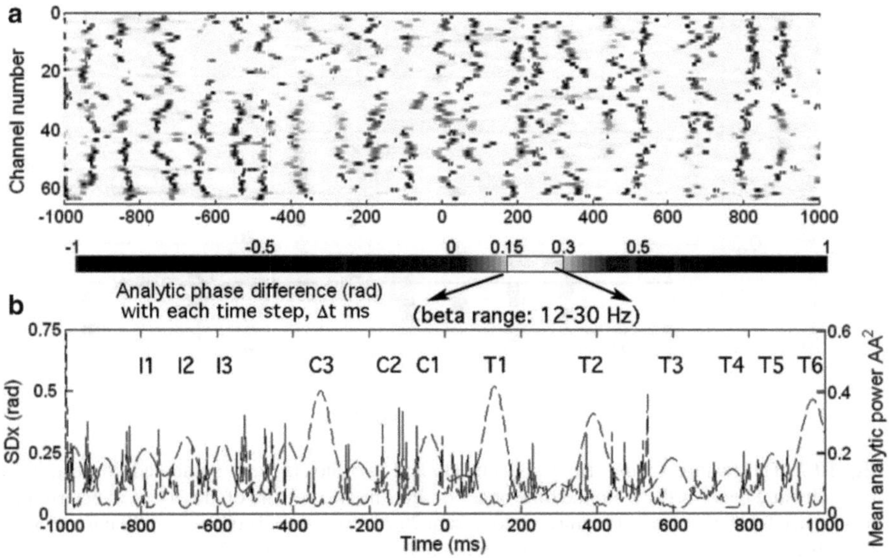

Fig. 10.6 An example is shown of the relations between analytic power and analytic phase differences. The irregularities are attributed to the multiple small avalanches that maintain criticality. (**a**) Successive analytic phase differences for each channel are plotted vs. time, *t*, in a 2-s segment of a trial. *Light areas* represent plateaus of small phase differences producing analytic frequencies within the applied temporal pass band. *Dark areas* represent phase jumps, where the value of analytic frequency was outside the spectral pass band, which implied that the phase and amplitude were undefined and indeterminate. (**b**) Successive values of the spatial standard deviation, $SD_X(t)$, of the analytic phase differences (*black spiky curve*) and the mean analytic power, normalized to the global z-score (*gray dash curve*). CS onset was at 0 ms (From Fig. 1 in Ruiz et al. 2010)

imposing the requirement for multimodal integration. A second set was acquired after the change that required the subjects to consider the cross-modal relation.

The marker used to locate potentially classifiable spatial AM patterns in the EEGs was provided by the analytic phase, φ(t). Band-pass filtering of the EEG and application of the Hilbert transform (Sects. 6.4.1 and 9.5; Freeman 2007) revealed episodes of spatially phase-locked oscillations (Fig. 10.6a). The epochs were quantified by the spatial standard deviation, $SD_X(t)$, of the analytic phase (B). During each episode of spatial coherence, the spatial mean of the analytic power rose to a maximum. The 64 peaks in analytic power gave a feature vector for each episode.

The search was begun in the beta range (12.5–25 Hz), on the premise that EEG bursts tended to recur (Fig. 9.9; Fig. 10.6) at rates in the theta-alpha range (Freeman et al. 2003a; Freeman 2009). The initial search gave a succession of feature vectors from the artifact-free EEGs of 20 reinforced trials (CS+) and 20 unreinforced (CS−) trials, which were numbered consecutively forwardly and backwardly from time of CS onset at $t=0$. Following initial classification results, the full width of the predicted range was explored by stepwise variation of the center frequency and bandwidth of the temporal filter and repeating the classification test to construct tuning

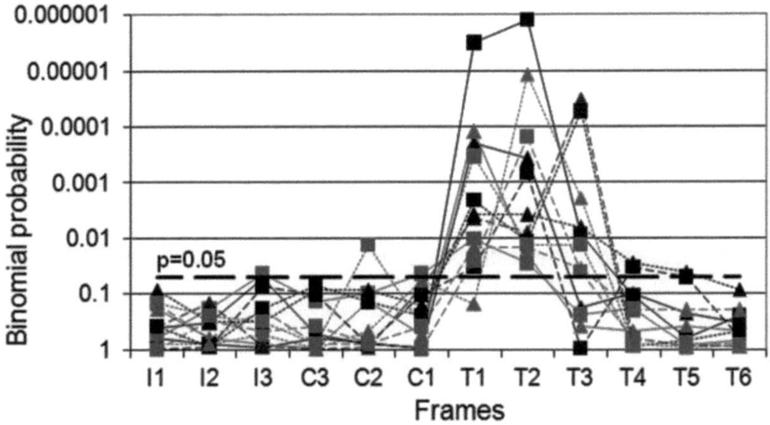

Fig. 10.7 The feature vectors from six subjects and two blocks were numbered consecutively forwardly and backwardly from the time of CS onset. The two centers of gravity were determined in 64-space for the clusters of feature vectors from 20 reinforced trials (CS+) and 20 unreinforced (CS−) trials. Each point representing the feature vector in 64-space of an AM pattern was classified as correct when its Euclidean distance to its own center of gravity was the lesser. Significance was estimated as previously (Barrie et al. 1996) by the binomial probability that the cumulative differences occurred by chance in a set of 40 trials. Significant separation of AM patterns was achieved by classifier-directed optimization for all six subjects, but only for the first three frames in the test period in the beta range. The % correct classification was further improved by applying a filter bank to the EEG before using the Hilbert transform (Fig. 9.10, Sect. 9.6.1) and repeating the classification in search of the optimal carrier frequency in each burst. This tuning procedure is comparable to optimizing the start time of each burst (Fig. 10.3d) by replacing each frame with a frame one or two window steps before and after the current step and repeating the classification in search of the optimal frame. *I* pretrial period, *C* control period, *T* test period. Trial duration = 6 s (From Fig. 3 in Ruiz et al. 2010)

curves (Sect. 6.4.3; Fig. 9.4a, Sect. 9.3.2; Fig. 10.3a, b). The procedure gave an optimal bandwidth of 7 Hz (15–22 Hz) about a center frequency (18.5 Hz) for the group of subjects.[4] Two other search parameters were then optimized: the thresholds for power or amplitude (Fig. 6.7a, Sect. 6.4.3) used to specify burst durations from the beginning and ending times of frames (Fig. 10.3b) and the choice of reference for recording and display (Fig. 10.9). Once significant percent-correct classification had been found for one block from a subject, each of these search parameters was explored over its range of sensitivity to optimize the degree of correct classification in the data set from other block. The procedure applied to data from each subject yielded significant degrees of separation of the CS+ and CS− clusters (Fig. 10.7) for

[4] Stephen O. Rice (Sect. 9.5) proved (Rice 1944) that the modal recurrence rate in Hz of beats in white noise passed through an ideal filter was proportional solely to the width of the pass band in Hz and was independent of the center frequency. We demonstrated that this proportionality held for brown and black noise as well. The constant was 0.641, which the pass band of 7 H predicted a burst rate of 4.5 Hz.

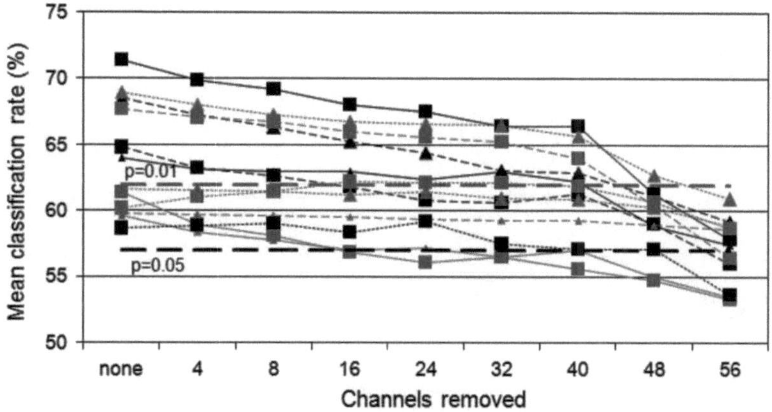

Fig. 10.8 The effect is shown of channel deletion on classification rate. As in all studies this far of ECoG, the classifying information in the EEG is spatially nonlocal and distributed with uniform density (From Fig. 5 in Ruiz et al. 2010)

the first two or three post stimulus frames for all subjects but only in the beta range. Further optimization of classification was achieved by using tuning curves to optimize the center frequency and pass band of each subject.

The data were tested to determine whether the classifying information might be concentrated in channels overlying the sensory areas. Groups of channels, four at a time, were randomly selected and deleted, while the classification procedure was repeated 80 times for each increment of removal. As in prior results with the ECoG (Fig. 8.7b, Sect. 8.4; Fig. 10.3c), the classificatory information for the EEG was distributed across the entire array, irrespective of local amplitude or variance (Fig. 10.8).

Each subject had a characteristic signature spatial pattern of amplitude that persisted through the duration of the study. The differences in AM pattern related to learning had the form of minor deviations widely scattered in the signature pattern (Fig. 10.5a). Re-referencing dramatically altered the appearance of the spatial images of analytic power (Fig. 10.9, left) and analytic phase (right) but had no effect on classification rates. The amplitude was biased toward low values in the vicinity of the reference site and high values at the greater distance. Most of the EEGs had relatively small deviations from the phase of the spatial ensemble average filtered signal. The exceptions were EEGs out of phase with the average, which were at sites located nearest the reference. Re-referencing to the ensemble average showed that the inverted signals had amplitudes lower than that of the reference, which inverted the shared waveform. The data support the conception of cortical neuropil operating as a continuous medium with intermittent condensation into spatially coherent narrow band oscillations (Pockett et al. 2009; Bressler and Menon 2010).

The EEG results show that the same form of macroscopic carrier of AM patterns holds at all levels from cortical columns to the entire neocortex in both hemispheres (Freeman et al. 2008). The methodology opens the way to identify and extract high-dimensional feature vectors for use in brain–computer interface technology,

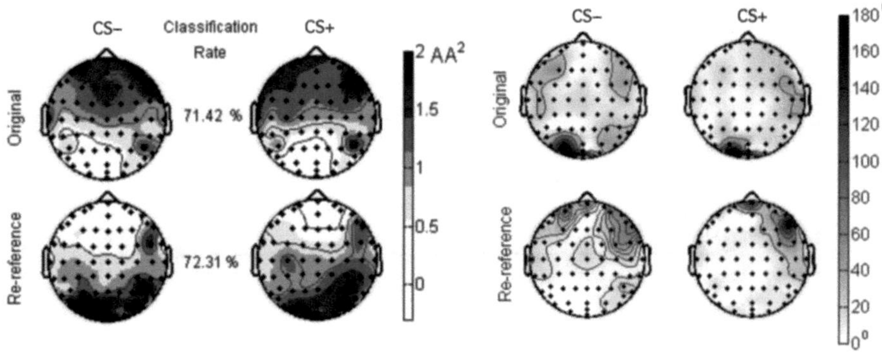

Fig. 10.9 The effects are shown of re-referencing the filtered EEG on the appearances of the AM patterns of the analytic power (*left*) and the PM patterns of the analytic phase (*right*) from block 1 of subject F. *Above*: the reference was located over the left occipital lobe adjacent to POz. *Below*: the reference was over the right frontal lobe. Re-referencing revealed a spatial trend (Freeman et al. 2003b) corresponding to the occipitofrontal alpha gradient. The analytic phase was referenced to the spatial average. All 64 signals were highly correlated, except for those from a small number of channels close to the reference channel, which showed phase reversal, because their signal amplitudes were less that the amplitude of activity at the reference channel (From Figs. 6 and 7, Ruiz et al. 2010)

for constructing and testing new models using differential equations (Freeman et al. 2012) and random graph theory (*neuropercolation*) (Kozma et al. 2005), and for cinematic display of pulsations and vortices in the ECoG (Freeman and Vitiello 2010). The spatially uniform density of classificatory information means that the placement of electrodes is arbitrary, provided that the locations are not changed in making comparisons between brain states or behavioral states.

The results raise challenges as well as answers. First among them is the absence of significant classification of bursts in the gamma range. However, recently another group of researchers (Brockmeier et al. 2012) has tackled the problem using the same data from the six subjects but with a better method of signal processing. Using a filter bank of wavelets in the gamma range coupled with the Hilbert transform to avoid the effects of phase dispersion, they have succeeded in demonstrating in three subjects' classification in the gamma range even stronger than that in the beta range. Yet other difficulties remain. One possibility is that occult EMG (Fig. 7.13, Sect. 7.7) obscured frequencies above 20 Hz (Whitham et al. 2007). Another is that gamma bursts and epsilon bursts may be more local, unlike beta bursts, and that higher density of spatial sampling will be necessary to extract them with spatial filtering (Sect. 7.7; Fig, 9.3, Sect. 9.3). In that case, PCA or ICA might yield good results. Yet another is that the temporal sampling rate (here 512 Hz) was too low to calculate the analytic signals in frequency ranges above beta. Improvements in sampling (more electrodes at higher spatial and faster digitizing) will materially help modeling as well as digital signal processing of EEG and ECoG alike.

Spatial imaging illuminates a deficiency in studies of the anatomical basis for scale-free dynamics. Qualitatively it is clear that layers I, V and VI of neocortex provide small numbers of very long connections that may suffice to explain long-range correlation distances (Fig. 8.4b; Fig. 10.9), especially when the cortex is on the knife-edge of bifurcation, but the evidence for the essential power-law distributions of axon lengths and transmission distance is fragmentary (Freeman and Breakspear 2007). The critical divergent–convergent topology of the lateral olfactory tract (Fig. 8.8a) has been verified by systematic microelectrode stimulation (Chap. 4 in Freeman 1975), but in the effort to document localization of microscopic networks, the statistical distributions of the synapses of large-scale neocortical projections are still to be fully documented. Further development of spatial ECoG and EEG imaging may encourage anatomists to extend the pioneering work of Sholl (1956), Bok (1959) and Braitenberg and Schüz (1998) on neocortical macrostatistics.

10.6 Summary

Low-level cognition involves operations that are closely related to sensory input and motor output. Examples are conditioned stimuli and responses in classical and instrumental conditioning. Spatial imaging of ECoG opens access to include also the operations of searching for desired sensory information by intermittent sampling (sniffing, glancing, harkening, touching). The microscopic sensory information carried by volleys of action potentials is transformed into macroscopic perceptual information carried by bursts of gamma oscillations (Figs. 8.3b, 9.6a, c and 10.1d, e), which imply the existence of limit cycle attractor landscapes that are initially formed by Hebbian and anti-Hebbian learning with and without reinforcement and modified thereafter by consolidation (Fig. 8.6, Sect. 8.4).

High-level cognition departs progressively from precise measures of input–output variables. Spatial imaging gives unprecedented access to the cortical correlates of the more complex operations. Four examples are given here: formation and use of abstract concepts, fusion of multiple unimodal images into a multimodal image, engaging in the use of language by extension of the study to human ECoG and extension insight learning to the EEG.

Spatial imaging makes it possible to correlate concepts with high-dimensional feature vectors and to model their formation and readout as macroscopic operations of cortical populations. Human neocortex uses the same information carrier as other animal cortices do: macroscopic bursts of oscillation that carry conceptual information in AM patterns. The bursts are located by maxima in the spatial standard deviation of the power in the wideband ECoG and by episodic high levels of coherence by phase locking.

The scale-free dynamics of neocortex may extend the macroscopic carrier into the spatial range of the EEG up to and including the entire neocortex operating as a unified system. The demonstration of noninvasive methodology for extracting high-dimensional feature vectors from normal human volunteers as the neural correlates of complex cognitive processes opens substantial new avenues for brain research.

References

Ackerman D (1990) A natural history of the senses. Random House, New York

Barrie JM, Freeman WJ, Lenhart M (1996) Modulation by discriminative training of spatial patterns of gamma EEG amplitude and phase in neocortex of rabbits. J Neurophysiol 76:520–539

Bok ST (1959) Histonomy of the cerebral cortex. Elsevier, Amsterdam

Braitenberg V, Schüz A (1998) Cortex: statistics and geometry of neuronal connectivity, 2nd edn. Springer, Berlin

Bressler SL, Menon V (2010) Large-scale networks in cognition: emerging methods and principles. Trends Cogn Sci 14:277–290

Brockmeier AJ, Hazrati MK, Freeman WJ, Principe J (2012) Locating global spatial patterns of waveforms during sensory perception in scalp EEG. Proc IEEE Conf EMBS #2098, San Diego, 28 Aug-01 Sept. http://cnel.ufl.edu/files/1340860427.pdf

Freeman WJ (1975) Mass action in the nervous system. Academic, New York

Freeman WJ (2007) Hilbert transform for brain waves. Scholarpedia 2(1):1338. http://www.scholarpedia.org/article/Hilbert_transform_for brain_waves

Freeman WJ (2009) Deep analysis of perception through dynamic structures that emerge in cortical activity from self-regulated noise. Cogn Neurodyn 3(1):105–116. doi:10.1007/s11571-009-9075-3

Freeman WJ, Barrie JM (2000) Analysis of spatial patterns of phase in neocortical gamma EEGs in rabbit. J Neurophysiol 84:1266–1278. http://soma.berkeley.edu/articles/BC.Phase.Cones.pdf

Freeman WJ, Breakspear M (2007) Scale-free neocortical dynamics. Scholarpedia 2(2):1357. http://www.scholarpedia.org/article/Scale-free_neocortical_dynamics

Freeman WJ, Burke BC (2003) A neurobiological theory of meaning in perception. Part 4. Multicortical patterns of amplitude modulation in gamma EEG. Int J Bifurcat Chaos 13: 2857–2866. http://repositories.cdlib.org/postprints/3345

Freeman WJ, Erwin H (2008) Freeman K-set. Scholarpedia 3(2):3238. http://www.scholarpedia.org/article/Freeman_K-set

Freeman WJ, Rogers LJ (2002) Fine temporal resolution of analytic phase reveals episodic synchronization by state transitions in gamma EEG. J Neurophysiol 87:937–945

Freeman WJ, Rogers LJ (2003) A neurobiological theory of meaning in perception. Part 5. Multicortical patterns of phase modulation in gamma EEG. Int J Bifurcat Chaos 13:2867–2887. http://repositories.cdlib.org/postprints/3346

Freeman WJ, Vitiello G (2009) Dissipative neurodynamics in perception forms cortical patterns that are stabilized by vortices. J Phys Conf Series 174(012011):1–24. http://www.iop.org/EJ/toc/1742-6596/174/1

Freeman WJ, Vitiello G (2010) Vortices in brain waves. Int J Mod Phys B 24(17):3269–3295. http://dx.doi.org/10.1142/S0217979210056025

Freeman WJ, Burke BC, Holmes MD (2003a) Aperiodic phase re-setting in scalp EEG of beta-gamma oscillations by state transitions at alpha-theta rates. Hum Brain Mapp 19(4):248–272. http://repositories.cdlib.org/postprints/3347

Freeman WJ, Burke BC, Holmes MD, Vanhatalo S (2003b) Spatial spectra of scalp EEG and EMG from awake humans. Clin Neurophysiol 114:1055–1060. http://repositories.cdlib.org/postprints/989

Freeman WJ, Gaál G, Jornten R (2003) A neurobiological theory of meaning in perception. Part 3. Multiple cortical areas synchronize without loss of local autonomy. Int J Bifurcat Chaos 13:2845–2856. http://repositories.cdlib.org/postprints/3344

Freeman WJ, Holmes MD, West GA, Vanhatalo S (2006a) Fine spatiotemporal structure of phase in human intracranial EEG. Clin Neurophysiol 117:1228–1243

Freeman WJ, Holmes MD, West GA, Vanhatalo S (2006b) Dynamics of human neocortex that optimizes its stability and flexibility. Int J Intell Syst 21:1–21. http://repositories.cdlib.org/postprints/2385

Freeman WJ, O'Nuillain S, Rodriguez J (2008) Simulating cortical background activity at rest with filtered noise. J Integr Neurosci 7(3):337–344. http://repositories.cdlib.org/postprints/3373

Freeman WJ, Kozma R, Bollobás B, Riordan O (2009) Chapter 7: Scale-free cortical planar network. In: Bollobás B, Kozma R, Miklös D (eds) Handbook of large-scale random networks. Bolyai mathematical studies, vol 18. Springer, New York, pp 277–324. http://www.springer.com/math/numbers/book/978-3-540-69394-9

Freeman WJ, Livi R, Opinata M, Vitiello G (2012) Cortical phase transitions, non-equilibrium thermodynamics and the time-dependent Ginzburg-Landau equation. Int J Mod Phys B (2011) arxiv:1110.3677v1 [physics.bio-ph]

Kay LM, Freeman WJ (1998) Bidirectional processing in the olfactory-limbic axis during olfactory behavior. Behav Neurosci 112:541–553

Kozma R, Freeman WJ, Erdí P (2003) The KIV model – nonlinear spatio-temporal dynamics of the primordial vertebrate forebrain. Neurocomputing 52:819–826. http://repositories.cdlib.org/postprints/1049

Kozma R, Puljic M, Bollobás B, Balister P, Freeman WJ (2005) Phase transitions in the neuropercolation model of neural populations with mixed local and non-local interactions. Biol Cybern 92(6):367–379. http://escholarship.org/uc/item/6rk1d0sj

Kozma R, Huntsberger T, Aghazarian H, Tunstel E, Ilin R, Freeman WJ (2008) Intentional control for planetary rover SRR2k. Adv Robot 21(8):1109–1127

Lehmann D, Pascual-Marqui RD, Michel M (2009) EEG microstates. Scholarpedia 4(3):7632. doi:10.4249/scholarpedia.7632

Ohl FW, Scheich H, Freeman WJ (2000) Spatial representation of frequency-modulated tones in gerbil auditory cortex revealed by epidural electrocorticography. J Physiol Paris 94:549–554

Ohl FW, Scheich H, Freeman WJ (2001) Change in pattern of ongoing cortical activity with auditory category learning. Nature 412:733–736

Ohl FW, Deliano M, Scheich H, Freeman WJ (2003) Early and late patterns of stimulus-related activity in auditory cortex of trained animals. Biol Cybern 88:374–379. doi:10.1007/s00422-002-0389-z

Ohl FW, Scheich H, Freeman WJ (2005) Neurodynamics in auditory cortex during category learning. Chapter 8. In: König R, Heil P, Budinger E, Scheich H (eds) The auditory cortex—a synthesis of human and animal research. Lawrence Erlbaum Associates, Mahwah, pp 429–444

Panagiotides H, Freeman WJ, Holmes MD, Pantazis D (2010) Behavioral states may be associated with distinct spatial patterns in electrocorticogram (ECoG). Cogn Neurodyn 5(1):55–66. doi:10.1007/s11571-010-9139-4

Pikovsky A, Rosenblum M, Kurths J (2001) Synchronization—a universal concept in non-linear sciences. Cambridge University Press, Cambridge

Pockett S, Bold GEJ, Freeman WJ (2009) EEG synchrony during a perceptual-cognitive task: widespread phase synchrony at all frequencies. Clin Neurophysiol 120:695–708. doi:10.1016/j.clinph.2008.12.044

Quian Quiroga R (2000) Obtaining single stimulus evoked potentials with wavelet denoising. Phys D Nonlinear Phenom 145:278–19

Quian Quiroga R, Kraskov A, Koch C, Fried I (2009) Explicit encoding of multimodal percepts by single neurons in the human brain. Curr Bio 19:1308–1313

Ramon C, Freeman WJ, Holmes MD, Ishimaru A, Haueisen J, Schimpf PH, Resvanian E (2009) Similarities between simulated spatial spectra of scalp EEG, MEG and structural MRI. Brain Topogr 22:191–196

Rice SO (1944) Mathematical analysis of random noise. Bell Syst Tech J 23:282–332

Ruiz Y, Li G, Freeman WJ, Gonzalez E (2009) Detecting stable phase structures on EEG signals to classify brain activity amplitude patterns. J Zhejiang Univ 10(10):1483–1491.

Ruiz Y, Pockett S, Freeman WJ, Gonzales E, Guang L (2010) A method to study global spatial patterns related to sensory perception in scalp EEG. J Neurosci Methods 191:110–118. doi:10.1016/j.jneumeth.2010.05.021

Sholl DA (1956) The organization of the cerebral cortex. Methuen & Co., London

Skarda CA (1999) The perceptual form of life. In: Núñez R, Freeman WJ (eds) Reclaiming cognition. Imprint Academic, Thorverton, pp 79–94

Tallon-Baudry C, Bertrand O, Peronnet F, Pernier J (1998) Induced gamma-band activity during the delay of a visual short-term memory task in humans. J Neurosci 18(11):4244–4254

Tass P, Kurths J, Rosenblum M, Weule J, Pikovsky A, Volkmann J, Schnitzler H, Freund H (1999) Complex phase synchronization in neurophysiological data. In: Uhl C (ed) Analysis of neurophysiological brain functioning. Springer, Berlin, pp 252–273

Whitham EM, Pope KJ, Fitzgibbon SP, Lewis T, Clark CR, Loveless S, Broberg M, Wallace A, DeLosAngeles D, Lillie P, Hardy A, Fronsko R, Pulbrook A, Willoughby JO (2007) Scalp electrical recording during paralysis: quantitative evidence that EEG frequencies above 20 Hz are contaminated by EMG. Clin Neurophysiol 118(8):1877–1888

Zelano C, Mohanty A, Gottfried JA (2011) Olfactory predictive codes and stimulus templates in piriform cortex. Neuron 72:178–187. http://www.sciencedirect.com/science/article/pii/S0896627311007318

Chapter 11
Synthesis

11.1 Introduction

According to neural network models, all sensory cortices operate on streams of incoming pulses, which are generated by sensory receptors distributed in the surfaces of the eye, ear, skin, and nose. The delivery of the receptor pulses is by axons that are arranged in degrees of topographic order, so that each receptor surface is mapped onto the cortical surface. Each stimulus to each cortex provides a packet of a particular kind of energy, whether chemical, electromagnetic, mechanical, thermal, etc., in time frames determined by the sources. Each receptor is tuned to a specific type and amount of energy. Prevailing theory holds that the energy in the stimulus carries a certain quantity of information, which is transduced by the tuned sensory receptor into the energy of a flow of receptor current. The axon converts the current into a pulse train, which transmits the information expressed in the location of the axon terminals and the firing rate. A stimulus given to a distribution of sensory neurons excites the formation of a spatiotemporal pattern of cortical pulses that represents the information delivered by the stimulus. The pulse pattern is driven by the cortical dendritic currents that are evoked by the afferent barrage. The currents also contribute to a local field potential (LFP) as they flow across the tissue resistance. The networks of neurons in each sensory cortex process the information by extracting the features of the stimulus, binding the features into categorical representations, and storing them for future recall.

This description of the network model of information processing has provided the basis for interpretation of the firing patterns of cortical neurons for the past five decades. It also led to the hypothesis that the spatial pattern of the dendritic potentials evoked by a stimulus could be extracted using the techniques that had been developed for multichannel EEG recording and analysis (Chap. 1). We tested the hypothesis by averaging the dendritic potential evoked by electrical stimulation of the topographically organized input path of the olfactory bulb (Fig. 8.8, Sect. 8.4). We found, as predicted, that the spatial AM and PM patterns were determined by the stimulus location. However, when we extended the analysis to potentials evoked by

W.J. Freeman and R. Quian Quiroga, *Imaging Brain Function With EEG: Advanced Temporal and Spatial Analysis of Electroencephalographic Signals*,
DOI 10.1007/978-1-4614-4984-3_11,

sensory stimuli on single trials without averaging, we failed to confirm the hypothesis in three aspects. First, the AM patterns lacked invariance with respect to the information specific to fixed stimuli. Instead, they changed with changes in the context, history, and experience of the subjects with the stimuli (Fig. 8.6). Second, the categorizing information was spatially nonlocal; no ECoG signal in an array had any more or less value for classification than any other (Fig. 8.7). Third, the time course of the cortical activity revealed segmentation into intermittent bursts of oscillation at gating frequencies in the theta and alpha ranges and with carrier frequencies in the beta and gamma ranges (Figs. 6.1 and 8.3), a generalization of theta-gamma coupling. We inferred that the AM and PM patterns of the carrier wave were determined not by the stimulus as revealed by averaging but by the properties of the cortex (Figs. 8.6 and 8.9) that depend on the mechanisms of learning.

The findings left us two options. One was to disregard ECoG evidence as irrelevant and epiphenomenal. The other was to construct a population model that would build on and extend the network model. We took the second path. We were encouraged to undertake the construction of the population model by support from three lines of evidence. The first was *anatomical*: the extreme density of neuronal packing in the neuropil, which can provide the richness of connections needed to store not just a representation of the stimulus but the memory that a subject has about a stimulus. The second was *physiological*: the capacity of cortex for self-organization of spatiotemporal patterns by means of exceedingly rapid changes in state, whether spontaneous or triggered by sensory input. The third was *phenomenological*: the subjective experience of immediate apprehension of the meaning of a conditioned stimulus through the flood of knowledge it evokes.

We base our population model on the nineteenth century conception that brains are thermodynamic systems that perform the work of cognition and control. The main physical work of cortex (Raichle and Mintun 2006) is done by synapses acting like chemical batteries that have high internal impedance, which are briefly switched on by action potentials and which drive ionic currents across the matching high transmembrane impedance at the trigger zones of axons (Fig. 6.2a, Sect. 6.2.1). The magnitude of energy expenditure in the resting and active states is not readily apparent in the electric and magnetic potentials because the sums of excitatory and inhibitory synaptic potentials subtract from each other and tend to zero. In contrast, the energies that the excitatory and inhibitory ionic currents require do not cancel; they add (Logothetis 2008). The distinction causes major disparities between the measures of blood flow or oxygen consumption and the electrical and magnetic measures of cortical energy dissipation (Freeman et al. 2009a). The disparities contribute to the concept of "dark energy" (Raichle 2006). Indeed the voracious appetite of brains for energy even at rest is apparent in the fact that the rate of metabolic energy dissipation by brains per unit mass is ten times greater than that of any other organ including the heart. We may reasonably suppose that constructing AM patterns is energy-intensive, so we propose a thermodynamic model of cortical function to explain AM pattern formation in cognition. Such a thermodynamic model might help to integrate the experimental evidence from the broad range of noninvasive measurements of cortical activity, including EEG, MEG, PET, BOLD, and fMRI (Freeman et al. 2009a).

The foundation for the science of thermodynamics was laid by a French military engineer, Sadi Carnot, who in 1824 published the first theory describing how steam engines do useful work. His formal representation of the transfer of energy from heat into work and vice versa, now known as the Carnot cycle, served as a kind of Rosetta stone that enabled physicists in the next 30 years to formulate the first and second laws of thermodynamics: the conservation of energy and the inevitable loss in every physical system of energy that can do useful work (*free energy*). That loss was designated as an increase in *entropy* (Sect. 3.4). In 1847 Hermann Helmholtz, an army surgeon turned neuroscientist, introduced thermodynamics into brain science by showing that the first law of thermodynamics held for muscle, nerve being too small. He proved that the action potential is an electrochemical wave propagating at a finite velocity. His replacement of the concept of animal spirit with that of nerve energy revolutionized neurobiology. The British essayist Herbert Spencer (1863) described the conservation of nerve energy as "an unquestionable truth that, at any moment, the existing quantity of liberated nerve-force, which in an inscrutable way produces in us the state we call feeling, must expend itself in some direction — must generate an equivalent manifestation of force somewhere." Charles Darwin (1872) continued: "This involuntary transmission of nerve force may or may not be accompanied by consciousness. Why the irritation of nerve-cells should generate or liberate nerve force is not known; but that this is the case seems to be the conclusion arrived at by all the greatest physiologists such as Mueller, Virchow and Bernard, and so on." Sigmund Freud in 1893 wrote: "[My] approach is derived from clinical observations of 'excessively intense' ideas in hysteria. … I have in mind the principle of neuronic inertia" (i.e., the conservation of energy) (Pribram and Gill 1976).

By 1900 this scientific foundation had collapsed (Freud 1985), largely due to the realization that *nerve energy* is not conserved, because the brain is an open system operating far from thermodynamic equilibrium. Neurophysiology remained at an impasse for 50 years until the 1950s, when a new foundation was constructed based on information theory (Freeman 2007; Freeman et al. 2012a). The concept of the flow of energy through the receptors into sensory cortices was replaced by the flow of information, which could be measured in the rates and intervals of pulse trains of neurons by treating the pulses as binary digits. Research based on this foundation has flourished both in experiment and theory (Friston 2009), owing to the mutual support between microscopic anatomy using the Golgi technique to describe neural networks, electrophysiology using the microelectrode to measure pulse trains, and Claude Shannon's information theory, despite Shannon's (1948) opposition to the use of his theory in semantics: "The fundamental problem of communication is to reproduce a message. Frequently the messages have meaning. … These semantic aspects are irrelevant to the engineering problem".

Considering that the forms of AM patterns appear more closely related to perceptual meaning than to sensory information, we propose to describe brains as systems that operate far from equilibrium, using the concepts of the "chemical morphogenesis" of spatiotemporal patterns that was pioneered in theoretical chemistry by Alan Turing (1952); "dissipative structures" that feed on metabolic free energy as introduced in nonequilibrium thermodynamics by Ilya Prigogine (1980);

"synergetics," the self-organization of patterns by circular causality with "order parameters" devised in theoretical physics by Hermann Haken (1983); the "spontaneous breakdown of symmetry" from quantum field theory (Vitiello 2001); and "random structures" formed by subcritical Hopf bifurcations in mean field models of probabilistic cellular automata (*neuropercolation*, Kozma 2007; Freeman et al. 2009b). Accordingly in Sect. 11.1 we introduce the history and basis for the thermodynamic model relying on phase transitions. In Sect. 11.2 we review the evidence for the model to explain how percepts form in sensory cortices, as revealed by macroscopic AM patterns. In Sect. 11.3 we apply the Carnot formalism to our experimental observations. In Sect. 11.4 we extend it to the Carnot vapor (Rankine) cycle to introduce criticality and phase transition. In Sect. 11.5 we describe the mechanism of readout of AM patterns, which returns the percept to the microscopic level of sparse coding for transmission between cortices. We relate the properties of concept cells (Quian Quiroga 2012) revealed by single-cell recording to the spatiotemporal AM patterns from ECoG recording. There and in Sect. 11.6, we speculate how thermodynamic and network models might be synthesized or exploited separately for describing mechanisms of higher cognitive processes.

11.2 Sensation to Perception by Phase Transition

The principal operation of sensory cortices is to use metabolic free energy to construct knowledge from information through the process of perception (Pribram 1991; Freeman 2001). We define information as a delimited set of facts or features that is sensed about something or someone, which we can relate to microscopic firing of neurons singly or in small networks in conformance with Shannon's theory. We define knowledge as an understanding of the interrelations of a mass of information that has been acquired by experience, which we can relate to macroscopic fields of ECoG and EEG oscillations. It is with difficulty that we relate knowledge to Shannonian information, owing to the unbounded richness of associations (Lucky 1989), so for this purpose we have adopted pragmatic information, $H_e(t)$ (Sect. 6.4.3) as a useful extension (Atmanspacher and Scheingraber 1990). The underlying issue that we face, the relation between material and mental variables, is undeniably an aspect of the philosophical mind-body problem, but we finesse the issue by relying on classifying our electrophysiological AM patterns with respect to the probabilities of performing conditioned responses by animal subjects or of reporting by human subjects. We do not need to know what our subjects are thinking. Our knowledge of what they are doing gives us already a sufficient grip (Merleau-Ponty 1942) to understand the correlation between patterns of EEG activity and behavior.

We identify two starkly differing states in sensory cortical macroscopic signals: a baseline state, in which its neurons are driven by sensory and centrifugal inputs (Gray and Skinner 1988), and an active state in which its neurons are driven by each other with such intensity that in effect they respond only to each other and not to extracortical inputs. Just as axons can be viewed as having baseline and active states, macroscopic signals show bistability in the normal process of cognition (Fig. 6.5) in having a receiving state with little macroscopic output and a transmitting state with no significant acceptance of input to the macroscopic pattern. Using the thermodynamic formalism, we can describe the baseline state of macroscopic signals as chaotic (taking the word chaotic in a broad sense), low-density receiving phase, that is sustained by uncorrelated firings of the neurons, and the active state as an orderly, high-density transmitting phase, in which the probabilities of firing of all of the neurons in a region of cortex are in varying degree phase-locked briefly in a narrow frequency band (Sect. 6.4.2). The pass band and center frequency appear to be determined by the same or related processes by which the Hebbian assembly selects the basin of attraction (Fig. 6.14, Sect. 6.9) leading to the AM pattern (Fig. 8.6, Sect. 8.4).

It follows immediately that there are two transitions between the two states of cortical signals. The seemingly random firing of neurons in the basal state condenses into an AM pattern in the EEG or ECoG, and then the order evaporates. The AM pattern suggests that a shimmering film of cortical activity briefly unifies a fraction of the variance of millions or even billions of neurons in a narrow spectral band. The texturing of such a knowledge-bearing film would be by sudden binding of the features of a stimulus with memories of all prior experience by the subject with that stimulus, which are stored in immense numbers of modified cortical synapses.

As discussed in Chap. 8, the process is seen in its simplest, prototypical form in the olfactory ECoG (Figs. 6.12 and 8.3a) with minimal preprocessing by neural networks at the input to the bulb. It begins with an intentional stance for search and an inhalation that brings molecules of an expected scent to the vast array of sensory receptors in the nose (schematized in Fig. 8.8a, Sect. 8.4). Each nostril has in round numbers 10^8 receptors with 10^3 types of receptor giving 10^5 receptors of each type. Each inhalation delivers molecules of an expected scent to perhaps 10^2 sensitive receptors but with a different subset on every inhalation. The problem for olfaction (and all senses) is to categorize the input to the type of receptor. The solution is to strengthen the excitatory connections between coexcited bulbar excitatory neurons but only during the presence of reinforcement (Sects. 8.4 and 8.5). Over a set of trials, the cumulative synaptic changes form a Hebbian assembly, in which excitation of any subset ignites the entire assembly by the process of spread that have similarities with what has been described mathematically as *percolation* (Kozma 2007; Freeman et al. 2009b).

As described in Sect. 8.4, the Hebbian assembly has multiple functions. It amplifies the microscopic input by reverberation (Amit 1995), generalizes to the category of CS that the input has selected, and abstracts by deleting useless information about which among the equivalent receptors actually received input. It converts the surge of input from the receptor cells to a burst of gamma oscillation (Fig. 8.3, Sect. 8.2; Fig. 8.8b). The oscillation is due to the negative feedback excitatory neurons that receive the input and the inhibitory interneurons. The tendency for oscillatory impulse responses is characteristic of the bulb (Fig. 8.5a, Sect. 8.3) and the prepyriform cortex (Fig. 6.13, Sect. 6.8), showing that they operate in a similar fashion as a narrow band pass filter. When struck by an impulse from background noise (or an electric shock), the activity of the loop rings at the characteristic frequency with an envelope that decays exponentially. However, the response of the bulb to the ignition of a Hebbian assembly is not a damped cosine with an exponential decay. The ringing at the characteristic frequency has an envelope with an exponential increase. The sign of the envelope exponent reverses from negative to positive (Fig. 9.13c). Simulation shows that the reversal is due to the positive feedback gain, k_{ee}, in the Hebbian assembly (Sect. 6.2.2) when the Hebbian synapse is enhanced by reinforcement learning. The small increase in positive feedback gain strongly increases the negative feedback gain, k_n, above unity (Fig. 6.10b, Sect. 6.5), when the sensory input of a CS ignites a Hebbian assembly (Fig. 8.8a, b). The result is a burst of gamma oscillation with each inhalation manifesting destabilization but only when the bulb receives a CS (which includes the accustomed background input). Then the singular operation of the assembly, which from sampling considerations we estimate comprises on the order of 0.1% of the mitral population, is to ignite the macroscopic transition that confines the entire bulb into the basin of the relevant attractor.

Formation of an AM pattern of each burst depends on whether the inhaled odorant mixture contains a CS as the key to a Hebbian assembly. The dependence requires the prior activation by the limbic system (Fig. 8.1) of a landscape of attractors in the bulbar dynamics (Fig. 6.14b) having a basin of attraction for each expected stimulus that constitutes attention. The landscape must include a basin for the background odorant mixture to which subjects have habituated, which gives the characteristic signature pattern for each subject (Sect. 8.4). The landscape must also include a catchall "I-don't-know" basin for novel odorants. That basin provides the unique and irreproducible chaotic activity that is necessary to form a new attractor (Skarda and Freeman 1987) during the subject's exploration by means of an orienting response (Fig. 8.3b, Sect. 8.2).

Owing to the massive divergence of $1{:}10^4$ from each neuron to others in regenerative feedback (Sect. 6.2.2), the feedback gain exceeding unity amplifies the signal with each reexcitation. Evaluation of the dependence of pulse density on the average wave density of the gamma oscillation shows that pulse density cannot increase beyond the limit imposed by the refractory periods (Q_m). What can increase is the number of the neurons in each mm^3 that are participating in the oscillation by time multiplexing the pulse density (Sects. 6.2.2 and 7.1). We postulate that geometric intensification with gain greater than unity results in asymptotic convergence to spatial saturation, in which all or nearly all of the cortical neurons participate in the high-density state. The most direct demonstration of low density in the resting state is by exhaustive sampling of the spike correlates of evoked potentials in the bulb, which show sparse, decaying pulse probabilities (Freeman 1974), compared with the ease of finding sustained pulse probability waves in the waking state that are correlated with ECoG bursts of oscillation (Fig. 6.11, Sect. 6.6). The difference between the two thermodynamic states is also seen by comparing the linear phase gradient of the bulbar evoked potential that is imposed by the extrinsic symmetry breaking from an electric shock (Fig. 8.5) with the conic phase gradient of the gamma burst resulting from spontaneous symmetry breaking (Fig. 8.9).

The requirements for preprocessing of sensory information are relatively few in the olfactory system: range compression, normalization, bias control, and spatial contrast enhancement (Freeman 2001). The preprocessing in vision, audition, and somesthesis is far more complex than in olfaction, as shown by the complexity of the neural networks in the six-layered neocortex specific to each modality. The similarities in mass action show that, after the sensory information is extracted, the integration by phase transition is the same in all sensory cortices in delivering a percept to the limbic forebrain. Our measurements of neocortical AM patterns show that each burst of coherent oscillation is large enough to encompass the entire extent of each sensory cortex (Fig. 8.4b, Sect. 8.3) and that it has the same statistical and spectral properties as those in olfaction, so we conclude that the same thermodynamic model holds for perception by all sensory cortices after the sensory preprocessing.

The destabilization means in dynamic terms that the cortex switches from a non-convergent attractor that maintains cortex in the region of criticality (Kozma et al. 2012) governing the background activity (Fig. 6.10, Sect. 6.7) to a limit cycle attractor (Fig. 9.13, Sect. 9.7.1). In thermodynamic terms, it means switching from a low-density state to a high-density state. Now we propose to push the thermodynamic model further by suggesting that at low density the activity is in a gas-like phase, while at high density it enters a liquid-like phase, so that we can model the switch as a phase transition. We conceive an energy barrier or threshold for onset of high density, above which a major change occurs by the coming into dominance of short-range forces among neurons and glia. The known mechanisms include electrical synapses (Bennett 2009), gap junctions (Hameroff 2009), ephapsis[1] (Anastassiou et al. 2011), and nonsynaptic diffusion neurotransmission (NDN, Bach-y-Rita 1995). Additionally, Freeman and Vitiello (2009) postulate that extremely short-range Coulomb and van der Waals forces may come into play that do not cause phase transitions and classifiable patterns, which clearly depend on long-range axo-synaptic transmission, but they may facilitate them.

The anatomical basis for the postulate is the extreme packing density of fine axonal, dendritic, and glial threads in the cortical neuropil. The physiological basis is that the firing rates of most neurons are well below carrier frequencies, so that a neural population sustains coherence by time multiplexing (Sect. 6.6). The psychological basis is the richness and intensity of experience in recognition of a stimulus. Each neuron may fire at random only once in a burst of 3–5 cycles. The disparity in rates suggests that short-range forces may facilitate achievement of maximal density rapidly, in less than 1/4 cycle of the carrier wave, and sustain it thereafter.[2]

[1] Ephapsis denotes the polarization of a neuron exerted by the extracellular ionic loop currents of neighboring neurons. Currents penetrating the axon of the neuron hyperpolarize it where they enter and depolarize it where they exit. Due to the high resistance of the membrane in comparison to the low shunting resistance of the extracellular compartment, the penetrating fraction is on the order of 0.1%. However, as cortex brings itself in criticality to the threshold for phase transition, that tiny fraction may come to dominate the interactions and precipitate high-density energy dissipation in the liquid-like state ('dark energy').

[2] We have not heretofore introduced the role of nonspiking neurons in neural feedback loops, owing to the complexity of the mechanisms. The best-studied instance is the dendrodendritic reciprocal synapses (Reese and Brightman 1965; Rall et al. 1966; Rall and Shepherd 1968) in the olfactory bulb between the mitral and internal granule cells (glutamate vs. GABA), which play a major role in the generation of bulbar gamma oscillations by negative feedback. It is the passive dendritic time constants (~6.25 ms) of the neurons in the four steps around the loop (25 ms) that determine the characteristic frequency of bulbar activity (40 Hz) (Freeman 1991).

We suggest that coordination of subthreshold oscillations may be facilitated by neural and glial embedding in the local fields of polar molecules (including water molecules, Freeman et al. 2012b). On this basis, we postulate that the liquid-like phase of cortical activity in perception differs radically from the gas-like recipient phase in sensation. We believe that the difference justifies our use of the term *phase transition* in the thermodynamic sense. We also suppose that this coupling is not easily formed and, once it has formed, is not easily broken. For the neural mechanism that makes and breaks AM patterns, we appeal to our evidence for singularity in the transition to a limit cycle attractor (Fig. 9.13, Sect. 9.6.1) at onset and the action of the null spike (Fig. 7.8d, Sect. 7.5) at termination.

11.3 Neurodynamics and Thermodynamics: The Carnot Cycle

We intend the thermodynamic model as an extension from the network model by upscaling to encompass the immense numbers of cortical neurons and their interconnections. In doing so, it is essential to identify the main state variables of thermodynamics with observable quantities in experimental and theoretical neuroscience. A theoretical framework for doing this is given by the Carnot cycle, which two centuries ago facilitated understanding how a steam engine worked. It may now play a similar role for how perception works. Carnot's model was an abstraction because the insights it yielded depended on fixing the four classical thermodynamic variables (mass, temperature, pressure, volume). He fixed mass and temperature, heated or cooled the mass by adding or removing energy, and showed how pressure and volume changed along isothermal isoclines. He then fixed the energy content and allowed the temperature to vary, giving adiabatic[3] isoclines. Two each of these isoclines formed a closed loop. His formalism enabled the next generation of physicists to discover and apply the first and second laws of thermodynamics. It is the example of changing one independent variable at a time that we follow in using the Carnot formalism to evaluate the brain energy required to create knowledge.

The original Carnot cycle was expressed in the relation between pressure and volume in an ideal gas. A modern form (Fig. 11.1a) is represented by a loop, in which the temperature in a system replaces pressure and entropy replaces volume. Entropy now serves also as an index for increasing disorder, or inversely as measure of increasing information (negentropy). An example is the use of chemical energy by a diesel engine to do useful work. There are four steps. The cycle can begin at any point, say (1) where the temperature is low and the entropy is high. In step 1–2, a piston applies mechanical free energy stored in a flywheel that compresses the fuel

[3] Adiabatic cooling is familiar in the drop of temperature when one ascends from the plains to the mountains, as the air expands without change in heat content and as sodium ions expand into the interior of an axon, cooling it during an action potential (Abbot 1960).

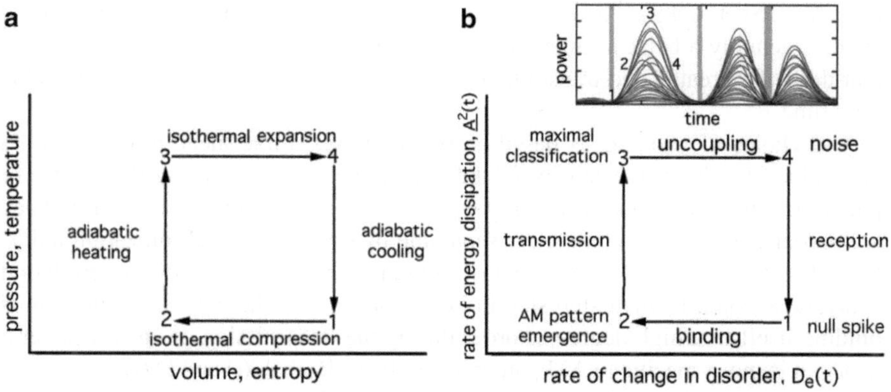

Fig. 11.1 (**a**) An ideal Carnot cycle with four isoclines is illustrated for a heat engine. (**b**) The variables representing temperature and entropy are replaced with indices derived from multichannel ECoG recording. The upper inset illustrates a sequence of three Carnot cycles in the superimposed filtered ECoGs from 64 electrodes (Fig. 6.5c). We start the cycle (1) at the minimal mean power, $\underline{A}^2(t)$, and maximal chaotic disorder, $D_e(t)$. At the other extreme, the power and information are maximal (the disorder is minimal (3)). The two states are connected by trajectories through intermediary states (2) and (4) that serve as markers for steps to distinguish overlapping processes described in the text. We can index the rate of useful work done in mobilizing and transmitting knowledge by the product of two variables: mean power, $\underline{A}^2(t)$, times the degree of increase in order (negentropy), $1/D_e(t)$, giving the information increase with each cycle (the area in the rectangle). This quantity is the pragmatic information, $H_e(t)$ (Fig. 6.7, Sect. 6.4.3). From Freeman et al. (2012a)

mixture, decreasing the entropy without changing the temperature (isothermal). In step *2–3*, an adiabatic process takes place; there is no heat exchange with the environment, but the temperature increases to maximum at fixed volume (3) with no change in entropy. In step 3–4, the mixture ignites, converting some of the chemical energy to mechanical energy as the piston volume increases (4) and converting the rest of it to heat. In step 4–1, the temperature drops to the starting value (1) and cools during an adiabatic process with no change in entropy. The formalism is useful because it separates the steps into four isoclines that form a closed loop, in which the area in the loop indexes the amount of work done by the system in a cycle. Furthermore, the formalism estimates the efficiency of the process by the ratio of the useful work obtained divided by the total energy dissipated in a cycle.

In the general terms of energy and entropy, we can introduce the formalism of the Carnot cycle into macroscopic cortical dynamics (Fig. 11.1b) without need for the classical state variables of pressure, temperature, volume, and mass, which brains closely regulate homeostatically at the macroscopic level, though they fluctuate at the microscopic level, especially temperature (Abbot 1960; Freeman et al. 2012a). Since brains operate far from equilibrium, we replace the static variables of energy and entropy with their rates of change. We replace temperature with a measured variable, which is the mean analytic power of AM patterns, $\underline{A}^2(t)$, and we replace entropy with the rate of change in the Euclidean distance, $D_e(t)$, with each digitizing step, Δt, between successive normalized feature vectors as a measure of disorder

(Fig. 6.6a, Sect. 6.4.2). The introduction of rates of change means that the Carnot cycle is actually a helix in time that is projected into the energy-entropy plane, so that the order created and transmitted as knowledge is cumulative and irreversible over time.

Our choice of variables is based on two premises. First, mean power, $\underline{A}^2(t)$, is determined by the pulse density of the generating neurons and by the degree of phase locking of the pulse probability waves. Both factors require free energy. They are partially though not completely separable by use of the synchronization index, $R_e(t)$ (Fig. 6.6b, Sect. 6.4.2), which does not require phase, frequency, or correlation analysis. Second, a small step size in $D_e(t)$ implies that the cortical population is binding itself in a high degree of order during the transmission of an AM pattern; hence, it is more specific to AM patterns than $R_e(t)$. Both $\underline{A}^2(t)$ and $D_e(t)$ are derived from the ECoG. They differ in that $\underline{A}^2(t)$ is the spatial mean of the analytic power at each time step, while $D_e(t)$ is the stepwise temporal difference of the feature vector after normalization of each frame (subtracting the frame mean and dividing by the spatial standard deviation of frame amplitude, $\underline{SD}_X$). One variable indexes the rate of consumption of free energy, and the other variable indexes the rate of decrease in order (increase in entropy). The area in the cycle indexes the rate of work done to create the order in each cycle. There is little to be said yet about cortical efficiency, other than stating that creating and utilizing knowledge are extraordinarily energy-intensive processes, and the cancellation of excitatory and inhibitory potentials is inherently wasteful.

We begin the idealized cycle at minimal power and maximal disorder (1), after a prior AM pattern has been quenched and the cortex is returned to its symmetric macroscopic baseline state. If a microscopic CS activates a mesoscopic Hebbian assembly or has already done so, symmetry is broken and a phase transition is initiated. The onset is often marked by a discontinuity in the analytic phase, $\phi(t)$. In the first step (1–2), phase coherence increases as shown by an increase in channel cross-correlation and observed power, as the assembly directs the cortex into the basin of its attractor. The reduction in degrees of freedom expresses the increase in order, which takes the particular form of the stable AM pattern (2). The high density of intracortical interaction precludes significant reception of further sensory input in the selected pass band. Maintenance of the pattern is assured by the balance between the energy drawn from ionic gradients and the energy dissipated by ionic flows, giving the oscillation a soliton-like stability. The free energy required for these operations is provided by the transmembrane ionic gradients of the neurons. The free energy drives dendritic currents and axonal pulses and is dissipated as heat. The sequestration of free energy in the coherence corresponds to a lower energy state, into which the cortex is "attracted".

In the second step (2–3), the density of the organized pulse cloud increases. Having been increased by phase convergence, the mean power increases by increasing dendritic current density to a maximum (3), as the AM pattern is transmitted at maximal pulse frequencies in the cortex through the output pathway (Fig. 8.8a, Sect. 8.4), which performs a spatial integral transformation (the Gabor transform and discussed in Sect. 8.6) into a holographic-like image with maximal resolution and classifiability (Fig. 9.2c, d, Sect. 9.3; Fig. 10.5a, Sect. 10.3), hence maximum information.

In the third and fourth steps (3–4 and 4–1), the analytic power decreases and the AM pattern terminates. The two concurrent processes of abatement of power and desynchroniation are not fully distinguishable, even with use of the synchronization index, $R_e(t)$; pulse sampling is required (Freeman 1974). The refractory periods of the axons weaken the interaction (reduce the feedback gain), so the average pulse firing rate decreases. The spatial distribution of characteristic frequencies about the central carrier frequency causes interference that manifests as beats as seen in Rayleigh noise (Fig. 6.4d; Sect. 9.3) but only in the macroscopic ECoG signals. The time interval between beats is solely determined by the width of the frequency distribution (Fig. 7.6, Sect. 7.4). Measurements of the spatial variance of the carrier frequencies (Fig. 6.5d, Sect. 6.4.1) place the predicted duration of bursts triggered by Hebbian assemblies within the range of theta wavelengths (Fig. 8 in Freeman 2009). We infer that only when the power in a burst has gone to zero in a null spike (Fig. 7.8d, Sect. 7.5) is the AM pattern terminated. This occurs after the cortex is released by the dissolution of the attractor landscape (Fig. 6.14). Following the beat, the same or another landscape may emerge, and the cortical populations in a new cycle can be sequestered in the same basin or in a differing basin of attraction.

The cortical processes of assembly ignition, burst formation, and AM pattern transmission all run thermodynamically downhill. The dissipation of free energy incurs what physiologists call an *oxygen debt*, meaning that the diminished ionic concentration gradients must sooner or later be restored (Fig. 6.1B). The debt is measured by blood oxygen level depletion (BOLD) using fMRI (Logothetis 2008; Freeman et al. 2009a). At the microscopic level, the energy dissipation during an action potential in axons is accompanied by cooling as sodium expands into the axoplasm (Abbot 1960). Cooling is overwhelmed after the pulse by heating as glucose is oxidized to provide the ATP that is required to operate the transmembrane ionic pumps. We infer that comparable fluctuations in temperature occur in the macroscopic cortex and that they are obscured by temporal and spatial smoothing (Freeman et al. 2009b). In the absence of direct measurements of temperature and thermal energy in respect to AM pattern genesis, what is important is that order is constructed in step 1–2 and transmitted in step 2–3 as AM patterns and that entropy and waste heat are discarded by the blood circulation in steps 3–4 and 4–1 as the oxygen debt is repaid.

11.4 Criticality and Phase Transitions: The Carnot Vapor (Rankine) Cycle

We extend the analogy using a generalization of the Carnot cycle that can include phase transitions. There are many variants to choose among; a convenient starting point is the conventional phase diagram (pressure vs. temperature) for the three main states of water, in which the phase boundary between gas and liquid ends at the *critical point*. For pressures and temperatures above that point, there is a domain of criticality (Sect. 6.9), in which gas and liquid states are intermingled in varying

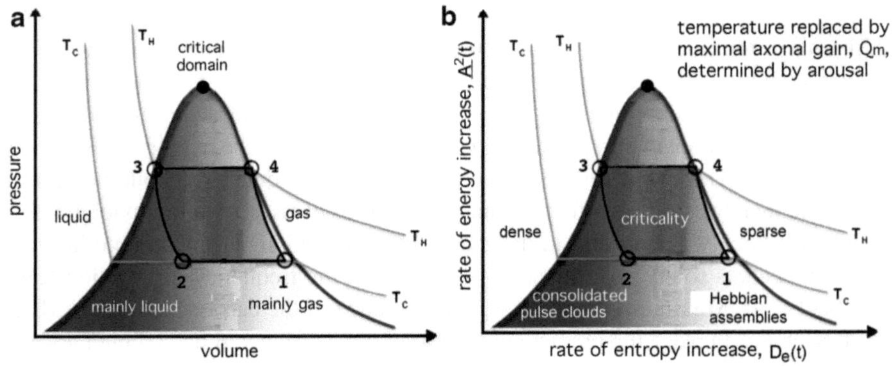

Fig. 11.2 (**a**) The four steps that comprise the generalized Carnot cycle are shown in a domain of criticality maintained by neural avalanches (Sect. 6.9). The gas and liquid phases coexist in varying degrees. In this formulation, energy is put in by heating in 2–3 and removed as waste heat in cooling 4–1. Adapted from Baratuci (2011) (**b**) We conceive the operation of cortex as using information from step 1 to select, construct, and transmit knowledge in step 3. High density of information appears in single neuron pulse trains in Hebbian assemblies selecting an attractor (1–2) and in readout neurons down sampling the AM pattern (3–4). The height of the cycle is determined by the degree of arousal as indexed by the maximum of axonal gain, Q_m (Fig. 6.10, Sect. 6.6). Free energy is derived from oxidative metabolism and is dissipated as heat in all four steps. From Freeman et al. (2012a)

degrees. Under a change of variables to the Carnot coordinates (pressure vs. volume), the critical domain appears as the shaded area. An example of the vapor cycle (Fig. 11.2a) illustrates the relation of pressure and volume along two isotherms for high and low temperatures, T_H and T_C. The four idealized steps of heat exchange have been embedded in the critical domain (Baratuci 2011). Beginning at minimal temperature and pressure (1), the vapor is condensed with no change in pressure (isothermal compression, 1–2), heated with increase in temperature (adiabatic compression, 2–3), evaporated at constant temperature and pressure with increased volume (isothermal expansion, 3–4), and cooled prior to disposal of waste heat with decreased temperature (adiabatic expansion).

In the model of cortical dynamics, the rate of energy dissipation as measured by the mean analytic power, $\underline{A}^2(t)$, replaces pressure. The rate of change in entropy replaces volume and in turn is replaced by the measure of AM pattern change, or inversely by the rate of increase in information (negentropy) as measured by $1/D_e(t)$. Temperature is unsuitable as a state variable for cognitive processing because in homeothermic animals it is homeostatically regulated. Instead the variable that most directly imposes order in the formation of cortical patterns is the intensity of synaptic interaction among all participating neurons, hence the order parameter, which is indexed by $H_e(t)$ and modeled as negative feedback gain, k_n (Sect. 6.2.2). The most important determinant of the loop gain is the normalized forward gain at trigger zones (Fig. 6.2a, Sect. 6.2.1) given by the function, $G_a(v)$, the sigmoid curve (Fig. 6.10, Sect. 6.6). This function has only one parameter, the magnitude of the upper asymptote, Q_m, which in the olfactory system is determined by the degree of arousal and motivation of intentional action (Fig. 8.1d). Inasmuch as arousal is a major factor in perception, we adopt Q_m as a measurable index to replace temperature and use it to symbolize the maximal intensity of global synaptic interaction.

We conceive that the ignition of a Hebbian assembly precipitates condensation into a coherent AM pattern (1–2). Three lines of evidence for condensation by a phase transition are the oft-noted phase discontinuity at onset (Fig. 9.8c, Fig. 9.9, Sect. 9.5), the jump to a new carrier frequency and AM pattern, and the accompaniment of the AM pattern by a conic phase gradient (Sects. 8.5 and 9.5c), for which the apex of the cone constitutes a singularity (Fig, 9.13c) that we propose may correspond to a site of nucleation (Fig. 8.9) (Freeman 1990). Condensation is followed by increased mean pulse firing rates (2–3) for the 3–5 cycles required for transmission (Fig. 9.7, Sect. 9.4.2). The subsequent fall in analytic power is due to lowering of mean firing rates by refractory periods (Fig. 6.11a, Sect. 6.5) and to decoherence upon evaporation of the condensed state (Fig. 6.5c) eventuating in the null spike (Fig. 7.8d, Sect. 7.5).

The Carnot formalism serves to organize many of the observed features of spatiotemporal patterns from high-density grid recording, and it provides physical meanings and measurable quantities to embody the concepts of free energy and entropy that are complementary to the symbolic meanings deriving from information theory (Friston 2009). More importantly the model clarifies and sharpens challenges for further investigation. The need is now brought into focus for distinguishing the contributions to $\underline{A}^2(t)$ of the mean pulse firing rates and the degree of coherence from phase locking in pulse probability waves, which is manifested in the modulation depths of statistical averages that ranges from 0% to 100% of the mean rates. Reliable statistical sampling methods with multiple microelectrodes are called for with which to estimate the pulse cloud density on single trials (Chap. 5), avoiding time averaging (Fig. 6.10, Sect. 6.6) (Freeman 1974).

We have posed the question whether the postulated liquid-like phase truly exists (Sect. 11.2). On the one hand, the cortical populations may oscillate between low and high density without convergence to attractors, as in the metastable state proposed by Kelso and Tognoli (2006) and chaotic itinerancy proposed by Tsuda (2001) (Sect. 6.4), which are in accordance with the ideal Carnot cycle in having no phase transitions (Fig. 11.1), but not with the evidence for discontinuities in the recordings. On the other hand, the properties of a liquid-like state that could differentiate it from mere high density are yet to be adequately defined. Is energy sequestered in the state of oscillation that is comparable to the latent heat of condensation? If so, is it released on evaporation similarly to the latent heat of water? Might the condensed phase impose some form of viscosity on intracortical communication and transmission? At first glance, the idea seems absurd. On second thought, the concept of internal friction implies only that energy is dissipated in the induction of synchronized oscillations, so that the generalized Carnot formalism (Fig. 11.2) describing the behavior of water might help to explain the appearance of vortices (Fig. 9.10, Sect. 9.5) in cinematic displays (Freeman 2011), often resembling miniature hurricanes (Fig. 9.12), some rotating clockwise, others counterclockwise, and most pulsing inwardly or outwardly for the several cycles of the bursts (Fig. 9.13) (Freeman and Kozma 2010). The vortices may be optical illusions from interference among overlapping phase gradients of multiple foci creating moiré. However, they are not to be dismissed out of hand, because they are consistent with the stationary phase gradients fitted with cones that accompany AM patterns (Fig. 9.5c). Moreover, the existence of vortices in association with phase transitions is predicted by modeling the cortical activity with concepts from many-body physics (Freeman and Vitiello 2009).

The role of neural viscosity might optimally be conceived as stabilization of the AM pattern within the frequency band set by the Hebbian assembly immersed in broadband noise. The prolongation was predicted by Rice (1950) and verified experimentally (Fig. 7.6d, Sect. 7.4; Fig. 9.7e, Sect. 9.5). Further speculation along these lines must await replication of our results with larger high-density spatial arrays and much faster temporal digitizing, because maximal resolution in the temporal, spatial, and spectral domains is essential for definitive exploration. The focus of the research should be to isolate the carrier of an AM pattern by an appropriate method of decomposition (Chap. 4; Sects. 6.3 and 7.3) and determine how close together might be the locations of the three manifestations of singularity: the null spike initiating the phase transition (Fig. 7.8, Sect. 7.5), the apex of the cone (Fig. 8.9, Sect. 8.5), and the center of rotation of the vortex (Fig. 9.10, Sect. 9.6.1). In theory the three points are predicted to coincide in space though not in time (Freeman and Kozma 2010). At present they usually do not. We propose that this is because we have not adequately decomposed overlapping bursts at different frequencies. Adequate tests will require substantial improvements in sampling and digital filter design.

11.5 Transmission, Reception, and Readout of Bursts

Transmission of the olfactory percept is by action potentials propagating on axons in the lateral olfactory tract (LOT) to excitatory axodendritic synapses on neurons in diverse targets in the basal forebrain, predominantly in the prepyriform or primary olfactory cortex (Fig. 8.8, Sect. 8.4). The divergent-convergent pathway performs a spatiotemporal integral transformation analogous to the Gabor transform of a lens or a holograph (Pribram 1991). The LOT delivers the same bulbar information to all targets by its multiple divisions, as in a hologram that can be broken into pieces, each piece having the entire output pattern at reduced resolution. The decision whether to accept or reject the transmitted data is local in accord with each local tuning (Traub et al. 1996). The prepyriform gamma bursts also contain classifiable AM patterns (Barrie et al. 1996), but the phase gradients and velocities of both the ECoG bursts and the evoked potentials are determined by the axons in the LOT. Maps of the amplitude and latency of the spread of ECoG oscillations (pp. 42–45 in Freeman 2000; Freeman and Barrie 2000) and ECoG phase (Freeman 1978) provide direct evidence that phase cones do not exist in prepyriform cortex, as they do in the olfactory bulb (Fig. 8.6) and the primary sensory neocortices (Fig. 9.5c).

The prepyriform cortex and intervening anterior olfactory nucleus have extensive connections recurrent to the bulb via the medial olfactory tract (Gray and Skinner 1988). Each of the three parts has its characteristic frequency, which is nonharmonic and incommensurate with the other frequencies (Kozma and Freeman 2001; Freeman and Erwin 2008). The positive excitatory feedback among them continuously increases the activity, while the thresholds and refractory periods continuously curb it but with no convergence to any characteristic frequency, so the three-way interaction results in the broad-spectrum nonconvergent background ECoG (Skarda and Freeman 1987). When the prepyriform cortex is deprived of bulbar input by cutting or inactivating the LOT, the cortex goes silent, while bulbar activity becomes periodic, revealing the characteristic frequency of its limit cycle attractor (Gray and Skinner 1988). We infer that the prepyriform cortex lacks the capacity for self-organization and that its AM and PM patterns are not emergent but driven by bulbar input.

The LOT transmits the knowledge that is stored in the bulb to the prepyriform cortex in pulse clouds. The dense neuropil of layers I and II (Fig. 8.2) sustains the knowledge as classifiable AM patterns in the liquid-like phase of pulse densities. The prepyriform output is transmitted by the deep pyramidal cells in layer III, which have widely radiating basal dendrites. We conjecture that they sample the cloud of layer II pulses that sustain the AM patterns by integrating over local patches, thereby spatially coarse-graining, and transducing prepyriform activity from macroscopic pulse densities to microscopic pulse frequencies through many-to-one convergence. By this operation, the prepyriform cortex returns the output of the olfactory system to the same microscopic level of sparse coding as the input to the bulb, but whereas bulbar pulse inputs on the PON convey sensory information, pulse outputs from the prepyriform to the brain would convey perceptual meanings.

It could be expected that the layer III pyramidal cells would project to more specialized neurons that could read and extract the conceptual meaning of a CS. Such neurons have been described in the human hippocampus and surrounding temporal cortex and have been named *concept* cells (Quian Quiroga et al. 2005; Quian Quiroga 2012), whose firing is correlated not with specific sensory stimuli but with a variety of stimuli that resemble, suggest, or are associated with a category such as a familiar person (e.g., "Jennifer Aniston cells"). In the mass action view, the layer III neurons transmitting the prepyriform output would correspond each to a dot in a pointillist painting or a pixel in a hologram. Each neuron would have an implicit representation of the meaning of a stimulus, but of course, they would not act in isolation, as a large portion of layer III would have to provide an adequate sample of the active population to convey a particular concept with its full context.

In preceding Chapters (Sects. 8.6, 9.7, and 10.3), we noted that a major target of prepyriform transmission in olfaction is the entorhinal cortex in the medial temporal lobe, which is the main gateway to the hippocampus. Concept cells were found in both these areas in humans, which also receive projections, through the parahippocampal and perirhinal cortices, from purely visual areas, such as the inferior temporal, auditory, and somatic areas via the cingulum, and the frontal lobe through the uncinate fasciculus. In Sect. 10.4 we described AM patterns from the inferotemporal ECoG (Fig. 10.5, Sect. 10.4) that were classified with respect to categories of cognitive behavior. In analogy to the primary sensory cortices, we propose that the specialized entorhinal neurons are members of Hebbian assemblies and that the specialization is by synaptic changes due to reinforcement learning by which assemblies form (Fig. 8.8, A, Sect. 8.4). Then the vigorous firing of a concept cell would manifest the ignition of a Hebbian assembly, which would select an attractor representing a particular concept. But in contrast to the classic idea of Hebbian assemblies and attractors, the state of the system is quite dynamic, as it may go from one attractor to another, representing different related concepts in sequences resembling "chaotic itinerancy" (Tsuda 2001) and "metastability" (Kelso and Tognoli 2006).

The existence of AM patterns involving the cat entorhinal cortex (Figs. 10.3c, Sects. 10.3 and 10.5, Sect. 10.4) indicates that at least in this species the entorhinal output is simultaneously microscopic and macroscopic. That inference generates two hypotheses for higher cognitive processing. One is that all further processing is done at the microscopic level by combinatorial dynamics that is executed by neural networks throughout the allocortex and neocortex, culminating in activation of neural networks in the hippocampus representing concepts and forming new memories (Quian Quiroga 2012) and in the motor cortex and basal ganglia for implementing intentional behaviors. The hypothesis is supported at the microscopic level by the finding that concept cells with very different cognitive correlates are found close together in entorhinal cortex or hippocampus (Quian Quiroga 2012), with no evidence of spatial organization as found in adjacent areas of inferotemporal cortex (Tanaka 2003).

The other hypothesis is that the transition from microscopic to macroscopic and back again occurs with every Rankine cycle in higher cortices. By this hypothesis, the action-perception cycle would minimally require three successive Rankine cycles (Fig. 9.2a, b): first, sensation to perception, second, perception to conception, and third, conception to action (Fig. 8.1), which may most quickly close the loop of the action-perception cycle (Merleau-Ponty 1942). Category cells having high information content mediate the transitions between levels in step 1–2 (Hebbian neurons) and step 3–4 (readout neurons) of each Rankine cycle. The hypothesis is supported at the macroscopic level by classifiable AM patterns in EC0Gs involving multiple cortical areas (Fig. 10.3) and in the scalp EEG (Fig. 10.7). Further support is from simulation of perceptual dynamics in an intentional robot (Kozma et al. 2003, 2008).

11.6 Future Developments in Cortical Thermodynamics of Perception

Nonequilibrium thermodynamics can be construed as central in the coordinated sciences that address the biological foundation of the mind. Our description is focused on the electric fields of potential generated by axons and dendrites, which draw on a virtually infinite reservoir of electromotive energy that is universally stored in transmembrane ionic gradients. Synapses and trigger zones borrow energy freely by creating an oxygen debt, and the debt is repaid independently at far slower time scales. Our main thrust has been to apply theory and techniques of digital signal processing to optimize the description and interpretation of the spatiotemporal patterns we can observe and measure at the surfaces of the cortex and scalp. We have left for others the study of pattern correlates with measurements of the oxygen debt (BOLD) and its repayment through oxidative metabolism (PET). We have left untapped the wealth of information on the pattern correlates with magnetic potentials (MEG), which may access cortical areas in sulci beyond the reach of EEG. We have said little about the psychological and behavioral correlates of patterns, other than passing reference to the action-perception cycle of Merleau-Ponty (1942). We have briefly opened new doors to cinematic display of dynamic space-time patterns of ECoG and EEG (Freeman 2011) to spectral decomposition of EEG patterns (Ruiz et al. 2010; Brockmeier et al. 2012), and to further enhanced extraction of information regarding pulse-wave relations of unit activity (Fig. 6.11, Sect. 6.6). We have laid the foundation for advanced mathematical modeling using volume conductor theory (Ramon et al. 2009), random graph theory and neuropercolation (Kozma 2007; Freeman et al. 2009), many-body physics (Freeman and Vitiello 2009), and nonequilibrium thermodynamics (Freeman et al. 2012b).

The salient hypothesis arising from this study posits the existence of a liquid-like phase of cortical neuropil, which may provide the neural basis for virtually instantaneous integration of an enormous quantity of information revealed in the experience of recall and recognition in memory retrieval. The AM and PM patterns of beta and gamma bursts provide the main evidence supporting the hypothesis. The evidence includes descriptions of mechanisms by which the patterns are formed and dissolved. The most controversial aspect of the hypothesis may be the existence of a singularity in the corticodynamics, which may appear in three forms: the null spike in analytic power, the apex of the phase cone, and the center of rotation of cortical vortices. The hypothesis requires that these points coincide spatially for each new transition from the sparse phase to the dense phase. At present they seldom do, we believe for the reason that the classifiable patterns are distorted and obscured by concomitant background avalanches that support the state of criticality. Adequate testing with substantial improvements in ECoG sampling and decomposition is crucial because the hypothesis will stand or fail depending on the outcome.

The experimental evidence for verifying and extending the thermodynamic hypothesis from allocortex to neocortex is deficient in several respects. There is as yet no direct anatomical mapping of divergent-convergent projections comparable to the LOT for the output of other sensory cortices, which could perform a Gabor transform. The evidence for power-law distributions of axon lengths needed to mediate global phase transitions is scanty (Freeman and Breakspear 2007). There is as yet no identification of the areas of cortex (or of equivalent volumes of neurons in the basal ganglia) that would be needed by the primary neocortical sensory areas to implement the readout mechanism that is provided for the bulb by the prepyriform cortex. The lack of evidence does not disprove the hypothesis; rather it attests to the difficulty of making the necessary anatomical observations and measurements (Sholl 1956; Bok 1959; Braitenberg and Schüz 1998). What is required is investment in the field of macroscopic neuroanatomy focused on evaluation of the degree of conformance of the connectivity distributions to the requirements for modeling phase transitions.

Additional anatomical data would be illuminating, along with further refinements for recording, measuring, and analyzing the scalp EEG. At present using standard clinical arrays and digitizing frequencies, we have achieved the first glimpse (Sect. 10.5) of the wealth of new data that await harvesting with larger and denser arrays, increased sampling rates by an order of magnitude, and new techniques for minimizing the EMG, which at present virtually precludes systematic investigation of the gamma and epsilon activity in the EEG (Whitham et al. 2007). The evidence provided for global AM patterns in the EEG (Fig. 10.7) that is comparable to that from the ECoG (Fig. 10.3) documents their scale-free nature, but too little is known about their other properties, especially their phase gradients, at present to support speculation on how the AM patterns form in EEG, what the range of their cognitive correlates might be, how they are read out, and whether they manifest pulse clouds that play an active role in objective behavior and subjective experience. Their accessibility for noninvasive measurement in normal human subjects engaged in normal cognition provides the best reason that we, the authors, have undertaken the task of preparing this introduction to spatiotemporal EEG imaging. We hope also that engineers might learn how to model perceptual and conceptual dynamics so as to compact electronic information into knowledge and retrieve it for use in the way that brains do.

11.7 Summary

We describe the brain as a thermodynamic system that has evolved to create information and store and use it in compact form as knowledge. Using water as an analogy, we conceive neurons as having two states or phases (possibly more), one sparse and gas-like and the other dense and liquid-like.

The condensation from sparse phase to dense phase requires the ignition of a Hebbian assembly, which provides the transition energy to raise cortical activity above a protective barrier that prevents preemption by noise. The following evaporation is predicated on a pervasive tendency to disorder imposed by the distribution of feedback gains in the cortical populations that disperses the characteristic frequencies of oscillation of local populations.

Recurrence of the two phase transitions at beat frequencies is modeled by the Carnot and Rankine cycles. A measure of mean analytic power in the ECoG, $\mathbf{A}^2(t)$, replaces pressure and energy. A measure of the rate of change in the spatial AM patterns, $D_e(t)$, replaces volume entropy. The inverse of entropy gives the rate of information increase. Temperature is replaced by a measure of the intensity of feedback interaction, which is dependent on the degree of arousal. The relations among these three variables are expressed in isoclines that form a closed loop.

The area in the loop is given by the product of the rate of energy dissipated times the rate of information increase, $H_e(t)$, which is the index of the pragmatic information that is formed in each cycle and added to the store of knowledge. An index of the efficiency of cortical dynamics is given by the ratio of coherent power to total power, $R_e(t)$, given by the ratio of the $SD_X(t)$ of the mean ECoG divided by the mean $\underline{SD}_X(t)$ of the set of filtered ECoGs.

The cortical phase transitions are incorporated by embedding the loop in a domain of criticality. The domain is modeled on the vapor cycle of the phase diagram for water, which operates in the area beyond the critical point marking the end of the phase boundary between gas and liquid.

Creation or recall of a memory by a stimulus occurs in the macroscopic dense level; readout is in the microscopic sparse level. The mobilized knowledge is manifested macroscopically in spatial AM patterns from ECoGs of association cortices and the scalp EEG and microscopically in the firings of concept cells in networks. We conclude that understanding cognition and cogitation requires measurements at both levels of dynamics.

References

Abbot BC (1960) Heat production in nerve and electric organ. J Gen Physiol 43:119–127

Amit DJ (1995) The Hebbian paradigm reintegrated: local reverberations as internal representations. Behav Brain Sci 18:617–657

Anastassiou CA, Perin R, Markram H, Koch C (2011) Ephaptic coupling of cortical neurons. Nat Neurosci 14:217–223

Atmanspacher H, Scheingraber H (1990) Pragmatic information and dynamical instabilities in a multimode continuous-wave dye laser. Can J Phys 68:728–737

Bach-y-Rita P (1995) Nonsynaptic diffusion neurotransmission and late brain reorganization. Demos, New York

Baratuci B (2011) Learn thermodynamics. http://www.learnthermo.com/T1-tutorial/ch06/lesson-E/pg17.php#

Barrie JM, Freeman WJ, Lenhart M (1996) Modulation by discriminative training of spatial patterns of gamma EEG amplitude and phase in neocortex of rabbits. J Neurophysiol 76:520–539

Bennett MVL (2009) Gap junctions and electrical synapses. In: Squire LR (ed) Encycopedia of Neurosci, vol 4. Elsevier/Academic Press, New York, pp 529–548

Bok ST (1959) Histonomy of the cerebral cortex. Elsevier, Amsterdam

Braitenberg V, Schüz A (1998) Cortex: statistics and geometry of neuronal connectivity, 2nd edn. Springer, Berlin

Breakspear M, Stam CJ (2005) Dynamics of a neural system with a multiscale architecture. Philos Trans R Soc Lond B Biol Sci 1457:1051–1074

Brockmeier AJ, Hazrati MK, Freeman WJ, Principe J (2012) Locating global spatial patterns of waveforms during sensory perception in scalp EEG. In: Proceedings of the IEEE conference on EMBS #2098, San Diego, 28 Aug–1 Sept

Darwin C (1872) The expression of emotion in man and animals. Murray, London

Freeman WJ (1974) Average transmission distance from mitral tufted to granule cells in olfactory bulb. Electroencephalogr Clin Neurophysiol 36:609–618

Freeman WJ (1978) Spatial properties of an EEG event in the olfactory bulb and cortex. Electroencephalogr Clin Neurophysiol 44:586–605. See Fig. 7 for phase from frontal prepyriform, Fig. 8 for phase from temporal prepyriform

Freeman WJ (1990) On the problem of anomalous dispersion in chaoto-chaotic phase transitions of neural masses, and its significance for the management of perceptual information in brains. In: Haken H, Stadler M (eds) Synergetics of cognition, vol 45. Springer, Berlin, pp 126–143

Freeman WJ (1991) The physiology of perception. Sci Am 264:78–85

Freeman WJ (2000) Neurodynamics: an exploration of mesoscopic brain dynamics. Springer, London. http://soma.berkeley.edu/books/BD/MesoBrainDyn.html

Freeman WJ (2001) How brains make up their minds. Columbia University Press, New York

Freeman WJ (2007) Three centuries of category errors in studies of the neural basis of consciousness and intentionality. Neural Netw 10:1175–1183

Freeman WJ (2009) Deep analysis of perception through dynamic structures that emerge in cortical activity from self-regulated noise. Cogn Neurodyn 3(1):105–116

Freeman WJ (2011) Understanding Perception Through Neural 'Codes'. In: Special Issue on "Grand Challenges in Neuroengineering", IEEE Trans Biomed Engin 58(7):1884–1890. doi:TBME-00851-2010.R1 http://soma.berkeley.edu/videos/?video=6

Freeman WJ, Barrie JM (2000) Analysis of spatial patterns of phase in neocortical gamma EEGs in rabbit. J Neurophysiol 84:1266–1278

Freeman WJ, Breakspear M (2007) Scale-free neocortical dynamics. Scholarpedia 2(2):1357. http://www.scholarpedia.org/article/Scale-free_neocortical_dynamics

Freeman WJ, Erwin H (2008) Freeman K-set. Scholarpedia 3(2):3238. http://www. scholarpedia.org/article/Freeman_K-set

Freeman WJ, Kozma R (2010) Freeman's mass action. Scholarpedia 5(1):8040. http://www.scholarpedia.org/article/Freeman%27s_mass_action, http://soma.berkeley.edu/videos/?video=6

Freeman WJ, Vitiello G (2009) Dissipative neurodynamics in perception forms cortical patterns that are stabilized by vortices. J Phys Conf Ser 174(012011):1–24. http://www.iop.org/EJ/toc/1742-6596/174/1 http://repositories.cdlib.org/postprints/3379

Freeman WJ, Holmes MD, West GA, Vanhatalo S (2006a) Fine spatiotemporal structure of phase in human intracranial EEG. Clin Neurophysiol 117:1228–1243

Freeman WJ, Holmes MD, West GA, Vanhatalo S (2006b) Dynamics of human neocortex that optimizes its stability and flexibility. Int J Intell Syst 21:1–21. http://repositories.cdlib.org/postprints/2385

Freeman WJ, Ahlfors SM, Menon V (2009a) Combining EEG, MEG and fMRI signals to characterize mesoscopic patterns of brain activity related to cognition. Int J Psychophysiol 73(1): 43–52. http://repositories.cdlib.org/postprints/3386

Freeman WJ, Kozma R, Bollobás B, Riordan O (2009b) Chapter 7. Scale-free cortical planar network. In: Bollobás B, Kozma R, Miklös D (eds) Handbook of large-scale random networks. Bolyai mathematical studies, vol 18. Springer, New York, pp 277–324. http://www.springer.com/math/numbers/book/978-3-540-69394-9

Freeman WJ, Kozma R, Vitiello G (2012a) Adaptation of the generalized Carnot cycle to describe thermodynamics of cerebral cortex. Proc IEEE World Congr Comp Intell WCCI/IJCNN, Brisbane, Australia. IEEE Press, pp. 3229–3236.

Freeman WJ, Livi R, Obinata M, Vitiello G (2012) Cortical phase transitions, non-equilibrium thermodynamics and the time-dependent Ginzburg-Landau equation. Int J Mod Phys B 26(6):1250035. doi:10.1142/S021797921250035X

Freud S (1985) The complete letters of Sigmund Freud to Wilhelm Fliess, 1887–1904. Belknap Press/Harvard University Press, Cambridge, MA

Friston K (2009) The free-energy principle: a rough guide to the brain? Trends Cogn Sci 13(7):293–301

Gray CM, Skinner JE (1988) Centrifugal regulation of neuronal activity in the olfactory bulb of the waking rabbit as revealed by reversible cryogenic blockade. Exp Brain Res 69:378–386

Haken H (1983) Synergetics: an introduction. Springer, Berlin

Hameroff S (2009) The "conscious pilot"—dendritic synchrony moves through the brain to mediate consciousness. J Biol Phys. doi:10.1007/s10867-009-9148-x

Kelso JAS, Tognoli E (2006) Metastability in the brain. Neural networks IJCNN'06, pp 363–368. doi:10.1109/IJCNN.2006.246704

Kozma R (2007) Neuropercolation. Scholarpedia 2(8):1360

Kozma R, Freeman WJ (2001) Chaotic resonance: methods and applications for robust classification of noisy and variable patterns. Int J Bifurc Chaos 10:2307–2322

Kozma R, Freeman WJ, Erdí P (2003) The KIV model—nonlinear spatio-temporal dynamics of the primordial vertebrate forebrain. Neurocomputing 52:819–826. http://repositories.cdlib.org/postprints/1049

Kozma R, Huntsberger T, Aghazarian H, Tunstel E, Ilin R, Freeman WJ (2008) Intentional control for planetary rover SRR2k. Adv Robot 21(8):1109–1127
Kozma R, Puljic M, Freeman WJ (2012) Thermodynamic model of criticality in the cortex based on EEG/ECoG data. In: Plenz D (ed) Criticality in neural systems. Wiley, New York
Logothetis NK (2008) What we can do and what we cannot do with fMRI. Nature 453:869–878. doi:10.1038/nature06976
Lucky RW (1989) Silicon dreams: information, man and machine. St. Martin's Press, New York
Merleau-Ponty M (1942) The structure of behavior (trans: Fischer AL, 1963). Beacon Press, Boston
Pribram KH (1991) Brain and perception: holonomy and structure in figural processing. Lawrence Erlbaum Associates, New Jersey
Pribram KH, Gill MM (1976) Freud's 'Project' re-assessed. Preface to contemporary cognitive theory and neuropsychology. Basic Books, New York, pp 356–359
Prigogine I (1980) From being to becoming: time and complexity in the physical sciences. WH Freeman, San Francisco
Quian Quiroga R (2012) Concept cells: The building blocks of declarative memory functions. Nat Rev Neurosci 13:587–597. doi:10.1038/nrn3251
Quian Quiroga R, Reddy L, Kreiman G, Koch C, Fried L (2005) Invariant visual representation by single neurons in the human brain. Nature 435:1102–1107
Raichle M (2006) Neuroscience. The brain's dark energy. Science 314(5803):1249–1250
Raichle M, Mintun M (2006) Brain work and brain imaging. Annu Rev Neurosci 29:449–476
Rall W, Shepherd GM (1968) Theoretical reconstruction of field potentials and dendrodendritic synaptic interactions in olfactory bulb. J Neurophysiol 31:884–915
Rall W, Shepherd GM, Reese TS, Brightman MW (1966) Dendrodendritic synaptic pathway for inhibition in the olfactory bulb. Exp Neurol 14:44–56
Ramon C, Freeman WJ, Holmes MD, Ishimaru A, Haueisen J, Schimpf PH, Resvanian E (2009) Similarities between simulated spatial spectra of scalp EEG, MEG and structural MRI. Brain Topogr 22:191–196
Reese TS, Brightman MW (1965) Electron microscopic studies on the rat olfactory bulb. Anat Rec 151:492–497
Rice SO (1950) Mathematical analysis of random noise—and appendixes. Technical Publications Monograph B-1589. Bell Telephone Labs Inc., New York
Ruiz Y, Pockett S, Freeman WJ, Gonzales E, Guang L (2010) A method to study global spatial patterns related to sensory perception in scalp EEG. J Neurosci Methods 191:110–118. doi:10.1016/j.jneumeth.2010.05.021
Shannon C (1948) A mathematical theory of communication. Bell Syst Tech J 27:379
Sholl DA (1956) The organization of the cerebral cortex. Methuen, London
Skarda CA, Freeman WJ (1987) How brains make chaos in order to make sense of the world. Behav Brain Sci 10:161–195
Spencer H (1863) Essays: moral, political, and aesthetic. Appleton, New York
Tanaka K (2003) Columns for complex visual object features in the inferotemporal cortex: clustering of cells with similar but slightly different stimulus selectivities. Cereb Cortex 13:90–99
Traub RD, Whittington MA, Stanford IM, Jefferys JGR (1996) A mechanism for generation of long-range synchronous fast oscillations in the cortex. Nature 383:421–424
Tsuda I (2001) Towards an interpretation of dynamic neural activity in terms of chaotic dynamical systems. Behav Brain Sci 24:793–810
Turing AM (1952) The chemical basis of morphogenesis. Phil Trans R Soc Lond B 237B:37–72
Vitiello G (2001) My double unveiled. John Benjamins, Amsterdam
Whitham E, Pope K, Fitzgibbon S, Lewis T, Clark C, Loveless S, Broberg M, Wallace A, DeLosAngeles D, Lillie P (2007) Scalp electrical recording during paralysis: quantitative evidence that EEG frequencies above 20 Hz are contaminated by EMG. Clin Neurophysiol 118(8):1877–1888

Index

W.J. Freeman and R. Quian Quiroga, *Imaging Brain Function With EEG: Advanced Temporal and Spatial Analysis of Electroencephalographic Signals*, DOI 10.1007/978-1-4614-4984-3, © Springer Science+Business Media New York 2013

F

G

MIX
Papier aus verantwortungsvollen Quellen
Paper from responsible sources
FSC® C105338

If you have any concerns about our products,
you can contact us on
ProductSafety@springernature.com

In case Publisher is established outside the EU,
the EU authorized representative is:
Springer Nature Customer Service Center GmbH
Europaplatz 3, 69115 Heidelberg, Germany

Printed by Libri Plureos GmbH
in Hamburg, Germany